Applications in High Resolution Mass Spectrometry

Applications in High Resolution Mass Spectrometry

Food Safety and Pesticide Residue Analysis

Edited by

Roberto Romero-González

Antonia Garrido Frenich

elsevier.com

Elsevier
Radarweg 29, PO Box 211, 1000 AE Amsterdam, Netherlands
The Boulevard, Langford Lane, Kidlington, Oxford OX5 1GB, United Kingdom
50 Hampshire Street, 5th Floor, Cambridge, MA 02139, United States

Notices
Knowledge and best practice in this field are constantly changing. As new research and experience broaden our understanding, changes in research methods, professional practices, or medical treatment may become necessary.

Practitioners and researchers must always rely on their own experience and knowledge in evaluating and using any information, methods, compounds, or experiments described herein. In using such information or methods they should be mindful of their own safety and the safety of others, including parties for whom they have a professional responsibility.

To the fullest extent of the law, neither the Publisher nor the authors, contributors, or editors, assume any liability for any injury and/or damage to persons or property as a matter of products liability, negligence or otherwise, or from any use or operation of any methods, products, instructions, or ideas contained in the material herein.

Library of Congress Cataloging-in-Publication Data
A catalog record for this book is available from the Library of Congress

British Library Cataloguing-in-Publication Data
A catalogue record for this book is available from the British Library

ISBN: 978-0-12-809464-8

For information on all Elsevier publications visit our website at
https://www.elsevier.com/books-and-journals

Working together
to grow libraries in
developing countries

www.elsevier.com • www.bookaid.org

Publisher: Nikki Levy
Acquisition Editor: Patricia Osborn
Editorial Project Manager: Jaclyn A. Truesdell
Production Project Manager: Lisa Jones
Designer: Alan Studholme

Typeset by TNQ Books and Journals

Contents

List of Contributors.. xi
Preface.. xiii

CHAPTER 1   HRMS: Fundamentals and Basic Concepts............... 1
Franciso Javier Arrebola-Liébanas, Roberto Romero-González,
Antonia Garrido Frenich
  1.1 Introduction (To High-Resolution Mass Spectrometry)............. 1
      1.1.1 Basic Concepts (Units and Definitions)......................... 1
      1.1.2 Low-Resolution Mass Spectrometry Versus
             High-Resolution Mass Spectrometry............................ 3
  1.2 Resolution and Mass Resolving Power................................... 6
  1.3 Accurate Mass Measurement: Exact Mass and Mass Defect...... 7
  1.4 Mass Calibration in High-Resolution Mass Spectrometry 9
  1.5 General Considerations.. 12
  Acknowledgments... 13
  References.. 13

CHAPTER 2   HRMS: Hardware and Software............................. 15
Juan F. García-Reyes, David Moreno-González, Rocío Nortes-
Méndez, Bienvenida Gilbert-López, Antonio Molina Díaz
  2.1 Introduction... 15
  2.2 Principles of High-Resolution Mass Spectrometry
        Analyzers... 16
      2.2.1 Time-of-Flight... 17
      2.2.2 Fourier Transform Ion Cyclotron Resonance............... 19
      2.2.3 Orbitrap... 21
  2.3 Time-of-Flight Mass Spectrometry: Instrument
        Configuration and Main Features 22
      2.3.1 Stand-alone Electrospray Ionization Time-of-Flight
             and Hybrid Quadrupole Time-of-Flight
             Instrumentation... 22
      2.3.2 Improvements of Current (Quadrupole) Time-of-Flight
             Instrumentation... 24
      2.3.3 Ion Mobility Quadrupole Time-of-Flight 29
      2.3.4 Hybrid Ion Trap Time-of-Flight 32
      2.3.5 Gas Chromatography—Time-of-Flight and Gas
             Chromatography—Quadrupole Time-of-Flight 32

2.4 Orbitrap Analyzers: Instrument Configurations and Main
Features...33
2.5 Acquisition Modes in High-Resolution Mass Spectrometry......39
   2.5.1 Data-Dependent Acquisition.. 39
   2.5.2 Data-Independent Acquisition.................................... 43
   2.5.3 Postacquisition Approaches.. 44
2.6 Databases and the Internet Resources for High-Resolution
Mass Spectrometry 44
Acknowledgments...50
References..50

CHAPTER 3   Analytical Strategies Used in HRMS 59
Aan Agüera, Ana Belén Martínez-Piernas,
Marina Celia Campos-Mañas
3.1 Introduction...59
3.2 Advantages of High-Resolution Mass Spectrometry
in Pesticide Analysis..60
   3.2.1 Selectivity in High-Resolution Mass Spectrometry:
Accurate Mass and Resolution in Qualitative
Analysis 60
   3.2.2 Improving Selectivity by Tandem Mass Spectrometry
Information 62
   3.2.3 Quantitative Performance... 65
3.3 Data Analysis Workflows in High-Resolution Mass
Spectrometry 68
   3.3.1 Qualitative Screening Method Validation..................... 73
   3.3.2 Nontarget Analysis 75
3.4 Conclusions...78
Acknowledgments...78
References..78
Further Reading...81

CHAPTER 4   Current Legislation on Pesticides 83
Helen Botitsi, Despina Tsipi, Anastasios Economou
4.1 Introduction...83
4.2 Pesticides..83
   4.2.1 Identity and Physicochemical Properties..................... 83
   4.2.2 Pesticides Classification.. 84
   4.2.3 Pesticide Metabolites and Transformation Products....... 85
4.3 Legislation 86
   4.3.1 Pesticides Authorization 86

4.3.2 Maximum Residue Limits..88

4.3.3 Monitoring Programs...90

4.4 Analytical Quality Control—Method Validation 92

4.4.1 Guidelines for Pesticide Residue Analysis 98

4.5 Mass Spectrometry in Pesticide Residue Analysis...............110

4.5.1 Mass Spectrometry Identification and Confirmation......110

4.5.2 Potential of High-Resolution Mass Spectrometry
in Pesticide Residue Analysis...................................115

References..119

CHAPTER 5 Advanced Sample Preparation Techniques for Pesticide Residues Determination by HRMS Analysis 131

Renato Zanella, Osmar D. Prestes, Gabrieli Bernardi,
Martha B. Adaime

5.1 Introduction...131

5.2 Matrix Effects and the Influence of Coextracted
Components 133

5.3 Sample Preparation Techniques for Pesticide Residue
Determination by Chromatographic Techniques Coupled
to High-Resolution Mass Spectrometry 135

5.3.1 Dilute-and-Shoot...137

5.3.2 QuEChERS Method..138

5.3.3 Matrix Solid-Phase Extraction 141

5.3.4 Solid-Phase Extraction 142

5.3.5 Solid-Phase Microextraction and Stir Bar Sorptive
Extraction 147

5.3.6 Microwave-Assisted Extraction..............................149

5.3.7 Pressurized Liquid Extraction................................150

5.4 Perspectives and Conclusions 151

References..154

CHAPTER 6 Applications of Liquid Chromatography Coupled With High-Resolution Mass Spectrometry for Pesticide Residue Analysis in Fruit and Vegetable Matrices........................... 165

P. Sivaperumal

6.1 Introduction...165

6.2 Applications of Pesticide Residue Analysis in Fruit
and Vegetable Samples by LC-HRMS..............................168

6.3 Optimized Sample Preparation and Chromatographic
         Conditions for Mass Analyzers...................................186
6.4 Analytical Method Validation..187
    6.4.1 Matrix Effect..187
    6.4.2 Evaluation of the Matrix Interferences by
          UHPLC-HR/MS...189
    6.4.3 Limit of Detection, Limit of Quantitation,
          Accuracy, and Precision 192
6.5 Accurate Measurement of Pesticide Residues in Fruit
         and Vegetable Samples..192
    6.5.1 Determination of Chlorine Isotope in Food
          Samples...192
    6.5.2 Determination of Carbon, Chlorine, and
          Bromine Isotope in Food Samples...........................195
6.6 Evaluation of Pesticide Residues in Fruit and Vegetable
         Samples..195
6.7 Conclusion..197
Acknowledgments...198
References..198

**CHAPTER 7   Application of HRMS in Pesticide Residue
               Analysis in Food From Animal Origin.................... 203**
               Roberto Romero-González, Antonia Garrido Frenich
7.1 Introduction..203
7.2 Instrumental Requirements.....................................204
7.3 Analytical Procedures: Extraction and Chromatographic
         Conditions...210
7.4 Quantitative and Qualitative Applications.....................215
7.5 Differences Between Low-Resolution Mass Spectrometry
         and High-Resolution Mass Spectrometry Analytical
         Methods..226
7.6 Overview and Future Perspectives..............................228
References..229

**CHAPTER 8   Recent Advances in HRMS Analysis of Pesticide
               Residues Using Atmospheric Pressure Gas
               Chromatography and Ion Mobility 233**
               Lauren Mullin, Gareth Cleland, Jennifer A. Burgess
8.1 Introduction..234
8.2 Atmospheric Pressure Gas Chromatography.......................236
    8.2.1 Introduction..236
    8.2.2 Background..236

8.2.3 Current Design 237

8.2.4 Ionization Mechanisms...238

8.2.5 Ionization Trends for Pesticides...............................239

8.2.6 Pesticide Screening Using Atmospheric Pressure Gas Chromatography With High-Resolution Mass Spectrometry 239

8.2.7 Improved Selectivity and Sensitivity With Atmospheric Pressure Gas Chromatography...............................245

8.2.8 Carrier Gas Flow Rate Increase 246

8.3 Time-of-Flight Mass Spectrometry 247

8.4 Ion Mobility Separation..249

8.4.1 Background and Theory 249

8.4.2 Application of Traveling Wave Ion Mobility Spectrometry to Pesticide Screening.........................251

8.4.3 Spectral Selectivity Enhancement..........................253

8.4.4 Protomer Visualization...255

8.5 Summary and Conclusion 258

Acknowledgments...261

References...261

CHAPTER 9 Direct Analysis of Pesticides by Stand-Alone Mass Spectrometry: Flow Injection and Ambient Ionization 265

E. Moyano, M.T. Galceran

9.1 Introduction..265

9.2 Flow Injection Analysis..268

9.2.1 FIA Method Development 273

9.2.2 Sample Preparation for FIA-MS-Based Methodologies...277

9.2.3 FIA-MS Methods Performance 279

9.2.4 FIA-MS Applications..282

9.3 Ambient Mass Spectrometry 283

9.3.1 Electrospray-Based Techniques..............................284

9.3.2 Plasma-Based Techniques 290

9.3.3 Ambient Mass Spectrometry Method Development......292

9.3.4 Sample Handling for Ambient Mass Spectrometry-Based Methodologies..298

9.3.5 Ambient Mass Spectrometry Method Performance.......302

9.3.6 Ambient Mass Spectrometry Applications 305

9.4 Final Remarks and Future Trends....................................307

Acknowledgments...309

References...309

CHAPTER 10 Identification of Pesticide Transformation Products in Food Applying High-Resolution Mass Spectrometry **315**
Imma Ferrer, Jerry A. Zweigenbaum, E. Michael Thurman

10.1 Introduction 315
10.2 Experimental 316
    10.2.1 Chemicals and Reagents 316
    10.2.2 Greenhouse Study 316
    10.2.3 Plant Extraction 316
    10.2.4 Liquid Chromatography/Quadrupole-Time-of-Flight Mass Spectrometry Analysis 317
    10.2.5 Mass Profiler Software 318
10.3 Imidacloprid Metabolites in Plants 318
    10.3.1 Accurate Mass Databases 319
    10.3.2 Mass Profiler Professional 326
    10.3.3 Metabolite Distribution and Mass Balance for Imidacloprid Metabolites 329
10.4 Imazalil Metabolites in Plants and Soil 329
    10.4.1 Chlorine Filter Approach 330
    10.4.2 Soil Metabolites 332
10.5 Propiconazole Metabolites in Plants and Soil 332
    10.5.1 Chlorine Filter Approach 332
10.6 Conclusions 334
References 334

Index 337

List of Contributors

Martha B. Adaime
Federal University of Santa Maria, Santa Maria, Brazil

Ana Agüera
CIESOL, Joint Centre of the University of Almería — CIEMAT, Almería, Spain

Franciso Javier Arrebola-Liébanas
University of Almería, Almería, Spain

Gabrieli Bernardi
Federal University of Santa Maria, Santa Maria, Brazil

Helen Botitsi
General Chemical State Laboratory, Athens, Greece

Jennifer A. Burgess
Waters Corporation, Milford, MA, United States

Marina Celia Campos-Mañas
CIESOL, Joint Centre of the University of Almería — CIEMAT, Almería, Spain

Gareth Cleland
Waters Corporation, Milford, MA, United States

Anastasios Economou
National and Kapodistrian University of Athens, Athens, Greece

Imma Ferrer
University of Colorado, Boulder, CO, United States

Antonia Garrido Frenich
University of Almería, Almería, Spain

M.T. Galceran
University of Barcelona, Barcelona, Spain

Juan F. García-Reyes
University of Jaén, Jaén, Spain

Bienvenida Gilbert-López
University of Jaén, Jaén, Spain

Ana Belén Martínez-Piernas
CIESOL, Joint Centre of the University of Almería — CIEMAT, Almería, Spain

Antonio Molina Díaz
University of Jaén, Jaén, Spain

David Moreno-González
University of Jaén, Jaén, Spain

E. Moyano
University of Barcelona, Barcelona, Spain

Lauren Mullin
Waters Corporation, Milford, MA, United States

Rocío Nortes-Méndez
University of Jaén, Jaén, Spain

P. Sivaperumal
National Institute of Occupational Health, Ahmedabad, Gujarat, India

Osmar D. Prestes
Federal University of Santa Maria, Santa Maria, Brazil

Roberto Romero-González
University of Almería, Almería, Spain

E. Michael Thurman
University of Colorado, Boulder, CO, United States

Despina Tsipi
General Chemical State Laboratory, Athens, Greece

Renato Zanella
Federal University of Santa Maria, Santa Maria, Brazil

Jerry A. Zweigenbaum
Agilent Technologies Inc., Wilmington, DE, United States

Preface

Each problem that I solved became a rule, which served afterward to solve other problems.

Rene Descartes

Currently, mass spectrometry is an essential tool in food safety and its output has reached an unprecedented level, being the most adequate detection system for the analysis of pesticide residues in food matrices. In the last few years, new analyzers, ionization methods, and analytical strategies have been developed to increase the scope of analytical methods as well as the reliability of the results.

Up to now, high-resolution mass spectrometry (HRMS) has been considered as a complementary tool of conventional triple quadrupole (QqQ) analyzers. However, time of flight and/or Orbitrap are replacing low-resolution mass spectrometry analyzers because they can increase the number of compounds simultaneously monitored, retrospective analysis can be carried out, and unknown compounds can be identified. All these tasks can be performed using one benchtop platform at suitable sensitivity. In addition, if HRMS analyzers are coupled with ultrahigh-performance liquid chromatography (UHPLC) the number of compounds analyzed in one single run can increase considerably. Furthermore, when ambient ionization techniques are used with HRMS analyzers, sample treatment could be minimized, increasing sample throughput.

The use of these HRMS analyzers opens a new scenario in pesticide residue analysis, and potential users should know the strategies that can be applied to get all the information that these analyzers could provide, as well as the pros and cons in relation to QqQ.

Because of the fast application and implementation of these new procedures, we consider that setting the main principles and strategies based on these approaches is necessary to provide a comprehensive view of the cited tools, being a cornerstone in the food safety field.

Thus, this book would provide a complete overview of the possibilities that HRMS could offer in pesticide residue analysis in food, as well as the suitable workflow needed to achieve the goals proposed by scientists. Chapters 1–4 provide an overview of HRMS, basic concepts, hardware, general approaches (target and nontarget analysis), and current legislation related to this topic. Chapters 5–7 are "applied chapters" where the main extraction procedures used, as well as the application of HRMS in pesticide residue analysis in several types of matrices, are described. Finally, Chapters 8–10 describe new advances on the use of HRMS such as gas chromatography (GC)–HRMS, ambient ionization techniques, and identification of transformation products. The content of each chapter is described in more detail as follows.

To facilitate the introduction to the topics presented in this book, Chapter 1 describes the principles of HRMS, explaining several concepts such as monoisotopic mass, resolution, mass accuracy, isotopic pattern, etc. Moreover, the differences between low- and high-resolution mass spectrometry are also discussed. In Chapter 2, the several analyzers that could be used in HRMS are described, indicating the differences between the HRMS systems from different vendors, as well as the software that, nowadays, is available to process all the data provided by these analyzers. Finally, online resources such as ChemSpider and MassBank are described. Target, nontarget, and unknown analysis strategies are explained in Chapter 3, describing how a database is prepared and the workflow commonly used for nontarget and unknown analysis. Chapter 4 provides a complete overview of the general requirements indicated by international guidelines (i.e., SANTE) regarding the use of HRMS in pesticide residue analysis, as well as the parameters that should be validated. A section describing pesticides legislation (indicating MRLs) is also included. Bearing in mind that theoretically, unlimited number of compounds could be determined by HRMS, Chapter 5 describes the development of generic extraction methods that allow the simultaneous extraction of a huge number of pesticides that are needed. Chapters 6 and 7 cover the main applications describing the use of HRMS coupled with UHPLC during the analysis of pesticide residue analysis in fruits and vegetables (Chapter 6) and in food from animal origin (Chapter 7). Although most of the current applications focused on pesticide residue applying HRMS use liquid chromatography as the separation technique, in the last few years, GC has also been utilized, because of the development of new ionization sources, such as atmospheric pressure gas chromatography or new couplings such as GC-Orbitrap. Therefore, Chapter 8 is devoted to this topic to highlight the potentiality of GC—HRMS in pesticide residue analysis as well as new approaches such as ion mobility. Because of the potentiality of HRMS, chromatographic step could be removed from the conventional analytical method, and rapid detection of the target compounds could be performed. This approach is interesting for pesticides that cannot be commonly analyzed by multiresidue methods (i.e., very polar pesticides), and the use of HRMS could simplify the analytical strategy. Chapter 9 is focused on this approach as well as in the use of ambient mass spectrometry, which allows for the direct analysis of samples without sample extraction and chromatographic separation, especially when HRMS analyzers are used. Finally, Chapter 10 is dedicated to the advantages that HRMS provides for the identification of pesticide transformation products.

The book is intended for researchers and professionals working with LC—MS, such as food chemists, analytical chemists, toxicologists, food scientists, and everyone who uses/needs this technique to evaluate food safety. Moreover, undergraduate students would also be interested.

Finally, it has been a great pleasure to thank all the authors of this book for their work. All of them are specialists and we really appreciate their effort and time. We also give special thanks to the editorial and production teams of the publisher,

Elsevier, especially Karen R. Miller, who started this adventure, and Jackie Truesdell, for her patience and help, allowing this work to come to fruition.

Roberto Romero-González
Antonia Garrido Frenich
Almería, Spain
November, 2016.

HRMS: Fundamentals and Basic Concepts

Franciso Javier Arrebola-Liébanas, Roberto Romero-González, Antonia Garrido Frenich

University of Almería, Almería, Spain

1.1 INTRODUCTION (TO HIGH-RESOLUTION MASS SPECTROMETRY)

1.1.1 BASIC CONCEPTS (UNITS AND DEFINITIONS)

Mass spectrometry (MS) is an analytical technique commonly used for qualitative and quantitative chemical analysis. MS measures the mass—charge ratio (m/z) of any analyte, of both organic and inorganic nature, which has previously been ionized. Only the ions are registered in MS, but the particles with zero net electric charge (molecules or radicals) are not detected. Therefore, MS does not directly measure mass, but it determines the m/z, being m the relative mass of an ion on the unified atomic scale divided by the charge number, z, of the ion (regardless of sign). The m/z value is a dimensionless number.

Because the mass of atoms and molecules is very small, the kilogram as standard international (SI) base unit cannot be used for its measurement. For that, a non-SI unit of mass, unified atomic mass unit (u) is used. At this point, in this introductory section, it is worth clarifying some basic terms (units and definitions) in MS according to the International Union of Pure and Applied Chemistry (IUPAC) recommendations (IUPAC, 1997; Murray et al., 2013).

The u also called Dalton (Da), is defined as 1/12th of the mass of one atom of ^{12}C at rest in its ground state, being $1\,u = 1\,Da = 1.660538921\,(73) \times 10^{-27}$ kg (number in parentheses indicates the estimated uncertainty). In this way, the mass of other atoms or molecules is expressed relative to the mass of the most abundant stable isotope of carbon, ^{12}C, and this value is dimensionless.

The z is defined as absolute value of charge of an ion divided by the value of the elementary charge of the electron (e) rounded to the nearest integer, being $e = 1.602177 \times 10^{-19}$ C. The m/z unit is the thomson (Th), although it is now a deprecated term, being $1\,Th = 1\,u/e = 1.036426 \times 10^{-8}$ kg/C. For that, use of the dimensionless term m/z is accepted in the literature, and this criterion will be followed throughout this book.

Other basic concepts that are commonly used in MS will be shortly described to clarify the meaning of these throughout the following chapters.

Applications in High Resolution Mass Spectrometry. http://dx.doi.org/10.1016/B978-0-12-809464-8.00001-4

- *Atomic mass*: The number that represents the element's mass based on the weighted average of the masses of its naturally occurring stable isotopes. For example, the integer atomic mass of bromine is 80 Da. This is because there are only two naturally occurring stable isotopes of bromine, ^{79}Br and ^{81}Br, which exist in nature in about equal amounts. When the *relative mass* (Mr) of an ion, molecule, or radical is reported, it is based on the atomic masses of its elements.
- *Nominal mass*: Mass of a molecular ion or molecule calculated using the isotope mass of the most abundant constituent element isotope of each element (Table 1.1) rounded to the nearest integer value and multiplied by the number of atoms of each element. Example: nominal mass of $H_2O = (2 \times 1 + 1 \times 16)$ u = 18 u.
- *Monoisotopic mass*: Exact mass of an ion or molecule calculated using the mass of the most abundant isotope of each element. Example: monoisotopic mass of $H_2O = (2 \times 1.007825 + 1 \times 15.994915)$ u = 18.010565 u. The exact mass of the common elements and their isotopes are provided in Table 1.1.
- *Exact mass*: Calculated mass of an ion or molecule with specified isotopic composition.
- *Mass defect*: Difference between the nominal mass and the monoisotopic mass of an atom, molecule, or ion. It can be a positive or negative value.
- *Relative isotopic mass defect (RΔm)*: It is the mass defect between the mono-isotopic mass of an element and the mass of its A+1 or its A+2 isotopic cluster (Thurman & Ferrer, 2010). For instance, RΔm for the pair ^{35}Cl:^{37}Cl is 0.0030 Da.
- *Average mass*: Mass of an ion or molecule weighted for its isotopic composition, i.e., the average of the isotopic masses of each element, weighted for isotopic abundance (Table 1.1). Example: average mass of $H_2O = (2 \times 1.00794 + 1 \times 15.9994)$ u = 18.01528 u.
- *Accurate mass*: Experimentally determined mass of an ion of known charge.
- *Mass accuracy*: Difference between the mass measured by the mass analyzer and theoretical value.
- *Resolution or mass resolving power*: Measure of the ability of a mass analyzer to distinguish two signals of slightly different *m/z* ratios.
- *Mass calibration*: Means of determining *m/z* values of ions from experimentally detected signals using a theoretical or empirical relational equation. In general, this is accomplished using a computer-based data system and a calibration file obtained from a mass spectrum of a compound that produces ions of known *m/z* values.
- *Mass limit*: Value of *m/z* above or below which *ions* cannot be detected in a *mass spectrometer*.
- *Mass number*: The sum of the protons and neutrons in an atom, molecule, or ion. If the mass is expressed in u, mass number is similar to nominal mass.
- *Most abundant ion mass:* The mass that corresponds to the most abundant peak in the isotopic cluster of the ion of a given empirical formula.

Table 1.1 Nominal, Isotopic, and Average Masses of Some Common Stable Isotopes

Element	Isotope	Abundance	Nominal Mass	Isotopic Mass	Average Mass
H	^1H	99.9885	1	1.007825	1.00794
	^2H	0.0115	2	2.014102	
C	^{12}C	98.93	12	12.000000	12.0110
	^{13}C	1.08	13	13.003355	
N	^{14}N	99.632	14	14.003074	14.00674
	^{15}N	0.368	15	15.000109	
O	^{16}O	99.757	16	15.994915	15.9994
	^{17}O	0.038	17	16.999131	
	^{18}O	0.205	18	17.999160	
F	^{19}F	100	19	18.998403	18.9984
Na	^{23}Na	100	23	22.989770	22.9898
Si	^{28}Si	92.2297	28	27.976927	28.0855
	^{29}Si	4.6832	29	28.976495	
	^{30}Si	3.0872	30	29.973770	
P	^{31}P	100	31	30.973762	30.9738
S	^{32}S	94.93	32	31.972072	32.0660
	^{33}S	0.76	33	32.971459	
	^{34}S	4.29	34	33.967868	
Cl	^{35}Cl	75.78	35	34.968853	35.4527
	^{37}Cl	24.22	37	36.965903	
Br	^{79}Br	50.69	79	78.918336	79.9094
	^{81}Br	49.32	81	80.916289	
I	^{127}I	100	127	126.904476	126.9045

1.1.2 LOW-RESOLUTION MASS SPECTROMETRY VERSUS HIGH-RESOLUTION MASS SPECTROMETRY

It should be noted that mass measurements in MS can be carried out at either low resolution (LRMS) or high resolution (HRMS). An LRMS measurement provides information about the nominal mass of the analyte (Dass, 2007), i.e., the *m/z* for each ion is measured to single-digit mass units (integer mass). However, exact mass is measured by HRMS, i.e., the *m/z* for each ion is measured to four to six decimal points (Ekman, Silberring, Westman-Brinkmalm, & Kraj, 2009). This is very useful to structure elucidation of unknown compounds for analytes having the same nominal mass, but with very small differences in their exact masses. As a result, by LRMS measurements it is not possible to differentiate between imazalil, $C_{14}H_{14}Cl_2N_2O$ ($14 \times 12 + 14 \times 1 + 2 \times 35 + 2 \times 14 + 1 \times 16 = 296$ u), and flunixin, $C_{14}H_{11}F_3N_2O_2$ ($14 \times 12 + 11 \times 1 + 3 \times 19 + 2 \times 14 + 2 \times 16 = 296$ u), pesticides. However, this would be possible by using exact mass measurements, imazalil $C_{14}H_{14}Cl_2N_2O$ ($14 \times 12 + 14 \times 1.007825 + 2 \times 34.968852 + 2 \times 14.003074 + 1 \times 15.994915 = 296.048317$ u) and flunixin $C_{14}H_{11}F_3N_2O_2$ ($14 \times 12 + 11 \times 1.007825 + 3 \times 18.998403 + 2 \times 14.003074 + 2 \times 15.994915 = 296.077262$ u).

High-resolution mass spectrometers have evolved from the 1960s with the introduction of double-focusing magnetic-sector mass instruments (Picó, 2015). Next, Fourier transform ion cyclotron resonance (FT-ICR), time-of-flight (TOF), and Orbitrap mass analyzers were also introduced in the market. Also, hybrid HRMS instruments, such as quadrupole TOF (Q-TOF), ion trap (IT)-TOF, linear trap quadrupole (LTQ)-Orbitrap, or Q–Orbitrap, have been developed. These last analyzers provide tandem (MS/MS) or MS^n spectra of high resolution, in addition to accurate monoisotopic mass measurements, of great applicability both for the confirmation of target compounds and the identification of unknown compounds (Lin et al., 2015). The TOF and Orbitrap analyzers, single or hybrid instruments, are the most widely used in the analysis of organic contaminants, such as pesticide residues (Lin et al., 2015; Picó, 2015).

Among the main characteristics that define the performance of a mass analyzer are (Dass, 2007; de Hoffmann & Stroobant, 2007; Mcluckey & Wells, 2001) mass range, speed, efficiency, linear dynamic range, sensitivity, resolution (or its mass resolving power), and mass accuracy. The mass range is that over which a mass spectrometer can detect ions or is operated to record a mass spectrum. When a range of m/z is indicated instead of a mass range, this should be specified explicitly. The speed or scan speed is the rate at which the analyzer measures over a particular mass range. Efficiency is defined as the product of the transmission of the analyzer by its duty cycle, where the transmission is the ratio of the number of ions reaching the detector and the number of ions entering the mass analyzer, and the duty cycle can be described as the fraction of the ions of interest formed in the ionization step that are subjected to mass analysis.

Linear dynamic range is considered as the range over which ion signal is linear with analyte concentration. Sensitivity can be expressed as detection sensitivity or abundance sensitivity; the first is the smallest amount of an analyte that can be detected at a certain defined confidence level, while the second is the inverse of the ratio obtained by dividing the signal level corresponding to a large peak by the signal level of the background at one mass-to-charge unit lower or higher. A summary of these characteristics of high-resolution mass analyzers is shown in Table 1.2. As it can be observed, in terms of resolving power and accuracy, the FT-ICR analyzer presents the best values, followed by the recently introduced tribrid Orbitrap analyzer. TOF and Q-TOF analyzers have worse values, although the FT-ICR analyzer comprises the worst sensitivity.

Last but not least, two key characteristics of high-resolution mass analyzers are resolution (or its mass resolving power) and mass accuracy, which will be treated in more detail in the following two sections.

Table 1.2 Comparison of the Characteristics of Some High-Resolution Mass Spectrometry Analyzers

Analyzer	Mass Range	Speed	Linear Dynamic Range	Sensitivity	Resolving Power (FWHM)	Accuracy (ppm)
Magnetic sector	10,000	~1 s	10^9	10^6–10^9	100,000	<1
FT-ICR	10,000	~1 s	10^3–10^4	10^3–10^4	1,000,000	<1
TOF	>300,000	Milliseconds	10^6	10^6	30,000	3–5
Q-TOF	10,000	~Milliseconds	10^3–10^4	10^6	30,000	3–5
IT-TOF		~0.1 s	10^3–10^4	10^5	100,000	3–5
Exactive Orbitrap	4000	0.1 s	>5000	10^6	100,000	<3
LTQ-Orbitrap	4000	0.1 s	>5000	10^6	100,000/240,000	<3
Q-Orbitrap	8000	0.05 s	10^5	10^6	240,000	<2
Tribrid Orbitrap	6000	0.05 s	10^5	10^6–10^7	500,000	<1

FT-ICR, Fourier transform ion cyclotron resonance; FWHM, full width at half maximum; IT-TOF, ion trap TOF; LTQ, linear trap quadrupole; Q-TOF, quadrupole TOF; TOF, time-of-flight. Adapted from Picó, Y. (2015). Advanced mass spectrometry. In Y. Picó (Ed.), Comprehensive analytical chemistry, Vol. 68. Amsterdam: Elsevier.

1.2 RESOLUTION AND MASS RESOLVING POWER

Resolution or resolving power is the capacity of a mass analyzer to yield distinct signals for two ions with a small m/z difference (de Hoffmann & Stroobant, 2007). Unfortunately, there is confusion about these two concepts and also between mass resolving power and resolving power in MS, because the definitions provided by different documents are not exactly the same.

Dass (2007) defines the mass resolution of a mass spectrometer as its ability to distinguish between two neighboring ions that differ only slightly in their mass (Δm). According to this definition, it is the inverse value of the resolving power, $RP = m/\Delta m$, where m is the average of the accurate masses, $(m_1 + m_2)/2$, of the two neighboring ions. Xian, Hendrickson, and Marshall (2012) define the resolution as the smallest mass difference, $m_2 - m_1$ or Δm, between two mass spectral peaks such that the valley between their sum is a specified fraction (e.g., 50%) of the height of the smaller individual peak. A similar definition is given by Marshall, Hendrickson, and Shi (2002), as the minimum mass difference between two equal magnitude peaks such that the valley between them is a specified fraction of the peak height.

The IUPAC recommendations (Murray et al., 2013) define resolution *as $m/\Delta m$*, where m is the m/z of the ion of interest. Although depending on the method of measurement of $\Delta(m/z)$, it is possible to differentiate between the two concepts (Murray et al., 2013; Price, 1991). On one hand, resolution, as 10% valley, is the $(m/z)/\Delta(m/z)$ value measured for two peaks of equal height in a mass spectrum at m/z and $m/z + \Delta(m/z)$ that are separated by a valley for which the lowest point is 10% of the height of either peak, i.e., the peaks are resolved when the valley between the two m/z values is 10% of the height of either one (Fig. 1.1). For peaks of similar height separated by a valley, let the height of the valley at its lowest point be 10% of the lower peak, and the resolution should be given for a number of values of m/z. This 10% valley definition for the resolution is used with magnetic-sector analyzers (Ekman et al., 2009).

On the other hand, resolution, as peak width, expresses the $(m/z)/\Delta(m/z)$ value for a single peak, where $\Delta(m/z)$ is the width of the peak at a height, which is a specified fraction (50, 5, or 0.5%) of its maximum peak height (Fig. 1.1). The used fraction is often 50%, and $\Delta(m/z)$ is named as full width at half maximum (FWHM). FT-ICR, TOF, and Orbitrap analyzers use this 50% valley definition for set resolution (Ekman et al., 2009).

In addition, there is controversy in the definition of mass resolving power and resolving power in MS (IUPAC, 1997). The definition of the first term is similar to the definition of resolution indicated earlier (Murray et al., 2013), i.e., as a dimensionless ratio between $m/\Delta m$. Resolving power in MS is the ability of an instrument or measurement procedure to distinguish between two peaks differing in the quotient m/z by a small increment and expressed as the peak width in mass units. However, both terms have been unified in the current IUPAC definition as a measure of the ability of a mass spectrometer to provide a specified value of mass resolution.

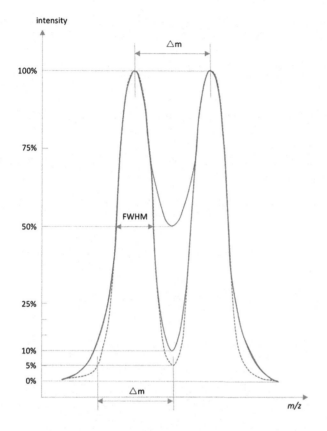

FIGURE 1.1

Methods of calculating mass resolving power.

Reprinted from Picó, Y. (2015). Advanced mass spectrometry. In Y. Picó (Ed.), Comprehensive analytical chemistry, *Vol. 68. Amsterdam: Elsevier, with permission from Elsevier.*

1.3 ACCURATE MASS MEASUREMENT: EXACT MASS AND MASS DEFECT

It is important to differentiate between accurate mass and exact mass. The first is the experimentally determined mass of an ion of known charge (Bristow and Webb, 2003; Sparkman, 2006) and it refers to a measured mass, while the second is the calculated mass of an ion or molecule with specified isotopic composition (Kim, Rodgers, & Marshall, 2006), and it refers to a calculated mass. Therefore, although an LR mass spectrometer can measure integer relative mass with high accuracy, the information obtained is not so complete as the measurement of accurate relative mass offered by HR mass spectrometers (Herbert & Johstone, 2003). The difference between the nominal mass and the monoisotopic mass of an atom, molecule, or ion, positive or negative value, is the mass defect.

The value of accurate mass measurement is illustrated in the following examples: (1) to distinguish compounds with the same integer nominal (molecular) mass in the same sample; (2) to determine the molecular formula or elemental composition for an unknown compound, which is helpful for its identification; and (3) to find out the fragmentation routes. As an example, Fig. 1.2 shows the fragmentation pattern for the flonicamid pesticide, which can be elucidated by HRMS.

In HRMS, the mass accuracy is the difference between the m/z value measured by the mass spectrometer and the theoretical m/z value. It can be reported as an absolute value; for instance, in millimass units (mmu) or millidalton (mDa):

$$\text{Mass accuracy(mmu)} = (m/z_{\text{measured}} - m/z_{\text{theoretical}}) \times 10^3$$

Also, it can be expressed as a relative value in parts per million (ppm):

$$\text{Mass accuracy(ppm)} = \left(\frac{m/z_{\text{measured}} - m/z_{\text{theoretical}}}{(m/z_{\text{theoretical}})} \right) 10^6$$

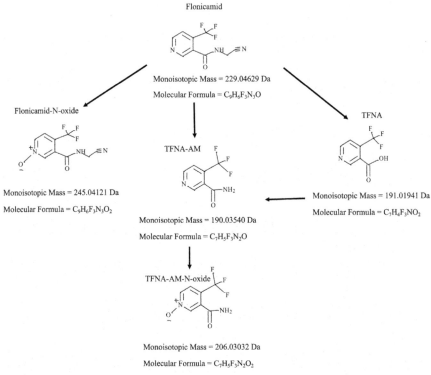

FIGURE 1.2

Fragmentation pattern for the flonicamid pesticide (TFNA: 4-trifluoromethylnicotinic acid; TFNA-AM: 4-trifluoromethilnicotinamide).

In general, an acceptable value of the measured mass should be within 5 ppm of the accurate mass (Gross, 1994). A key point to minimize error in accurate mass measurement is ensuring that the target ion is completely free of interfering ions, because these ions shift the mass of the target peak.

In general, high mass resolution and high mass accuracy depend on each other, because the latter tends to improve as the former is improved. HR allows to separate neighboring ions, and accurate mass can deliver molecular formulas (Gross, 2011). Therefore, it is important to note that HR alone does not equally imply measuring the accurate mass.

1.4 MASS CALIBRATION IN HIGH-RESOLUTION MASS SPECTROMETRY

Mass calibration is a relevant process in every mass spectrometer for a proper representation of ions in the m/z axis. It also results in a very important fact in HRMS where not only high mass resolution but also high mass accuracy is critical (Gross, 2011). For that, typically, mass reference compounds with a compilation of well-known m/z values are needed (Busch, 2004, 2005).

Calibration is frequently performed in an automatic or semiautomatic way by the mass spectrometer software when the list of ions of those mass calibration compounds are correlated with experimentally obtained m/z values. It is called external mass calibration if the mass calibration is stored in a calibration file for further measurements and the mass calibration standard is not used during acquisition of experimental mass spectra. Frequency of recalibration has influence on mass accuracy of the mass analyzer. The selection of the mass calibration compound depends on the ionization method and, of course, the mass analyzer used. However, (1) they should yield sufficient regularly spaced abundant ions across the entire scan range; (2) the reference ions should have negative mass defects to prevent overlap with typical compounds containing C, H, N, and O; and (3) they should be readily available, chemically inert, and sufficiently volatile. Some of the most common calibration standards and their masses and relative abundances can be found in literature (Dass, 2007). For example, perfluorokerosene (PFK) is often established as a mass calibration standard in electron ionization (EI). PFK provides numerous fragment ions that may be used up to m/z 700–1100 depending on the type of mixture used (commercially available from low to high boiling grades). Also, perflourotributylamine (FC-43) is also proposed as mass calibration standard thanks to its characteristics ions up to 614 in an EI spectrum (Sack, Lapp, Gross, & Kimble, 1984). When a high-mass calibration (i.e., up to 3000 u) is required, triazines and a mixture of fluorinated phosphazenes called Ultramark can be used as reference calibrants. For electrospray instruments, the most typical calibration standards are CsI, poly(-ethylene glycol) (PEG), poly(ethylene glycol) bis(carboxymethyl ether), poly(ethylene glycol monomethyl ether), and poly(propylene glycol). MALDI users also have several reference compounds available, such as α-CHCA matrix (dimer + H$^+$),

4-hydroxy-3-methoxycinnamic acid (trimer + Na$^+$), angiotensin I and II, bradyki-nin, substance P, desArg1-bradykinin, gramicidin, and autodigestion products of trypsin.

As an alternative, internal mass calibration can be performed. For that, the mass calibration standard is introduced using a second inlet system into the ion source, for instance, as a volatile standard. As an alternative, it can be mixed with the analyte before analysis. This last option presents more limitations than the use of alternative inlet systems.

Typical mass accuracy obtained by internal mass calibration used to be better than those obtained by external calibration. Some examples are 0.1−0.5 ppm with FT-ICR, 0.5−1 ppm with Orbitrap, 0.5−5 ppm with magnetic sector, or 1−10 ppm with TOF analyzers.

Fast atom bombardment instruments are sometimes internally calibrated with good mass accuracy by using the matrix peaks for a mass calibration but it is preferred a mixture of the standard with the analyte (matrix) if unwanted reactions are not observed and proper solubility of the analyte and standard. One typical mass calibration standard used is PEG with an average molecular weight of 600 u (PEG 600). In this sense, the reproducibility of mass calibration after several scan cycles is improved because of affection of magnets by hysteresis.

MALDI mass calibration can be compromised if thick sample layers are used with on-axis TOF instruments. However, orthogonal acceleration TOF analyzers present better results. In some cases, for example, in the analysis of synthetic poly-mers, the formation of evenly spaced oligomer ions can be used as internal mass calibration (Dienes et al., 1996). In the case of TOF analyzers, the conversion from a measured flight time to mass requires a mass calibration. The computer makes the calculations using proper algorithms once the values of flight time for a few calibrant masses are known. Sometimes, a second calibration step is requested to achieve enough accuracy (Ferrer & Thurman, 2009). It must be carefully controlled changes in flight distances or accelerating potentials to obtain mass accu-racies of 1 ppm, but again, the internal mass calibration can correct such instrument factors with an automatic data processing carried out at the same time as that of the analysis of the sample. Generally, a mass calibration per day or week is enough to obtain a proper accuracy of m/z for many TOF instruments, but it is adequate to use internal mass calibration, especially when long analyses are performed (Chernush-evich, Loboda, & Thomson, 2001).

FT-ICR mass analyzers with superconducting magnets are frequently very stable for many days of use in normal applications. In this case, a mass accuracy better than 1 ppm can be achieved in a wide mass range (Rodgers, Blumer, Hendrickson, & Marshall, 2000).

Some HR mass spectrometers such as double focusing systems (DFS) can be operated by multiple ion detection (MID) mode where the intensities of some ions typical of a target analyte can be continuously monitored to increase sensitivity, precision, and selectivity of the method. A monitoring window is selected if the instrument is coupled to a chromatographic inlet device (gas or liquid chromato-graph). For data acquisition, the magnet gets blocked in one mass and electric scans are carried out modifying the acceleration voltage. Each scan suffers a rectification of the mass calibration at the same time that experimental data are obtained. This technique is called lock-mass technique and improves mass accuracy increasing reli-ability of mass spectrometric data obtained. Resolution of the instrument can also be recalculated on each scan. It can also be improved doubling the calibration masses in a technique called Lock-plus-cali mass technique. This scan-to-scan mass calibra-tion processed in the background improves confidence of the analytical data. For this internal mass calibration, the calibration standard is leaked continuously from the reference inlet system into the ion source. Two ion masses are selected from the reference substance: one mass that is below the analyte target masses and another one above the analyte target masses. The lowest mass is called "lock mass" and the highest is named "calibration mass." The magnet is locking the magnet at the start of each MID process and performing a mass calibration based on the lock mass. All analyzer jumps to the calibration and target masses by fast electrical jumps of the acceleration voltage. It provides a fine calibration in only a very few milliseconds (Thermo Fisher Scientific Inc., 2007). Fig. 1.3 shows the typical sequence of steps during a MID process in DFS instrument.

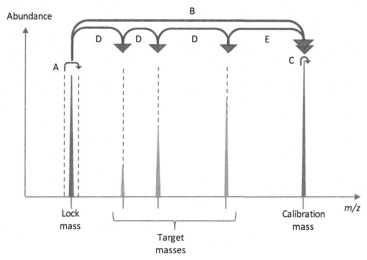

FIGURE 1.3

Internal mass calibration and target mass detection during a multiple ion detection (MID) process in a double focusing system (DFS) instrument: A, magnet locking and lock mass sweep. Mass calibration and resolution determination; B, electrical jump to calibration mass; C, calibration mass sweep and mass calibration; D, electrical jumps to target masses; E, electrical jump to calibration mass for mass calibration.

1.5 GENERAL CONSIDERATIONS

Empirical determination of a molecular formula for a substance can be very useful in organic MS. HRMS can be an alternative to traditional chemical methods based on the tedious, slow, and often inaccurate process of breaking down of a known weight of a molecule into its constituent elements and weight of them (Herbert & Johstone, 2003). It should be mentioned that the mass of an electron is very small compared with masses of any element, and therefore, frequently, the mass of M^+ is considered the same as that of M. Therefore, an HR mass spectrometer can be used to measure relative atomic, molecular, or fragment ion masses with high accuracy. An acceptable value of the measured mass should be within 5 ppm of the accurate mass (Gross, 1994).

It is essential that the ion of interest is properly resolved from all other neighboring ions because any interfering ion would introduce an error in mass measurement. Therefore, a high resolving power is very important for accurate mass measurements. For small molecules, a resolving power higher than, i.e., 10,000 is not critical. For example, a resolving power of 770 would be enough for mass-resolving $C_2H_4^{+\bullet}$ (28.031300) and $CO^{+\bullet}$ (27.994915). Nevertheless, a resolving power of at least 5500 is needed to separate $C_{13}H_{16}-C_2H_4^{+\bullet}$ and $C_{13}H_{16}-CO^{+\bullet}$ ions, even if the difference between both masses is 0.036385 u. Another difficulty is that the possible elemental composition becomes quite larger as the mass increases.

Full scan can be one of the simplest acquisition modes to obtain an accurate mass value when an internal mass calibration is carried out. Accurate masses of all ions can be determined in one single chromatographic run, and elemental composition can be achieved with an adequate accuracy of less than 5 ppm. Peak-matching mode is a more accurate mass measurement technique with a typical accuracy lower than 0.3 ppm. For that, only one ion is determined at a time, for example, using the MID technique described earlier. It is recommended that the mass of the mass calibration ion must be within 2% of the unknown mass to give the highest accuracy. A combination of slow scanning of the accelerating voltage and computer programs that improve signal averaging, smoothing, and peak centroiding improves mass accuracy (Hammar, Pettersson, & Carpenter, 1974).

A nominal mass may result from several combinations of elements, but only one composition can match an accurate mass. Nevertheless, more possible combinations can occur with an increased number of atoms in a molecule. A list of potential molecular formulas for various masses has been proposed (Beynon & Williams, 1963). Diverse algorithms and computer programs are available for elemental composition; many of them are available online.

Finally, it should be mentioned that $R\Delta m$ is also a valuable tool for confirmation purposes. The sum of $R\Delta m$ of all isotopes of the molecule is defined as the isotopic mass average (IMA), and Padilla-Sánchez et al. (2012) used this approach for the reliable identification of small pesticides such as ethephon and N-acetyl-glufosinate, obtaining an experimental $IMA_{(A+2)}$ value of −0.00286 Da for ethephon, which was within the theoretical interval of −0.0029 ± 0.0001 Da.

ACKNOWLEDGMENTS

The authors gratefully acknowledge Andalusian Regional Government (Regional Ministry of Innovation, Science, and Enterprise) and FEDER for financial support (Project Ref. P-12-FQM 1838).

REFERENCES

Beynon, J. H., & Williams, A. E. (1963). *Mass and abundance tables for use in mass spectrometry*. Amsterdam: Elsevier.

Bristow, A. W. T., & Webb, K. S. (2003). Intercomparison study on accurate mass measurement of small molecules in mass spectrometry. *Journal of the American Society for Mass Spectrometry, 14*(10), 1086–1098.

Busch, K. L. (2004). Masses in mass spectrometry: Balancing the analytical scales. *Spectroscopy, 19*(11), 32–34.

Busch, K. L. (2005). Masses in mass spectrometry: Perfluors and more. Part II. *Spectroscopy, 20*(2), 76–81.

Chernushevich, I. V., Loboda, A. V., & Thomson, B. A. (2001). An introduction to quadrupole-time-of-flight mass spectrometry. *Journal of Mass Spectrometry, 36*(8), 849–865.

Dass, C. (2007). *Fundamentals of contemporary mass spectrometry*. Hoboken, New Jersey: John Wiley & Sons, Inc.

Dienes, T., Salvador, J. P., Schürch, S., Scott, J. R., Yao, J., Cui, S., et al. (1996). Fourier transform mass spectrometry—advancing years (1992-mid. 1996). *Mass Spectrometry Reviews, 15*(3), 163–211.

Ekman, R., Silberring, J., Westman-Brinkmalm, A., & Kraj, A. (Eds.). (2009). *Mass spectrometry. Instrumentation, interpretation, and applications*. Hoboken, New Jersey: John Wiley & Sons, Inc.

Ferrer, I., & Thurman, E. M. (2009). *Liquid chromatography time-of-flight mass spectrometry — principles, tools, and applications for accurate mass analysis*. Hoboken: Wiley-Interscience.

Gross, J. H. (2011). *Mass spectrometry. A textbook* (2nd ed.). Berlin Heidelberg: Springer.

Gross, M. L. (1994). Accurate masses for structure confirmation. *Journal of the American Society for Mass Spectrometry, 5*(2), 57.

Hammar, C. G., Pettersson, G., & Carpenter, P. T. (1974). Computerized mass fragmentography and peak matching. *Biomedical Mass Spectrometry*, (1), 397–411.

Herbert, C. G., & Johstone, R. A. W. (2003). *Mass spectrometry basics*. Boca Ratón: CRC Press.

de Hoffmann, E., & Stroobant, V. (2007). *Mass spectrometry. Principles and applications* (3rd ed.). Chichester: John Wiley & Sons Ltd.

IUPAC. *Compendium of chemical Terminology*, 2nd ed. (the "Gold Book"). Compiled by A. D. McNaught and A. Wilkinson. Blackwell Scientific Publications, Oxford (1997). XML on-line corrected version: http://goldbook.iupac.org (2006) created by M. Nic, J. Jirat, B. Kosata; updates compiled by A. Jenkins.

Kim, S., Rodgers, R. P., & Marshall, A. G. (2006). Truly "exact" mass: elemental composition can be determined uniquely from molecular mass measurement at ~0.1 mDa accuracy for molecules up to $\sim$ 500 Da. *International Journal of Mass Spectrometry, 251*(2−3), 260−265.

Lin, L., Lin, H., Zhang, M., Dong, X., Yin, X., Qub, C., et al. (2015). Types, principle, and characteristics of tandem high-resolution mass spectrometry and its applications. *RSC Advances, 5*, 107623−107636. Royal Society of Chemistry.

Marshall, A. G., Hendrickson, C. L., & Shi, S. D. H. (2002). Scaling MS plateaus with high-resolution FT-ICRMS. *Analytical Chemistry, 74*(9), 252A−259A.

Mcluckey, S. A., & Wells, J. M. (2001). Mass analysis at the advent of the 21st century. *Chemical Reviews, 101*(2), 571−606.

Murray, K. K., Boyd, R. K., Eberlin, M. N., Langley, G. J., Li, L., & Naito, Y. (2013). Definitions of terms relating to mass spectrometry (IUPAC Recommendations 2013). *Pure and Applied Chemistry, 85*(7), 1515−1609.

Padilla-Sánchez, J. A., Plaza-Bolaños, P., Romero-González, R., Grande-Martínez, A., Thurman, E. M., & Garrido-Frenich, A. (2012). Innovative determination of polar organophosphonate pesticides based on high-resolution Orbitrap mass spectrometry. *Journal of Mass Spectrometry, 47*, 1458−1465.

Picó, Y. (2015). Advanced mass spectrometry. In Y. Picó (Ed.), *Comprehensive analytical chemistry* (Vol. 68). Amsterdam: Elsevier.

Price, P. (1991). Standard definitions of terms relating to mass spectrometry. A report from the committee on measurements and standards of the American Society for Mass Spectrometry. *Journal of the American Society for Mass Spectrometry, 2*(4), 336−348.

Rodgers, R. P., Blumer, E. N., Hendrickson, C. L., & Marshall, A. G. (2000). Stable isotope incorporation triples the upper mass limit for determination of elemental composition by accurate mass measurement. *Journal of the American Society for Mass Spectrometry, 11*(10), 835−840.

Sack, T. M., Lapp, R. L., Gross, M. L., & Kimble, B. J. (1984). A method for the statistical evaluation of accurate mass measurement quality. *International Journal of Mass Spectrometry and Ion Processes, 61*(2), 191−213.

Sparkman, O. D. (2006). *Mass spec desk reference* (2nd ed.). Pittsburgh: Global View Publishing.

Thermo Fisher Scientific Inc.. (2007). [Online], Available from http://tools.thermofisher.com/content/sfs/brochures/high-resolution-mid-data-acquisition-for-target-compound-analysis-with-the-dfs-gcms-system.pdf.

Thurman, E. M., & Ferrer, I. (2010). The isotopic mass defect: A tool for limiting molecular formulas by accurate mass. *Analytical and Bioanalytical Chemistry, 397*, 2807−2816.

Xian, F., Hendrickson, C. L., & Marshall, A. G. (2012). High resolution mass spectrometry. *Analytical Chemistry, 84*, 708−719.

HRMS: Hardware and Software

2

Juan F. García-Reyes, David Moreno-González, Rocío Nortes-Méndez,
Bienvenida Gilbert-López, Antonio Molina Díaz

University of Jaén, Jaén, Spain

2.1 INTRODUCTION

Undoubtedly, the establishment of electrospray ionization (ESI) in the late 1980s (Fenn, Mann, Meng, Wong, & Whitehouse, 1989, p. 64; Yamashita & Fenn, 1984, p. 4451) and its enormous potential for biomedical applications prompted the development and growth of liquid chromatography—mass spectrometry (LC—MS) instrumentation. Several manufacturers launched LC—MS instrumentation in the mid-1990s with either ion trap or triple quadrupole (QqQ) analyzers, which established itself as the gold standard for quantitative trace analysis. Meanwhile, the use of high-resolution LC—MS instrumentation with time-of-flight (TOF) analyzers was commercially available from 1996 with the development of first hybrid quadrupole time-of-flight (Q-TOF) instruments. Micromass commercialized the first Q-TOF mass spectrometer in 1996 (Q-TOF-1), which was furnished with ESI source, quadrupole mass filter, hexapole collision cell, orthogonal acceleration, and microchannel plate detector. It featured a resolving power of 5000. Unfortunately, these instruments were originally applicable only to qualitative purposes such as the determination of molecular mass and elemental composition through accurate mass analysis. There were several issues (linear dynamic range due to detector saturation and scarce robustness) that limited the use of such instrumentation for quantitative purposes.

The last two decades have witnessed an unprecedented growth of mass spectrometric instruments of increasing high-mass resolving power. Such ability enabling accurate mass measurements permits the determination of elemental compositions for small molecules up to 500—1000 Da in mixtures containing thousands of chemical components (Marshall & Hendrickson, 2008, p. 579). There has also been an outstanding growth in the use of high-resolution instrumentation for quantitative application. This fact has been supported by relevant improvements of instrument performance in terms of resolution, sensitivity, and speed. Among them, in 2004, several vendors (Agilent, Bruker, Waters) offered updated and upgraded versions of TOF instruments, with improvements on electronics to prevent detector saturation, temperature control of flight tube, and continuous accurate mass calibration systems (dual sprayer, MassLock, etc.), which broadened the range of applications and enable more versatile and user-friendly operation. In 2005, the

Applications in High Resolution Mass Spectrometry. http://dx.doi.org/10.1016/B978-0-12-809464-8.00002-6

orbital ion trapping analyzer (Orbitrap), an alternative concept of high-resolution instrumentation—probably one of the major breakthrough in mass spectrometry in the last decades—was launched by Thermo Fisher Scientific. Its advantaging features, particularly related to resolving power, have fostered lately the development of high-performance TOF analyzers with enhanced performance and solutions to challenge the current Orbitrap lead.

Besides TOF and Orbitrap, high-resolution mass spectrometry (HRMS) can be also carried out using Fourier transform ion cyclotron resonance (FT-ICR) first commercialized in 1983 (Nier, Yergey, & Gale, 2015) and magnetic sector instruments. Provided the current literature and the number of instruments sold and manufacturers offering these technologies, it seems that the interest for these technologies is diminishing. In gas chromatography—high-resolution mass spectrometry (GC—HRMS), for years GC has been coupled to HR magnetic sector instruments, mostly for dioxin analysis. Yet, GC—HRMS is the reference instrumentation for analysis of many persistent organic pollutants (Eljarrat & Barceló, 2002, p. 1105; Hernández et al., 2012, p. 1251; Van Babel & Abad, 2008, p. 3956) particularly dioxins, including polychlorinated dibenzo-p-dioxins and polychlorinated dibenzofurans. However, recently there has been a growing interest in GC—HRMS with TOF and Orbitrap mass analyzers (Mol, Tienstra, & Zomer, 2016, p. 161; Peterson, Balloon, Westphall, & Coon, 2014, p. 10,044; Peterson, Mcalister, Quarmby, Griep-Raming, & Coon, 2010, p. 8618; Van Mourik, Leonards, Gaus, & De Boer, 2015, p. 259) partly associated with the introduction of atmospheric pressure chemical ionization (APCI) sources (Van Babel et al., 2015, p. 9047), which ultimately may replace magnetic sector instruments within the next decade.

In this chapter, an overview of the more commonly used HRMS instrumentation viz. those capable of featuring resolving power above 10,000 (full width at half maximum, FWHM) is provided, including orthogonal acceleration/reflectron TOF mass spectrometers and Fourier transform high-resolution mass spectrometers (Orbitrap and FT-ICR instruments) coupled with either LC or GC. Along with the ability to provide accurate mass measurements, these instruments offer several hardware combinations yielding hybrid instruments [coupled with ion traps and/or quadrupoles and even ion mobility (IM) spectrometry modules] providing interesting possibilities in terms of acquisition mode to face different qualitative and quantitative applications. On the other hand, the combination of accurate mass measurements with available databases (often free on the Web) permits unprecedented capability to scrutinize samples and their content in an automated fashion.

2.2 PRINCIPLES OF HIGH-RESOLUTION MASS SPECTROMETRY ANALYZERS

In this section, the principles of the main HRMS analyzers used nowadays are briefly outlined. A thorough explanation of the physical principles and the involved technical aspects related to TOF, Orbitrap, and FT-ICRMS instruments is provided elsewhere (Fjeldsted, 2016, p. 19; Hu, Noll, Li, Makarov, & Hardman, 2005, p. 430; Perry, Cooks, & Noll, 2008, p. 661).

2.2.1 TIME-OF-FLIGHT

TOF is a relatively mature technology (Fjeldsted, 2016, p. 19). Actually, the concept of TOF mass analysis was first proposed by Stephens in 1946 (Stephens, 1946, p. 691), and subsequently developed by Cameron and Eggers and referred to as a "velocitron" in 1948 (Cameron & Eggers, 1948, p. 605). The basic principle of TOF instrument relies on the velocity-dependent separation and subsequent detection of ions as they travel through a flight tube. Charged ions produced are accelerated into TOF tube and are allowed to drift along the path. Provided that velocity is dependent on mass-to-charge (*m/z*) ratio, the lighter ions will fly faster and reach the detector earlier. There are two main types of TOF analyzers: linear and of orthogonal acceleration. Whereas the use of linear time-TOFs was associated with pulsed ionization methods such as laser beams with matrix-assisted laser desorption/ionization (MALDI) experiments, the latter is the most commonly used configuration nowadays. Fig. 2.1 shows the scheme of an orthogonal acceleration TOF. It consists of

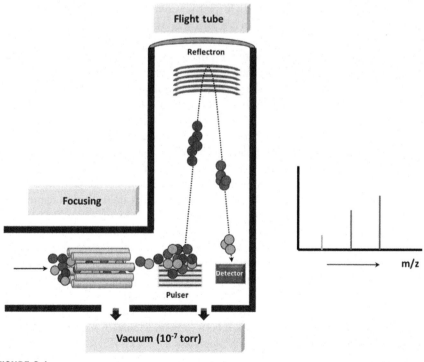

FIGURE 2.1

Scheme of an orthogonal acceleration time-of-flight instrument.

an ionization interface, which is followed by an ion transfer section (ion optics) and focusing region. The continuous ion beam emerging from the ion source is sharpened within the ion optic path and sent toward the pulser. A high-voltage pulse is applied to the pusher plate to orthogonally accelerate pockets of ions into the flight tube.

Wiley and McLaren (1955, p. 1150) realized the importance of space focusing. They found that small differences on the position of the ions created in the ion source and the distances to the detector affected resolving power dramatically. Ions created at the back of the ion source volume had a longer distance to travel to the detector than those that were created closer to the front of the ion source volume. They demonstrated that with the appropriate increase in voltage at the back pusher plate within the ion source it was possible to significantly reduce the loss of resolution because of the origin of the ions within the source. This advance was connected to the early development of the first TOF commercial instrument offered by Bendix in the mid-1950s (Fjeldsted, 2016, p. 19; Wiley, 1956, p. 817).

Therefore, during TOF analysis, it is important that all ions receive an equal acceleration to, as much as possible, reduce spreading in time. The focusing of the ion beam before reaching the pusher and the following orthogonal acceleration become critical to ensure equal starting conditions for all ions regardless of their individual masses and thermal energy. Thus, the initial spatial and ion energy spread among the ions are minimized. Mamyrin (1966) proposed the use of an ion mirror (referred to as a reflectron) in the flight path to compensate for the spread in the initial energy of the ions. Mamyrin showed that when a proper reflecting field was established, ions with greater energy proportionally travel further into the reflectron before being reflected with the effect that the arrival time of the ions at the detector shows a greater tolerance to the effects of their initial energy spread. Therefore, a careful design of the reflectron permits a compensation of speed differences, as caused by the remaining energy difference among the accelerated ions. In addition, the implementation of the reflectron doubled the flight path length to achieve longer flight times and therefore increased resolution.

TOF instruments are of pulsed nature. The use of ion packets is central to TOF analyzers. Initially, the creation of ion packets was generated by employing a pulsed ionization source such as an electron ionization filament or a beam deflector. In the latter approach, a pulsed ion packet was created through deflection by gating, or momentarily steering, the ion beam into the TOF analyzer. In this sense, the concept of orthogonal acceleration was proposed by Dawson and Guilhaus (1989, p. 155). A continuous beam of ions was momentarily pushed in a direction perpendicular to the direction of the continuous ion beam. The packet of orthogonally directed ions enters the TOF tube that is therefore oriented perpendicular to the continuous ion beam rather than in-line. With this configuration, the energy spread is substantially minimized. The introduction of the orthogonal TOF geometry was of paramount importance for adapting TOF to atmospheric pressure ionization sources, such as ESI, and the subsequent growth in TOF applications in LC/MS.

The detector—located at the end of the flight path—records the arrival time and the number of incoming ions. The square of the flight time is proportional to the *m/z* of the detected ion. Hence, masses (*m/z*) can simply be calculated after an instrument calibration process. Because of the fast ion flight times, a single push cannot be used to record a spectrum. Hundreds of consecutive pushes are summed to produce a single, averaged spectrum instead (corresponding to one chromatographic data point). A single pulser push accelerates only the ions passing above the surface of the pusher plate. The next push should not be started before the heaviest ion accelerated by the preceding push has reached the detector. This results in duty cycles clearly below 100%. The very short ion flight times within the flight tube put extreme importance on the frequency of the detection device and the following signal processing. In the past, fast time-to-digital conversion (TDC) detectors were only able to measure the flight time, but could not count the number of the simultaneously incoming ions at a given time. Such devices have been replaced by analog-to-digital conversion (ADC) detectors, which unlike TDC devices can account for the number of ions simultaneously hitting the detector plate, thus enabling the expansion of the dynamic range to up to four decades. Yet this point remains as one of the main weaknesses of TOF detectors; they undergo saturation from a threshold intensity, which not only affects the measured signal amplitude but also can result in a relevant mass shift of the ions being affected by saturation.

2.2.2 FOURIER TRANSFORM ION CYCLOTRON RESONANCE

FT-ICR was first described by Marshall and Comisarow in 1974 (Comisarow & Marshall, 1974, p. 282; Comisarow & Marshall, 1996, p. 581). Yet, it is the instrument providing the highest performance in terms of resolving power. In fact, the resolving power of either a magnetic sector or a TOF depends on the path length during the experiment, 7 m being the path length of the highest resolution magnetic sector and between 1 and 2.5 m in current TOF analyzers (leaving aside multireflec-tron TOF instruments). In contrast, an ion with *m/z* 1000 at a standard 9.4 T magnet exhibits cyclotron frequencies of 144,346 Hz and thus, has an effective path length higher than 9 km in 1 s ($2\pi \times 0.01$ m $\times$ 144,346 Hz = 9070 m).

FT-ICR enables the accurate determination of *m/z* of ions based on the cyclotron frequency of the ions in a fixed magnetic field (Comisarow & Marshall, 1996, p. 581; Marshall & Hendrickson, 2008, p. 579). The ions are trapped in a Penning trap (a magnetic field with electric trapping plates) where they are excited (at their resonant cyclotron frequencies) to a larger cyclotron radius by an oscillating electric field orthogonal to the magnetic field. After the excitation field is removed, the ions are rotating at their cyclotron frequency in phase (as a "packet" of ions). These ions induce a charge (detected as an image current) on a pair of electrodes as the packets of ions pass close to them. The resulting signal is called a free induction decay, transient, or interferogram that consists of a superposition of sine waves. The signal is digitalized and subjected to discrete fast Fourier transformation to yield a spectrum of ion cyclotron frequencies, which may then be converted to a spectrum of *m/z* (Marshall, Hendrickson, & Jackson, 1998, p. 1). The expression that describes the rotation of the ions at a cyclotron frequency [v_c (Hz)] in the spatially uniform magnetic field (*B*) is:

$$v_c = ezB/2\pi m$$

in which *B* is the magnetic field, *m* is the mass, and *e* is the elementary charge. At room temperature, typical ion cyclotron orbital radii are at the submillimeter level. Considering that ions with a certain *m/z* rotate with random phase, it is necessary to resonantly excite the ions with an oscillating or rotating electric field to yield a spatially coherent packet of ions of each given *m/z*. The motion of the different ion packets gives rise to a time-domain signal consisting of the difference in current induced on a pair of opposed electrodes (Fig. 2.2).

Nowadays, only one vendor commercializes a portfolio of FT-ICR instruments (Bruker Daltonics, Bremen, Germany). Up to three different models are offered including a hybrid quadrupole FT-ICR 7-Tesla (T) (solariX XR) and 12-T or 15-T high-field solariX XR with resolving power up to 600,000 at *m/z* 400 with 1 s transient (and up to 10,000,000 resolving power depending on magnet and acquisition time). To enable a flexible range of experiments, these instruments permit a wide range of fragmentation techniques including in-source collision-induced dissociation (CID), CID in collision cell, electron-capture dissociation (ECD), electron-transfer dissociation (ETD), and sustained off-resonance irradiation (SORI) CID in cell. In the past, Thermo also offered a hybrid linear ion trap (LTQ)-FT-ICRMS, but it was withdrawn in 2014, with the emergence of high-field Orbitrap, which nearly matched the performance of FT-ICR systems. The applications foreseen for this type of instrumentation are petroleomics (Cho, Ahmed, Islam, & Kim, 2015, p. 248; Nikolaev, 2015, p. 421; Schwemer, Rüger, Sklorz, & Zimmermann, 2015, p. 11,957) and state-of-the art proteomics research (Marshall & Chen, 2015, p. 410; Nicolardi, Bogdnov, Deelder, Palmblad, & Van

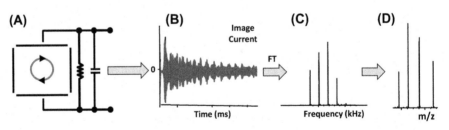

FIGURE 2.2

Summary of Fourier transform mass spectrometry based detection of ions in Fourier transform ion cyclotron resonance mass spectrometry: (A) excited ion cyclotron rotation, (B) differential image-current signal from opposed detection electrodes, (C) frequency-domain signal, and (D) mass spectrum obtained by calibrated frequency-to-*m/z* conversion. A typical acquisition of 1 s compatible with liquid separation permits resolving power above 400,000 (depending on the magnet used). Higher acquisition time permits increased values of resolving power above 1,000,000. For details, see Comisarow and Marshall (1996, p. 581) and Marshall and Hendrickson (2008, p. 579).

Der Burgt, 2015, p. 27,133), being the application of FT-ICR in pesticide testing scarcely addressed as the requirements and maintenance cost of such sophisticated instrumentation are not met by standard routine laboratories (Gilbert-López et al., 2013, p. 419).

2.2.3 ORBITRAP

The Orbitrap mass analyzer is the first high-performance mass analyzer that employs trapping of ions in electrostatic fields, together with a sophisticated ion injection process, which enables high resolution, mass accuracy, and excellent sensitivity for addressing numerous analytical applications in both research and routine analysis. The principle of Orbitrap is based on the Kingdon ion trap described in 1923 (Kingdon, 1923, p. 408). It was a trapping device consisting of a charged wire stretched along the axis of an enclosed metal can. The wire establishes an electrostatic field within the can, and ions that possess sufficiently high tangential velocity orbit the wire, rather than directly colliding with it (Eliuk & Makarov, 2015, p. 61). It was first commercially introduced in 2005 (Eliuk & Makarov, 2015, p. 61). It provides a very convenient balance between the advantages of a benchtop instrument featuring high resolution approaching FT-ICR performance, whereas the instrument size, laboratory space requirements, and regular operational maintenance are more comparable to Q-TOF instruments.

A thorough explanation of the mass analysis process is detailed elsewhere (Eliuk & Makarov, 2015, p. 61). This mass analyzer is based on the confinement of ions in an electrostatic potential well (Makarov, 1999, US 5886346; Makarov, 2000, p. 1156; Makarov & Hardman, 2006, US 6998609 B2) created between two carefully shaped electrodes: an inner coaxial spindled-shaped electrode and an outer (barrel-like) electrode, which is actually composed of two symmetrical halves electrically isolated from each other, set out for two purposes: establishment of the ion trapping fields and as receiver plates for image-current detection. The electrodes are precisely machined so that the electrostatic attractions of the ions to inner electrode are finely balanced by centrifugal forces, which cause the ions to orbit around the spindle. Previously, ions are first accumulated on an external injecting device (C-trap) that traps the ions in gas-filled quadrupole being then injected tangentially into the mass analyzer in short pulses (Fig. 2.3). Besides, an axial field causes the ions to oscillate harmonically along the spindled-shaped electrode. The outer electrodes allow differential image-current detection. These image currents produced by the oscillating ions are detected, followed by a fast Fourier transform (FT) to convert the time-domain signal to frequency domain and then to *m/z* spectrum. Here the resolution is directly proportional to number of harmonic oscillations detected. The resolving power can be enhanced by increasing the gap between inner and outer electrodes providing higher field strength for a given voltage. As the maximum acquisition time is limited in Orbitrap, the resolution power is not as high as FT-ICR. However, commercial Orbitrap analyzer provides a nominal resolution power as high as 500,000 at FWHM (Table 2.1).

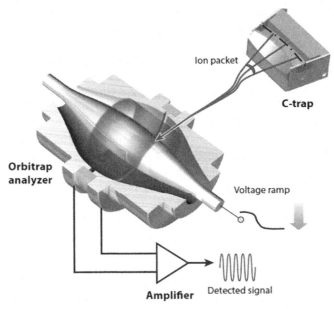

FIGURE 2.3

Cross section of the C-trap ion accumulation device and the Orbitrap mass analyzer with an example of an ion trajectory. During the voltage ramp, the ion packets enter the Orbitrap mass analyzer forming rings that induce current which is detected by the amplifier.

Reprinted with permission from Thermo Fisher Scientific, Copyright © 2016.

2.3 TIME-OF-FLIGHT MASS SPECTROMETRY: INSTRUMENT CONFIGURATION AND MAIN FEATURES

2.3.1 STAND-ALONE ELECTROSPRAY IONIZATION TIME-OF-FLIGHT AND HYBRID QUADRUPOLE TIME-OF-FLIGHT INSTRUMENTATION

The Micromass Q-TOF-1 (1996) was the first commercially available Q-TOF instrument launched in 1996, featuring a resolving power of 5000 (Morris et al., 1996, p. 889). Later, other small manufactures including Perceptive, Analytica or Bradford, Sensar, and JEOL introduced ESI-TOFs. In 1999, Micromass—part of Waters—updated the Q-TOF system (Q-TOF-2) with an increase in the resolving power up to 10,000 and presented a stand-alone TOF instrument (Waters LCT). A similar instrument was also commercialized by Applied Biosystems (now Sciex) around 2000 (Applied Biosystems Mariner ESI-TOF). In 2001, Waters upgraded the performance of Q-TOF (Q-TOF Ultima) using W-type flight path providing a resolving power of 17,500. After that, in 2002 Applied Biosystems (now Sciex) launched the Q-STAR XL, its first Q-TOF, which was updated in 2008 (Q-STAR Elite) with improvements in detector design and increased resolving power.

Table 2.1 Overview of the Technical Specifications and Characteristics of Time-of-Flight (TOF) and Orbitrap

Mass Analyzer Type	Manufacturer	Instrument Name	Resolving Power (FWHM Defined at m/z)	Mass Accuracy		m/z Range	Acquisition Speed (Hz)
				Internal Calibration	External Calibration		
Q-TOF	Bruker Daltonics	MicroOTOF-Q II	20,000 (m/z 922)	<2	<5	50–20,000	20
		MaXis impact	40,000 (m/z 386)	<1	<3	50–20,000	50
		MaXis 4G	60,000 (m/z 1222)	<0.6	<2	50–20,000	30 (MS), 10 (MS/MS)
	Waters	XEVO G2 Q-TOF	22,500 (m/z 956)	<1	–	20–20,000	30
		Synapt G2-S HDMS	50,000 (m/z 956)	<1	–	20–16,000	30
	Agilent	6500 Q-TOF series	42,000 (m/z 922)	<1	–	50–100,000	50
	Sciex	TripleTOF 4600	30,000 (full-range)	<0.5	<1	5–40,000	100
		TripleTOF 5600	35,000 (full-range)	<0.5	<2	5–40,000	100
		TripleTOF 6600	40,000 (full-range)	<0.5	<2	5–40,000	100
IT-TOF	Shimadzu	LC-MS-IT-TOF	10,000 (m/z 1000)	3	5	50–5000	10
Orbitrap	Thermo scientific	Exactive		<1	<5	50–4000	12 at RP 17,500
Q-Orbitrap		Q-Exactive	140,000 (m/z 200)	<1	<5	50–4000	12 at RP 17,500
LTQ-Orbitrap		Orbitrap Elite	240,000 (m/z 400)	<1	<3	50–4000	8 at RP 17,500
Tribrid-Orbitrap		Orbitrap Fusion Lumos tribrid	500,000 (m/z 200)	<1	<3	50–6000	18 at RP 17,500

FWHM, full width at half maximum; HDMS, high-definition mass spectrometry; IT-TOF, ion trap time-of-flight; LTQ-Orbitrap, linear ion trap Orbitrap.

Reproduced from Royal Society of Chemistry with permission, Copyright 2015 [Lin, L., Lin, H., Zhang, M., Dong, X., Yin, X., Qu, C. et al. (2015). Types, principle, and characteristics of tandem high-resolution mass spectrometry and its applications. RSC Advances, 5(130), 107623–107636].

The second generation of TOF instruments emerged by 2004, enabling a significant enhancement of both linearity and robustness. Three major vendors (Waters, Agilent Technologies, and Bruker) offer either completely new or completely transformed instruments. Waters first came up with both LCT TOF Premier and Q-TOF Premier using 4 GHz TDC detection and programmable enhanced dynamic range, which decreases the ion beam intensity to reduce detector saturation, extending the quantitation capabilities of the instrument. These TOF instruments featured continuous accurate mass calibration (MassLock). Both Bruker and Agilent introduced ESI-TOF instruments in 2004 with 10,000 resolving power, being the equivalent Q-TOF instruments launched in 2005 (micrOTOF-Q) and 2006 (Agilent 6510 Q-TOF), respectively, with resolving power of c. 15,000 in both cases. Follow-up products (microOTOF-Q II operating at 2 GHz and 6520 Q-TOF operating at 4 GHz) enabled resolutions of about 20,000. These two instruments used ADC instead of TDC, featuring—according to manufacturers—a dynamic range of up to five decades. A comprehensive explanation on the details of each technology is given elsewhere (Fjeldsted, 2003, 2016, p. 19). An example of the typical hardware configuration of a current Q-TOF instrument is illustrated in Fig. 2.4, where the schematics of Q-TOF Agilent 6550 is shown including a dual ion funnel ion sampling and a two-stage ion mirror (Fjeldsted, 2016, p. 19).

2.3.2 IMPROVEMENTS OF CURRENT (QUADRUPOLE) TIME-OF-FLIGHT INSTRUMENTATION

The improvement of TOF instrumentation can be attributed to different aspects, which are interrelated so that it is difficult to isolate the contribution of each individual factor to the overall performance.

Ionization step. First of all, ion production and atmospheric sampling is a key aspect on the overall sensitivity. Significant progress has been made from early electrospray experiments in the 1980s. The use of a nebulizing gas was proposed to increase the efficiency of ion production. To accelerate droplet size reduction, heated gases were then used (Turbo V source, Sciex). Then, different manufacturers came up with similar heated ESI based sources, such as the Jet Stream from Agilent, first implemented in 2008 Q-TOF 6530. Jet Stream (Agilent) consists of an enhanced ESI technology employing concentrically applied heated gas providing an average fivefold sensitivity increase compared with nonthermally enhanced ESI design (although this may vary depending on each individual compound).

Ion sampling. Major efforts have also been carried out for the improvement of ion sampling in the mass spectrometer. For this purpose, a balance between sampling orifice/inlet capillary diameter and vacuum demands must be achieved. Additionally, the ion transfer step has also been improved with more efficient ion guide technologies such as the ion funnels (Agilent) and other ion guide devices [QJet ion guide (Sciex) or Stepwave (Waters)], permitting an increase in the ions entering the mass spectrometer with the subsequent improvement of sensitivity.

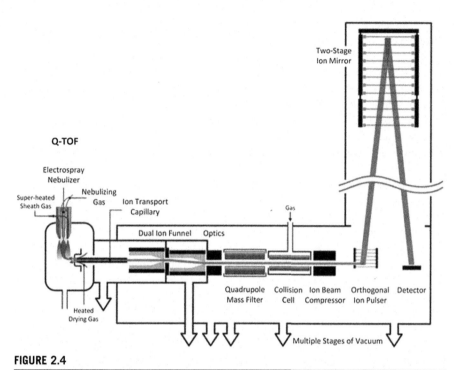

FIGURE 2.4

Key elements of modern Quadrupole time-of-flight (Q-TOF) mass spectrometer.

Reprinted with permission from Elsevier, Copyright © 2016 [Fjeldsted, J.C. (2016). Advances in time-of-flight mass spectrometry. In S. Pérez, P. Eichhorn, & D. Barceló (Eds.), Comprehensive analytical chemistry. *Applications of time-of-flight and Orbitrap mass spectrometry in environmental, food, doping and forensic analysis (p. 19). Amsterdam: Elsevier].*

The use of radio frequency (RF) fields exerts a force on ions over a wide range of pressure and mass. Based on this principle, the QJet ion guide designed by Sciex improved efficiency in separating ions from the noncharged species (neutrals), improving the focusing of captured ions into the mass spectrometer. Another approach (StepWave) was presented by *Waters* and first implemented in 2008 in the Synapt G2 hybrid instrument based on the use of a set of ion guides to shift the ion beam axis offset to the atmospheric pressure sampling inlet. RF and direct current (DC) potential are applied onto a set of ion concentric rings with overlapping apertures, so that the ions are shifted off the main axis into smaller off-axis concentric apertures, which focus the ion beam and direct it to the subsequent mass analysis stages while neutral gas passes along the principle axis being diverted to the exhaust.

Based also on the use of stacked ring RF ion guides, R.D. Smith et al. showed that ion confinement could be achieved at pressures in the range from 0.1 to 30 torr (Kelly, Tolmachev, Page, Tang, & Smith, 2010, p. 294), when constructing a set of ring electrodes having a linear decrease in the inner diameters in the last section of the stack (the so-called ion electrodynamic funnel), which could be used individually or as a pair in succession (*dual ion funnel*) with the second having a lower operating pressure than the preceding funnel. These RF ion funnel technologies enable highly efficient ion capture and may provide up to a 10-fold increase in the atmospheric sampling efficiencies and overall sensitivity (Fjeldsted, 2016, p. 19). The use of multiplexed (hexabore) capillaries has also been proposed to increase ion sampling efficiency (Kelly et al., 2010, p. 294).

Resolving power and flight path length. Unlike Orbitrap, the resolving power in TOF analyzer increases with *m/z* values. Several solutions have improved the overall performance of commercial TOF instrumentation particularly in terms of resolving power, attaining values up to 50,000 or even higher. Most of the improvements should be attributed to both ion optics (focusing of ions before TOF analysis) and detection electronics including the increase in the detector operation frequency of ADC TOF detectors (from 1 to 5 GHz) (Fjeldsted, 2016, p. 19). Besides, there is another obvious strategy that increases resolution (increasing the flight path length), although at the expense of sensitivity. This can be accomplished by using either longer flight tubes (up to 1.5 m) or multiple reflection by placing a second mirror between the ion pusher and detector, reflecting the beam twice, thus doubling the effective flight path length [such as *W optics* mode in Waters, *N-optic* design (Sciex) or folded path technology enabling up to 64 reflections (LECO)]. Ion mirror (reflectron) has also been improved through the use of two-stage mirrors, which compensate ion energy spread from the ion pusher. Nevertheless, reflectrons not only double the flight path length, but also are responsible for a relevant loss of recorded ion abundances. Hence, users of such multireflectron instruments have to come across a compromise between sensitivity and selectivity. Last, but not least, flight tube temperature control in the system also reduces the internal energy spread of the ions with the same mass, so that an increase in resolution is also achieved.

Ion detection. Ion detection and digitization have also undergone significant improvements in the last decade. It seems that most of the manufacturers have found that multichannel plate—based detectors with ADC are the best choice to avoid detector saturation and lack of linearity (Fjeldsted, 2016, p. 19). However, the use of very fast TDC conversion and a multianode detector is also used in instruments offering excellent performance such as Sciex TripleTOF 5600 Q-TOF (Fig. 2.5) (Andrews, Simons, Young, Hawkridge, & Muddiman, 2011, p. 5442). A detailed discussion on this matter is described by Fjeldsted (2003, 2016, p. 19).

Accurate mass measurements and mass calibration. Mass accuracy is central to HRMS. This fact is particularly important in TOF instruments, as they are generally less robust than FT-MS instruments in terms of the stability of accurate mass measurements and the subsequent accuracy (ppm mass errors). This is a key parameter, as small changes in the experimental conditions (such as laboratory temperature) usually affect the actual time measurements of ions TOF, provided the change in drift tube path length due to temperature changes (small thermal expansion or contraction of the flight tube length). A high-performance temperature

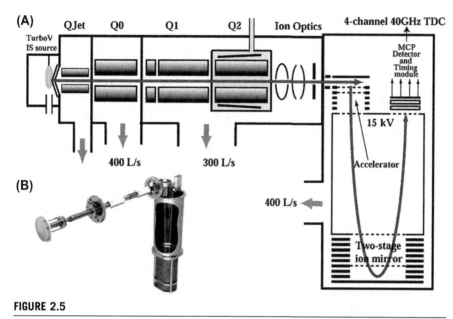

FIGURE 2.5

(A) Schematic of a Quadrupole time-of-flight instrument from Sciex (TripleTOF 5500) using multianode time-to-digital conversion (TDC) detection. (B) An image of the machined TripleTOF MS instrument platform.

Reproduced with permission from American Chemical Society, Copyright © 2011.

control assembly of the drift tube is mandatory to assess mass accuracy performance of TOF instruments.

For this purpose, mass axis calibration is an important process for obtaining reliable results. Mass calibration consists of comparing the observed experimental mass measurement of one or two known ions with their (theoretical) exact mass. Calibration may be internal (i.e., the reference masses are for ions of known elemental composition in the same mass spectrum of the analyte) or external (i.e., reference masses from a mass spectrum of another analyte acquired under similar conditions). Internal calibration is typically at least twofold more accurate than external (Marshall et al., 1998, p. 1). Accurate mass correction is usually performed in full-scan (FS) data, as the ions selected are available in all acquired spectra through the run. However, when recording accurate mass spectra in MS/MS mode with precursor ion selection, the lack of such calibrating ions hampers the accuracy of product ion spectra.

In the first generation of TOF instruments, external mass calibration was applied before the chromatographic run. This approach did not deliver results better than 5−10 ppm relative mass error because of the significant drifting of instrumental status (e.g., flight tube temperature). To overcome this situation, various strategies have been proposed to provide continuous accurate mass calibration in commercial

TOF instruments, allowing them to feature routine relative mass errors in the range of 1−3 ppm, when using mass calibration devices such as MassLock (Waters), the dual sprayer solution (Agilent Technologies), or similar technologies presented by other manufacturers such as the TwinSprayer (Sciex), which is an independent calibrant delivery path proposed in the recently commercialized compact Sciex Q-TOF (Sciex X500R Q-TOF) dedicated for small molecule applications (Fig. 2.6). These assemblies provide a compensating mechanism to keep flight times constant for a given fixed mass (reference solution). First, Waters proposed the discontinuous periodical infusion of a lock mass solution by a switching device that alternatively permitted the flow from two independent ESI sprayers (the reference solution and the analytical sample) (Eckers et al., 2000, p. 3683; Wolff et al., 2001, p. 2605). This strategy is not flawless because it introduces mass signals—unrelated to the sample—into both the spectrum and the data file, which are used to calibrate the entire data set at the end of the run. Likewise, Agilent proposed the use of a dual sprayer assembly, where a calibrant solution was continuously sprayed—orthogonal to the analytical sprayer—and introduced in the mass spectrometer (Fig. 2.6). This solution provided excellent results even enabling the measurement of the mass of an ion using twin fragment ions (Ferrer & Thurman, 2005, p. 3394). However, although they are orthogonal, there might be some interaction between the analytical and reference sprayers, which may yield signal suppression if the concentration/abundance of the reference solution analytes is not controlled. As mentioned previously, the newest Sciex Q-TOF (Sciex X500R Q-TOF) makes use of an

(A)   **(B)**

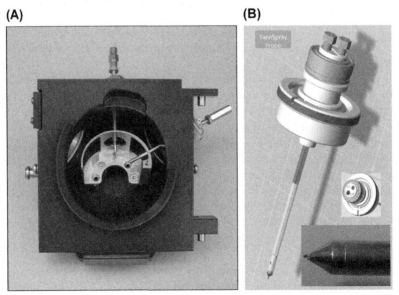

FIGURE 2.6

Examples of continuous calibration systems for mass calibration in time-of-flight instrumentation: (A) Agilent MSD TOF (2004); (B) Sciex X500R (2016).

Reprinted with permission from Sciex and Agilent Technologies, Copyright © 2016.

independent calibrant delivery path embedded in the analytical sprayer (Fig. 2.6), which also provides excellent accuracy figures, thanks also to the sophisticated temperature control of the flight tube.

2.3.3 ION MOBILITY QUADRUPOLE TIME-OF-FLIGHT

A relatively recent development with great potential in the detailed analysis of complex samples is the introduction of gas-phase IM as an additional dimension of separation within LC–MS. The combination of IMS with MS (IM-MS) incorporates shape and size as additional orthogonal dimensions to chromatography and mass spectrometry. It takes advantage of the differences in drift times of charged species in an inert gas with size and shape [in quantitative terms expressed as the collisional cross section (CCS)] being the key determinants of their mobilities (Beucher, Dervilly-Pinel, Prevost, Monteau, & Le Bizec, 2015, p. 9234; Fjeldsted, 2016, p. 19). As this separation takes place in the millisecond timescale, it is compatible (at the expense of some sensitivity) with both chromatographic separation (peak widths of a few seconds) and TOF ion detection system (ion flight times in the microsecond range and ion recording on the detector surface in the nanosecond scale). This approach proves very valuable in the separation of isomers that are undistinguishable solely based on accurate mass measurements. With slight differences in their CCS and thus drift times, such isomers can be resolved in this additional third dimension (fourth with MS/MS experiments) of LC–HRMS instrumentation (Beucher et al., 2015, p. 9234).

The use of IM is not new, although in the past IM-MS experiments were restricted to homemade instruments (Lapthorn, Pullen, & Chowdhry, 2013, p. 43). The progress and developments in hybrid IM mass spectrometry instrumentation in the last decade have prompted the commercial availability of various instruments from different vendors opening the door for a wider acceptance of this technique. The first commercial IM-mass spectrometry instrument became available around 2006 (Giles et al., 2004, p. 2401; Thalassinos et al., 2004, p. 55), the so-called Synapt high-definition mass spectrometry system (Waters) (Fig. 2.7), which was based on the use of high-transmission traveling-wave ion mobility (TWIM) cell. A follow-up version enabled a resolution of 40 (Pringle et al., 2007, p. 1). This quadrupole ion mobility separation time-of-flight mass spectrometry (Q-IMS-TOF MS) consists of a quadrupole unit followed by high-performance IM separation. It consists of three traveling wave–enabled stacked ring ion guides (TWIGs), namely the trap TWIGs, ion mobility separator (IMS), and the transfer TWIG. The trap TWIG has the function of trapping and releasing ions into the IMS, while the transfer TWIG functions to deliver the ions into the TOF analyzer (Fjeldsted, 2016, p. 19). This module also permits fragmentation of ions within the trap and in transfer TWIG. Ions are separated as they pass through IMS based on their size, charge, and cross section. Q-IMS-TOF provides excellent data-dependent (DDA) and data-independent acquisition (DIA), which are useful in metabolite profiling to detect and identify even low-level metabolites from endogenous matrices associated with high background noise.

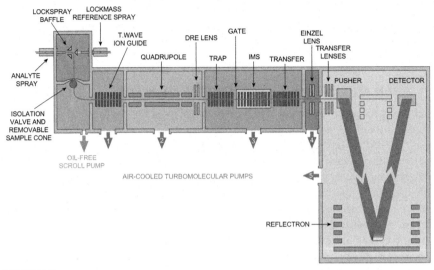

FIGURE 2.7

A schematic diagram of the Synapt high-definition mass spectrometry system. All of the turbomolecular pumps have 220 L/s pumping speed (model EXT255H, Edwards, Crawley, UK) and are backed by a 35 m 3/h scroll pump (model XDS35i, Edwards, Crawley, UK).

Recently, a new commercial IMS-Q-TOF instrument has been launched (Kurulugama, Imatani, & Taylor, 2013) based on the use of a uniform low-field drift tube and ion funnel sampling and focusing that allows high-performance IM (RP of c. 80) keeping also high sensitivity (Fig. 2.8) (Kurulugama et al., 2013). The separation is carried out based on the different drag force that ions undergo in the drift tube while they move through it because of the electric field applied. The countercurrent drug force is due to the collision of the ions with the buffer gas molecules. The drug force experienced by the ions depends on their collision cross sections (a function of size and shape, electric charge, and mass), so that ions with larger cross sections are slowed more easily by collisions with the buffer gas in the drift tube.

Differential ion mobility (DMS) has also been coupled in Sciex QTOF instruments (Sciex SelexION available in TripleTOF 5600+ and 6600) as a means to increase selectivity, enable isobaric separations, and reduce chemical background noise. DMS is—according to some authors—also known as High-field asymmetric waveform ion mobility (FAIMS) (Guevremont, 2004, p. 3; Kanu, Dwivedi, Tam, Matz, & Hill, 2008, p. 1), which has also been implemented with LC—TOFMS instruments using a chip operated at ambient pressure (Brown et al., 2010, p. 9827; Brown et al., 2012, p. 4095). A thorough explanation on the fundamentals and theory of separation for both FAIMS and DMS is available elsewhere (Kolakowski & Mester, 2007, p. 842).

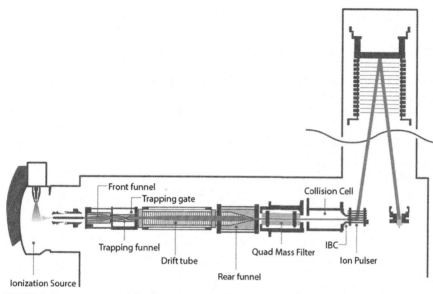

FIGURE 2.8

Schematic diagram of the Agilent 6560 ion mobility quadrupole time-of-flight (IM-Q-TOF) mass spectrometer (Kurulugama, Darland, Kuhlmann, Stafford, & Fjeldsted, 2015, p. 6834). It comprises of a uniform field drift tube coupled to a quadrupole TOF mass spectrometer. Ions generated in the source region are carried out into the front ion funnel through a single-bore glass capillary. The trapping funnel accumulates and release ions into the IM drift tube. Ions are separated in the 78-cm-long drift cell, which is typically operated at a 19 V/cm drift field. Ions exiting the drift tube enter the rear ion funnel, which compresses the ion beam before transmission to the high-resolution Q-TOF analyzer.

Reproduced with permission from Royal Society of Chemistry, 2015.

The applications of LC—IMS-Q-TOF in small molecule applications are relatively scarce despite this technique being available for nearly one decade. Limited literature is currently available in routine small molecule studies such as food safety, environmental, or forensics applications (Beucher et al., 2015, p. 9234; Far, Delvaux, Kune, Eppe, & De Pauw, 2014, p. 11,246; Fjeldsted, 2016, p. 19; Goscinny, Joly, De Pauw, Hanot, & Eppe, 2015, p. 85; Smith et al., 2013, p. 76; Stephan et al., 2016, p. 6545). The results obtained are generally superior in terms of selectivity, as no interfering compounds occur at the same retention time, exact mass, and drift time, yielding cleaned mass spectra or extracted ion chromatograms, although at the expense of some loss of sensitivity and overall system ruggedness. Besides LC sample introduction, supercritical fluid extraction coupled to ion mobility high-resolution mass spectrometry (SFC-Q-TWIM-TOFMS) (Beucher et al., 2016) using a Synapt G2 with TWIM coupled with Q-TOFMS mass analysis has also been proposed.

2.3.4 HYBRID ION TRAP TIME-OF-FLIGHT

The ion trap time-of-flight (IT-TOF) (Shimadzu) is a relatively new type of hybrid tandem mass spectrometer that combines the unique MS^n capability of quadrupole ion trap (QIT) with high resolution/accurate mass for both MS and MS^n modes provided by the TOF analyzer. Therefore, accurate mass measurements of the different MS/MS stages may provide the actual composition of each fragment to foster structure elucidation, as it may also be carried out in a hybrid LTQ Orbitrap. The main technological challenges are the efficient introduction of ions into the QIT and the simultaneous ejection of trapped ions to the TOF.

In the LC−MS-IT-TOF, ions are injected into the ion trap in pulses that are produced through the action of a combination of the skimmer, octapole, and lens optics (compressed ion injection). In creating these pulses, the ions are first accumulated in the octapole for a fixed period of time, with more ions accumulated at longer accumulation times for higher signal intensity where needed. The value for the ion accumulation time can be freely set in the method file. However, when the automatic sensitivity control feature is activated, the ion accumulation time in the octapole is adjusted automatically and instantaneously during the analysis to effectively avoid saturation of the detector (Shimadzu, 2009). Thus, the octapole acts as an ion gate, holding and controlling ions before release into the ion trap itself. The ion trap performs two functions: pulses ions into the TOF mass analyzer at a precise start time and achieves high CID efficiency experiments as pulsed argon gas eliminates CID energy mass dependence. Studies using LC−IT-TOF for pesticide screening or the determination of other contaminants in food are relatively scarce (Li, Zhang, Ma, Li, & Guo, 2015, p. 316; Zhan, Xing, Sun, Ting, & Chew, 2015).

2.3.5 GAS CHROMATOGRAPHY−TIME-OF-FLIGHT AND GAS CHROMATOGRAPHY−QUADRUPOLE TIME-OF-FLIGHT

The development of GC-compatible APCI and atmospheric pressure photoionization interfaces and their combination with HRMS has revitalized and raises back the interest in GC−MS, given its attractive features for nontarget analysis. The use of soft ionization techniques promotes the formation of molecular ions as compared to the commonly observed extensive fragmentation under EI conditions and thus results in an increase in sensitivity at detecting the molecular ion. In contrast, the use of a hybrid instrument with precursor ion isolation and dedicated fragmentation in a collision cell is required for unambiguous confirmation with MS/MS experiments.

In 2011 LECO began commercialization of Pegasus GC−HRT instrument with folded flight path technology enabling high resolution with high spectral production frequency (200 Hz) and three resolution operational modes: unit mass, high resolution (>25,000 FWHM), and ultrahigh-resolution (featuring R = 50,000 although with limited mass range) based on up to 64 reflections—up to 20 m effective path length

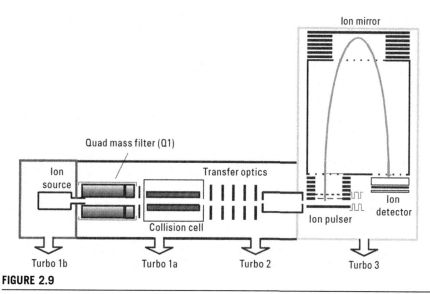

FIGURE 2.9

Ion source, ion optics, and mass filter from the Agilent gas chromatography (GC)
—quadrupole time-of-flight mass spectrometer.

Reproduced with permission from Agilent Technologies, Copyright 2016.

(Tranchida, Franchida, Dugo, & Mondello, 2016, p. 524). These features—including
scan speed—map well against the requirements of comprehensive gas chromatography systems (GC × GC) and may find several applications (Zimmermann et al.,
2014). Agilent also commercializes a dedicated GC−Q-TOF instrument with electron
impact ionization, which has been applied to pesticide testing in fruits and vegetables
(Belmonte et al., 2015, p. 2162; Zhang et al., 2012, p. 39) (Fig. 2.9).

2.4 ORBITRAP ANALYZERS: INSTRUMENT CONFIGURATIONS AND MAIN FEATURES

Different Orbitrap instruments (Thermo Scientific) have come out in the last decade.
A detailed summary of each of the variants commercialized is provided in Table 2.2
(Martins, Bromirski, Prieto Conaway, & Makarov, 2016, p. 3). The main features are
described, organized from the simplest configuration (stand-alone Orbitrap) to more
complex instrument architectures (tribrid Orbitrap) rather than by the date of
commercialization.

Stand-alone Orbitrap (Exactive). Until the introduction of the Exactive (stand-
alone Orbitrap) in 2009 (Bateman, Kellmann, Muenster, Papp, & Taylor, 2009, p.
1441), the relatively high cost and complexity of Orbitrap hybrid platforms (LTQ-
Orbitrap) limited their implementation in routine laboratories such as food safety,
environmental monitoring, and clinical and toxicological testing. The Exactive is

Table 2.2 Overview of Orbitrap Mass Spectrometry (as of Year 2015)

Family	Name	Launch Year	Front End	Orbitrap Analyzer	Other Analyzers	Fragmentation Methods	Other Features
Research-Type Instruments							
LTQ-Orbitrap	Classic (Makarov et al., 2006, p. 2113)	2005	Capillary skimmer	**Standard 3.5 kV**	LT, 2 detectors	CID	Discovery version (2007)
	XL (Olsen et al., 2007, p. 709)	2007	As above	As above	As above	CID, **HCD, (ETD)**	MALDI version 2008 (Stupat et al., 2009, p. 1451)
	Velos (Olsen et al., 2009, p. 2759)	2009	**S-lens**	As above	Dual-pressure LT	As above	–
	Elite (Michalski et al., 2012)	2011	As above	**High-field 3.5 kV**	As above	As above	**eFT signal processing** (Lange, Damoc, Wieghaus, & Makarov, 2014, p. 16)
Orbitrap Fusion	Classic (Senko et al., 2013, p. 11,710)	2013	As above	**High-field 5 kV**	QMF, LT	CID, HCD, (ETD)	**(Internal calibration)**
	Lumos	2015	**HCCT and ion funnel**	As above	Segmented QMF	As above	As above
Routine-Type Instruments							
Exactive	Classic (Bateman et al., 2009, p. 1441)	2008	Capillary skimmer	Standard 5 kV	None	(HCD)	–
	Plus and plus EMR	2012	S-lens	As above	None	As above	eFT, **EMR version** (2013)
Q-Exactive	Classic (Rose et al., 2012, p. 1084)	2011	As above	As above	**QMF**	HCD	eFT
	Plus	2013	As above	As above	**Segmented QMF**	As above	eFT, **inject prefiltering**
	Focus	2014	As above	High-field 5 kV	QMF	HCD	eFT
	HF (Scheltema et al., 2014, p. 3698)	2014	As above	As above	As above	As above	As above
Q-Exactive GC	Classic (Silcock et al., 2015)	2015	**In vacuum EI/CI**	As above	As above	As above	As above

CID, collision-induced dissociation; eFT, enhanced Fourier transform; EMR, enhanced mass range; ETD, electron-transfer dissociation; fETD, front-end electron-transfer dissociation; HCCT, high-capacity transfer tube; HCD, higher-energy collision-induced dissociation; LT, linear trap; LTQ, linear ion trap Orbitrap; QMF, quadrupole mass filter. **Bold** denotes first appearance of a feature; references belong to the first publications devoted to the corresponding instrument.
Reproduced from Elsevier, with permission, Copyright 2016 [Martins, C.P.B., Bromirski, M., Prieto Conaway, M.C., & Makarov, A.A. (2016). Orbitrap mass spectrometry: Evolution and applicability. In S. Pérez, P. Eichhorn, & D. Barceló (Eds.), Applications of time-of-flight and Orbitrap mass spectrometry in environmental, food, doping and forensic analysis (p. 3). Amsterdam: Elsevier].

a relatively compact benchtop instrument, which only consists of a stand-alone Orbitrap analyzer (omitting the ion trap using in hybrid LTQ-Orbitrap platforms). Thus, it does not have the ability to perform ion/mass selection. Because the total ion population is collected in the C-trap and injected into the Orbitrap analyzer, an automated gain control mechanism was implemented for reproducible ion injection.

The Exactive permits high-resolution/accurate mass (HR/AM) screening of known and unknown compounds with high sensitivity—approaching QqQ instruments in the MRM mode—and selectivity (high resolution up to 100,000 and mass accuracy lower than 5 ppm without internal calibration). Thanks to the high intrascan dynamic range (over four decades of magnitude) and fast polarity switching (both positive and negative HR spectra at 70,000 within a second), the development of (quantitative) screening methods based on exact mass and RT of compounds with sensitivity approaching MRM methods is relatively straightforward. FS screening also allows retrospective examination of data based on a posteriori hypothesis of additional compounds of interest, because of the FS data acquisition rather than strictly targeting specific ions of interest as it is done with standard QqQ analyses. Besides FS, the Exactive can also be operated in pseudo-MS/MS mode without precursor ion isolation, using a high-energy collision cell (HCD), which was optional—not available in all Exactive units. This mode of analysis, referred to as all-ion fragmentation (AIF) is similar to analogous techniques available on instruments sold by Agilent, Sciex, or Waters (all-ion mode, MS^{ALL}, and MS^E respectively). This mode is really convenient as it permits parallel acquisition of FS of intact molecules plus the characteristic fragmentation of all species, although at the expense of a lack of the selectivity provided by the absence of precursor ion isolation step.

Recently, two additional versions of Exactive were launched: Exactive Plus and Exactive Plus EMR. The first one included the same Orbitrap analyzer used in Q-Exactive, thus providing a resolving power of up to 140,000. The EMR stands (enhanced mass range) version permits acquisition over the range m/z 300 to m/z 20,000 and is intended for applications addressing structural studies of high–molecular weight biomolecules.

Hybrid quadrupole Orbitrap (Q-Exactive). To perform precursor ion isolation and fragmentation experiments (similarly to the equivalent Q-TOF or QqQ instruments), the Exactive mass spectrometer was combined with a front quadrupole with the mission to act as a mass filter and to isolate targeted precursor ions that were subsequently fragmented in the HCD cell (similar to that used in the stand-alone Orbitrap) and mass analyzed in the Orbitrap (Fig. 2.10). The so-called "Q-Exactive" instrument came out in 2011 (Rose, Damoc, Denisov, Makarov, & Heck, 2012, p. 1084). To achieve faster acquisition rates, an advanced processing technique [enhanced Fourier Transform (eFT)] for transforming the transient detection signal produced by the Orbitrap mass analyzer was first implemented with the Q-Exactive. Its use combined with fast quadrupole isolation and HCD fragmentation provided improved data quality and acquisition rates. The HCD cell or the C-trap is filled with ions while the previous MS/MS detection cycle is ongoing. High resolution could be acquired at a rate up to 12 Hz, thus avoiding the problem of low speed for accurate mass MS/MS common to hybrid LTQ-Orbitraps.

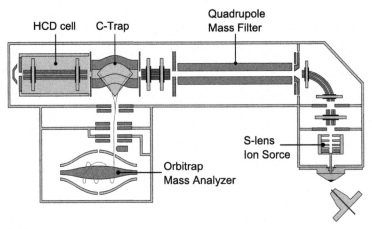

FIGURE 2.10

Scheme of a hybrid quadrupole-Orbitrap. *HCD cell*, high-energy collision-induced dissociation.

Reproduced with permission from Thermo Scientific, Copyright 2016.

Recently, three new versions of Q-Exactive have been launched: an upgraded version of the classic Q-Exactive with enhanced ion optics and transmission (advanced quadrupole technology and advanced active beam guide) (Q-Exactive Plus); an entry-level and more affordable version of Q-Exactive for routine laboratory applications (Q-Exactive Focus), with a resolving power up to 70,000 (*m/z* 200), a scan rate of 12 Hz at 17,500 (*m/z* 200), and full MS/MS capability; and finally an advanced version for high-end high-throughput proteomics applications using the high-field Orbitrap analyzer (Q-Exactive HF) featuring a resolving power of 240,000 (at *m/z* 200 and 1.5 spectra per second) (Table 2.3).

Table 2.3 Acquisition Speed and Resolution of Q-Exactive, Orbitrap Elite, and Q-Exactive HF

Q-Exactive (2011)		Orbitrap Elite (2011)		Q-Exactive-HF (2014)	
Resolution Setting *m/z* 400 (FWHM)	Spectra per Second (Hz)	Resolution Setting *m/z* 400	Spectra per Second Hz	Resolution Setting *m/z* 200 (FWHM)	Spectra per Second (Hz)
12,500	12	15,000	7.7	15,000	18
25,000	7	30,000	6.9	30,000	12
50,000	3	60,000	4.0	60,000	7
100,000	1.5	120,000	2.3	120,000	3
		240,000	1.2	240,000	1.5

FWHM, *full width at half maximum.*

Hybrid linear ion trap Orbitrap (LTQ-Orbitrap). The hybrid linear ion trap (LTQ-Orbitrap) was the first commercial instrument to incorporate an Orbitrap mass analyzer (Fig. 2.11). This configuration provides flexibility to undertake different experiments, including MS and MS^n spectra using either the Orbitrap for high resolution (up to 120,000 FWHM at m/z 400) and mass accuracy or the ion trap. One of the most commonly used operation mode is high FS HR/AM acquisition using data-dependent MS/MS scan in the ion trap. The LTQ-Orbitrap instrument features scan rates in the 4–5 Hz range in MS/MS with nominal mass detection or 3 Hz MS/MS spectra with accurate mass (Makarov et al., 2006, p. 2113). In addition, to obtain structural information not possible with CID fragmentation, additional fragmentation techniques such as ETD reagent ion source and HCD were implemented.

High-Field (LTQ) Orbitrap (Orbitrap Elite). The high-field version of the Orbitrap analyzer was first introduced in 2011 as a part of the Orbitrap Elite instrument (Michalski et al., 2012) (Fig. 2.11). Later, it was also utilized in a high-end version of Q-Exactive (Q-Exactive HF) intended for advanced proteomics applications and also in the tribrid Orbitrap Fusion. The high-field Orbitrap effectively doubles the operating frequency of the previous version, and combined with the eFT technique it permitted high resolving power of 24,0000 at m/z 400 for a 768-ms transient (acquisition time) yielding about fourfold increase in

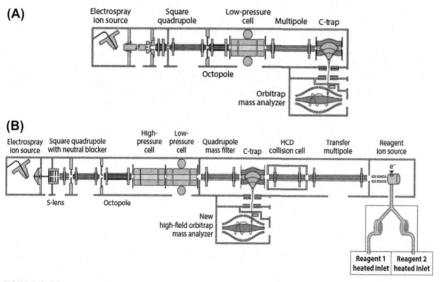

FIGURE 2.11

(A) Schematic of the linear ion trap (LTQ)-Orbitrap mass spectrometer with traditional ion trap followed by Orbitrap mass spectrometer. (B) Schematic of the Orbitrap Elite mass spectrometer with S-lens, dual-pressure linear ion trap, high-energy collision-induced dissociation (HCD) cell, electron-transfer dissociation (ETD) option, and high-field Orbitrap mass analyzer.

Reproduced with permission from Thermo Scientific, Copyright 2016.

resolving power at the same transient length compared to the earlier standard Orbitrap. Thus, the high-field Orbitrap can be operated at higher acquisition rates for the same resolution, ultimately enabling a faster acquisition cycle and/or increasing the number of MS/MS analyses in a given run. The high-field Orbitrap Elite instruments with appropriate modifications and acquisition times of c. 3 s were reported to provide resolving power above 1,000,000 for m/z 300—400 (Denisov, Damoc, Lange, & Makarov, 2012, p. 80). In addition, the increase in resolving power/acquisition rate provides significant advantages in quantitative experiments such as SILAC.

Tribrid (LTQ/Q-Orbitrap). The tribrid Orbitrap instrument (Orbitrap Fusion and Orbitrap Fusion Lumos) consists of a quadrupole mass filter, an Orbitrap, and a linear ion trap mass analyzers (Fig. 2.12). The high-field Orbitrap used features 500,000 resolving power at m/z 200. The system is conceived so that operation can be fully parallelized, maximizing the use of the ion current for the execution of complex acquisition modes, because of its ability to simultaneously isolate ions with one analyzer and separately detect ions in the two remaining analyzers (Senko et al., 2013, p. 11,710). For instance, precursor ion mass isolation can be performed with either the quadrupole or the ion trap while fragmentation can be generated by HCD, CID, ETD, or the novel electron-transfer and higher-energy collision dissociation fragmentation type at any level of MS^n. In addition, these precursor and fragment ions can be detected in either the Orbitrap or ion trap analyzers, providing a plethora of possibilities to conduct complex experiments to elucidate chemical

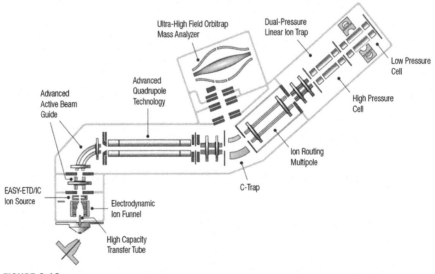

FIGURE 2.12

Quadrupole/Orbitrap/ion trap tribrid mass spectrometer architecture (Orbitrap Fusion Lumos) mass spectrometer.

Reproduced with permission from Thermo Scientific, Copyright 2016.

structures. This instrument is designed particularly for complex bioanalysis applications rather than small molecule analysis. Unfortunately, its relatively high cost—considering the standard prizes for benchtop HRMS instruments for small molecule application—dramatically reduces the implementation of such systems for food safety or environmental testing applications.

GC–Q-Orbitrap. The use of Orbitrap analyzer for GC analysis was first reported in 2010 based on a hybrid linear ion trap-Orbitrap LC instrument (Peterson et al., 2010, p. 8618; Peterson et al., 2014, p. 10,044). Later, a dedicated GC–quadrupole Orbitrap (GC–Q-Orbitrap) with electron impact ionization was conceived instrument (Peterson et al., 2010, p. 8618; Peterson et al., 2014, p. 10,044) using a high-field Orbitrap analyzer, offering the possibility of four modes of operation including FS high-resolution MS, selected ion monitoring (SIM) with quadrupole isolation, "AIF" MS/MS with beam-type CID in an HCD cell, and dedicated MS/MS high-resolution spectra with precursor ion isolation in the quadrupole and fragmentation in the HCD cell. At a resolution of 60,000, the Exactive GC system has a scan speed of c. 7 Hz (Thermo Scientific, 2016). The scan rate increases up to 18 Hz at a resolution of 12,500 (m/z 272), and up to 23 Hz (32 ms transient) at a resolution of 10,000 (m/z 200), providing low relative mass errors typically within 2 ppm (Thermo Scientific, 2016), which anticipates a broad range of application in the small molecule research field including metabolomics (Peterson et al., 2014, p. 10,044) or food safety testing (Mol et al., 2016, p. 161). An example is shown in Fig. 2.13 for the peak of hexachlorobenzene and the effect of resolution/scan rate on the number of points defining a chromatographic peak with c. 6-second width.

2.5 ACQUISITION MODES IN HIGH-RESOLUTION MASS SPECTROMETRY

In recent years, the traditional acquisition in HRMS workflow has evolved toward allowing a wide range of tools to identify or quantify compounds. This evolution has been made possible by improvements in HRMS instrumentation. As stated before, the introduction of new hybrid instruments including Q-TOF, Q-IMS-TOF, IT-TOF, Q-Orbitrap, etc. has allowed change in how mass spectrometry users work. A brief summary of the different tools is shown in Fig. 2.14. As could be observed, two types of data acquisition could be employed, namely data dependent and data independent.

2.5.1 DATA-DEPENDENT ACQUISITION

According to Mann et al. DDA is mode of data collection in MS/MS in which a fixed number of precursor ions whose m/z values were recorded in a survey scan (FS single-mass) are selected in real time using predetermined rules and are subjected to a second stage of mass selection in an MS/MS analysis (Fig. 2.14) (Mann,

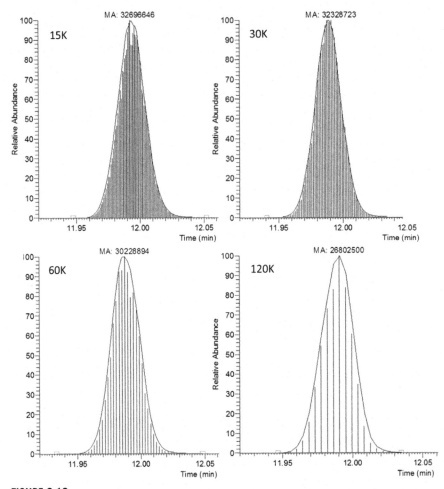

FIGURE 2.13

Gas chromatography (GC)–Q-Orbitrap analysis of hexachlorobenzene (63 pg on column) automated gain control (AGC) target set at 1e6. Scan rate versus mass resolving power. Extracted ion chromatograms for m/z 283.80,945 (hexachlorobenzene). Sticks represent intensities for individual scans (no smoothing applied), MA: peak area. For details, see (Mol et al., 2016, p. 161).

Hendrickson, & Pandey, 2001, p. 437; Murray et al., 2013, p. 1515). After acquiring the product ion mass spectra, the system returns back to the survey scan.

Precursor ion scan. Precursor ion scan (PIS) is one of the most used in the single-step MS/MS experiment, in which a second mass analyzer is set at the mass of the selected product ion; the first mass analyzer is scanned from that mass upward, resulting in a mass spectrum that contains signals for all the precursor

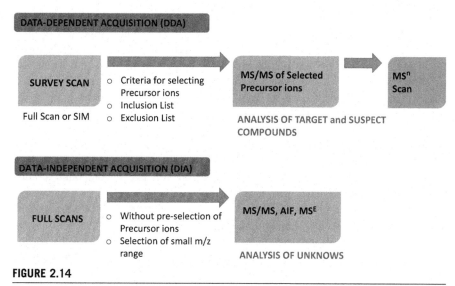

FIGURE 2.14

Data-dependent and data-independent acquisition strategies in high-resolution mass spectrometry.

ions that dissociate to the selected product ion. This mode could be carried out by the different hybrid mass spectrometers available including Q-TOF and Q-Orbitrap. However, this is limited because these MS/MS experiments have to be fixed previously. So, it was necessary to have a suspect compound. Thus, the analyst must have a more comprehensive understanding of system studied.

Ion intensity-dependent. In this DDA acquisition method, the user has to fix an ion intensity threshold. Then, the compounds, which have reached this ion intensity threshold, will be subjected to MS/MS experiments. It can be concluded that there is no need for any previous knowledge of the m/z values of the precursors (Ma & Chowdhury, 2013, p. 1285). However, the main drawback of this tool is the sensitivity of the analytes. It would be very complicated to obtain MS/MS spectra of low-level metabolites using intensity-dependent acquisition because it will mainly select high-intensity endogenous ions in biological matrices for MS/MS and MS^n acquisition (Ladumor et al., 2016, p. 199).

Pseudo neutral loss-dependent acquisition. Certain compounds under the action of CID could present a specific neutral loss. Thus, the MS/MS acquisition is triggered by this fragmentation in pseudo neutral loss−dependent acquisition. DDA is based on the acquisition of two consecutive FS. The first is performed at low collision energy (CE) (i.e., 5 eV), followed by another scan with a higher CE ramping (i.e., 20−40 eV). Then, the specific m/z differences of ion pairs (neutral loss) between consecutive low and high collision energy full-scan MS are obtained. The MS/MS acquisition will be triggered when the exact neutral falls within the specified mass tolerance window (Ladumor et al., 2016, p. 199; Ma & Chowdhury, 2013, p. 1285).

List dependent. In this approach the user has to build an inclusion list of accurate masses of target or suspect compounds. The different compounds of this list present a very specific mass tolerance range, which has been fixed by the user. Then, the precursor ions are predicted with the help of in silico tools such as MetaSite (Molecular Discovery), MetabolitePilot (AB Sciex), and MetWorks (Thermo Electron Corporation) in FS mode at real time. Finally, the instrument will obtain the product ion (MS/MS) spectrum of the ion of interest. This approach has proven to be a powerful tool for acquiring product ion spectra for low-level metabolites in complex biological matrices, because, the number of false positives due to the matrix is reduced (Ma & Chowdhury, 2013, p. 1285). However, the development of an accurate mass inclusion for every compound could considerably reduce the laboratory-throughput.

Isotope pattern—dependent. Isotope pattern data—dependent MS/MS has been developed for selective data acquisition of analytes that display a distinct isotope pattern. This acquisition uses some elements such as Cl and Br. It is well known that these elements have unique isotopic patterns that can be easily recognized in their mass spectra. Thus, the software is programmed to detect any ion with unique isotopic pattern in the full MS survey scan (Ma, Wen, Ruan, & Zhu, 2008, p. 1477). Once a listed metabolite ion is found in the survey scan, MS/MS acquisition of the metabolite is automatically triggered. DDA based on isotope pattern—dependent could be extended to compounds that contain synthetically incorporated isotopes (e.g., $^{13}C-$, $^{2}H-$, $^{18}O-$, $^{15}N-$, etc.) or the radiolabeled compound ($^{14}C-$) with a distinct $^{12}C/^{14}C$ isotopic pattern (Ma & Chowdhury, 2013, p. 1285).

Parallel reaction monitoring. Parallel reaction monitoring (PRM) is the closest scan mode to working with a QqQ mass analyzer using SRM. A list of targeted precursor ions, retention times, and CEs are used, providing the most sensitive and selective quantitation results in very complex matrices. PRM is carried out by Q-Exactive instruments (Law & Lim, 2013, p. 551). In PRM mode a precursor ion list has to be developed by the user. Then a full fragment-ion spectrum of each precursor ion is recorded continuously throughout the entire LC separation. Although precursor ion spectrum is not registered in standard PRM mode, one can still use an alternative setting to acquire a full MS scan or SIM spectrum at an expense of duty cycle (termed targeted DDA or directed DDA) (Schmidt, Gehlenborg, & Bodenmiller, 2008, p. 2138) to conduct MS^{1}- and MS^{2}-based analyses. The main advantage of PRM is the use of ultrahigh-resolution Orbitrap mass analyzer that is able to separate interferences from the true signals, thus significantly enhancing the selectivity of the method compared to the conventional SRM approach (Gallien, Duriez, Demeure, & Domon, 2012, p. 148). The high resolving power of the Orbitrap mass analyzer provides a significant improvement on the selectivity of measurements. Thus, low limits of quantification and detection are generally reached in spite of the high complexity of the samples, especially for the highest values of transient time. This is explained by the theoretical increase in S/N ratio in proportion to the square root of transient time in the FT spectra (Rauniyar, 2015, p. 28,566) and by the discrimination of nearly isobaric product ions generated from coisolated precursors in complex samples. However, the main handicap of this acquisition mode is that the number of precursor ions is limited because this number depends on the duty cycle or transient length of the (Orbitrap) mass analyzer and the chromatographic conditions. The use of different tools including parallelization (up to 10 precursor windows at one time), time-scheduling, and relaxation on the Orbitrap resolving power could increase the number of monitored precursor ions.

2.5.2 DATA-INDEPENDENT ACQUISITION

It should be clear that DDA modes are highly dependent from prior information collected by the analyst (Arnhard, Gottschall, & Pitterl, 2015, p. 405), because, the precursor ions should be previously fixed and analyzed by MS/MS. DIA has now become a powerful alternative for quantification and identification proposes (Doerr, 2014, p. 35). This acquisition mode could be considered as a simple and generic mode, which is based on nonspecific CID. Therefore, the MS/MS spectra are obtained in a nonselective manner.

Full scan. Until relatively recently, the analysis of unknown compounds was carried out by TOF or Orbitrap mass spectrometer. FS or full-spectrum mode is the most widely used acquisition data mode in HRMS. In FS, the mass spectra are continuously acquired within a fixed period of time (mostly ≤ 1 s) (Niessen & Falck, 2015, p. 1). Additionally, the range of m/z studied could be selected. However, FS is not adequate for confirmation propose of unknown compounds. To overcome this problem, a very interesting tool is in-source CID fragmentation (Abranko, García-Reyes, & Molina-Díaz, 2011, p. 478), where a *pseudo*-MS/MS experiment takes place in an intermediate pressure section of the mass spectrometer, between the atmospheric pressure source and the high vacuum of the mass spectrometer. Ions are generated in ionization chamber, being introduced in the vacuum region. Then, ions could be accelerated under various voltage conditions, prompting collisions with surrounding species, which can produce sufficient energy to yield (diagnostic) fragment ions. Thus, it would be possible to obtain MS/MS spectra using a TOF instruments.

PIS without precursor ion isolation. This acquisition mode was devolved for metabolite using with Q-TOF or Orbitrap technology [MS^E (Waters), all-ion mode (Agilent), MS^{ALL} (Sciex), and all-ion fragmentation (Thermo Scientific)]. In this approach, no precursor ion isolation is undertaken. Thus, two FSs have to be applied, one at low CE and the other one at a higher CE range. The mass spectra obtained from low CE provide intact molecular ion information. On the other hand, the mass spectra from the high CE contain MS/MS information (Ma & Chowdhury, 2013, p. 1285). However, the interpretation of the obtained product ion spectra is a challenging task, because of the presence of fragment interferences from coeluting components and background matrices. So, it was necessary to obtain good chromatographic performance to obtain peaks without a fragment ions derived from the interference. In Orbitrap, AIF works in the same way of MS^E. In this case, low-energy spectrum is obtained from the C-Trap and the high-energy mass spectrum is obtained from the HDC cell in a Q-Orbitrap instrument (Geiger, Cox, & Mann, 2010, p. 2252).

Nonselective MS/MS spectra. This strategy has been developed by SCIEX. It is a very useful way to obtain MS/MS spectra TripleTOF 5600 system. This DIA called sequential window acquisition of all theoretical fragment-ion spectra (SWATH) is an alternative approach that combines a high-specificity DIA method with a novel targeted data extraction strategy to mine the resulting fragment ion data sets. The SWATH strategy is based on a recurring cycle of a survey scan and a Q1 isolation strategy. At the beginning, a survey scan with low CE covers the user-defined mass range (Q1 set to full transmission). The mass range then is consecutively scanned using predefined Q1 narrow ion windows (20 Da), applying a range of CEs to produce product ion spectra, which are analyzed by the TOF analyzer (Roemmelt, Steuer, Poetzsch, & Kraemer, 2014, p. 11,742). In a wide mass range the Q1 window could be stepped, obtaining FS composite MS/MS spectra at each step, with an LC compatible cycle time. To collect this large amount of data, it was necessary that instrument has a high scan speed (100 Hz). The main advantage of SWATH is that the user need not select a list of specific target compounds, because this step will be carried in a postacquisition step (Ladumor et al., 2016, p. 199).

2.5.3 POSTACQUISITION APPROACHES

There are several postacquisition algorithms to ease the interpretation of mass spectra data. Among them, mass defect filter (MDF) has been developed to improve the detection of common and uncommon compounds. MDF mode could filter out most of interferences in a complex matrix. To erase this undesirable information, the software establishes a mass defect range of the common compounds studied. The interference ions, for which m/z is outside of this mass range, will not be considered. It should be noted that the use of HRMS is mandatory because the m/z difference between interference and analytes could be quite small. This approach has been successfully applied to resolve metabolite ions and isobaric interferences when they were separated by greater than ~ 50 mDa in a typical mass range of 200–1000 Da (Ladumor et al., 2016, p. 199; Zhang, Zhang, Rayb, & Zhu, 2009, p. 999). MDF has been used as a postacquisition data processing method. In fact, most of commercial brands have available software with this tools. On the other hand, this approach has been applied for real-time acquisition using a Q-TOF (TripleTOF) instrument (Bloomfield & Le Blanc, 2009).

2.6 DATABASES AND THE INTERNET RESOURCES FOR HIGH-RESOLUTION MASS SPECTROMETRY

Up to date, the analysis of untargeted compounds has been carried out in multiple steps (Tautenhahn et al., 2012, p. 826). First, mass spectrometry data for each of the samples are acquired. Then, the different unknown peaks have to be selected by the differences found between a group of samples. These compounds are identified manually in a database by m/z ratios of the peaks of interest. Finally, to obtain an unequivocal identification, tandem mass spectrometry (MS/MS) data from the sample could be compared with MS/MS data obtained from bibliography or a commercially available standard. Thus, untargeted workflow could be quite daunting. The use of free-access mass spectrometry database is an alternative to facilitate identification of compounds in the untargeted workflow. There are more and more complementary mass spectrometry data compilations (Oberacher, 2013, p. 312). In this section an overview of the most relevant database is shown.

MassBank. MassBank was the first mass spectra database of small chemical compounds for life sciences (<3000 Da) (MassBank, 2016). This repository was designed to share the obtained mass spectral data among the scientific community (Horai et al., 2010, p. 703). Thus, MassBank data could be useful for the chemical identification and structure elucidation of chemical compounds detected by mass spectrometry. Among other features, MassBank supplies (1) tandem mass spectra acquired on broad range of mass spectrometers as well as different ion sources; (2) a friendly interface, which could be used to find mass spectrum by chemical name or molecular formula; and (3) an easy way to search similar spectra on a neutral loss-to-neutral loss basis.

Chemspider. ChemSpider is a free online service that delivers information of more than 57 million structures from hundreds of data sources (Chemspider, 2016). These sources include data from government databases, commercial chemical supplier catalogs, academic and commercial website. Thus, it could be considered as the richest single source of structure-based chemistry information. For identification purposes, ChemSpider is useful to obtain MS and MS/MS spectra from known compounds. The spectra can only be found by systematic name, trade name, SMILES, InChI, or CSID. However, it is not possible to obtain information of compounds, which have not been partially identified before.

METLIN Metabolomics Database. This web-based database has been developed by the Scripps Research Institute (California, USA) (METLIN, 2016). This database is designed to provide, among others, structural and physical data on known endogenous metabolites, drugs, and pesticides and their metabolites and high-accuracy mass spectrum data from known compounds and their derivatives (Smith et al., 2005, p. 747). In addition, a wide MS/MS spectra collection has been developed. This information has been obtained on an Agilent 6510 Q-TOF instrument by collecting compound-specific reference spectra at three different CEs. The mass spectrometer is operated in both ESI positive and negative modes. Currently, there are more than 700,000 high-resolution MS/MS spectra available. It should be noted that this number is continuously expanding.

As an example, an unknown compound search is shown in Fig. 2.15. In METLIN screen data search the experimental m/z has to be selected. Additionally, the user could indicate the mass error or ionization mode, or delimit the search between either the molecular ions or several adduct ions (Fig. 2.15A). Then, several compounds were displayed according to the mass error (Fig. 2.15B). Finally, the user can select the MS spectra or MS/MS spectra at four different CEs. So, the tandem MS data that METLIN provides are therefore particularly valuable to researchers who are in the early stages of metabolite identification.

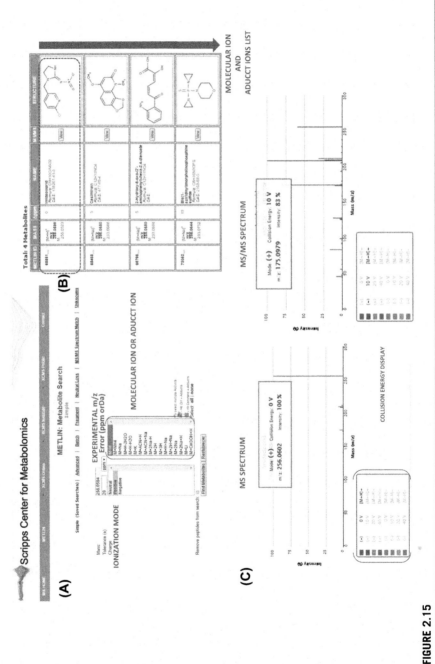

FIGURE 2.15

Basic search for an unknown compound using METLIN database: (A) m/z settings; (B) list of candidates; (C) MS and MS/MS spectrum with different collision energies.

A new database namely isoMETLIN has been developed through METLIN (Cho et al., 2014, p. 9358; IsoMETLIN, 2016). This database could facilitate the identification of metabolites that have been isotopically labeled. Thus, in the field of pesticide analysis it could be a powerful tool, as the number of isotope labeled species is gradually increasing. This database allows to search all computed isotopologues derived from METLIN on the basis of m/z values and specified isotopes of interest, such as ^{13}C or ^{15}N. In addition, isoMETLIN provides experimental MS/MS data on hundreds of isotopomers.

The Spectral Data Base for Organic Compounds. The Spectral Data Base for Organic Compounds (SDBS) (2016) is a free online site developed by National Institute of Advanced Industrial Science and Technology in Japan. SBDS contains six different types of organic compounds including an electron impact mass spectrum (EI-MS), a Fourier transform infrared spectrum (FT-IR), a ^1H nuclear magnetic resonance (NMR) spectrum, a ^{13}C NMR spectrum, a laser Raman spectrum, and an electron spin resonance (ESR) spectrum. This database has more than 25,000 MS spectra, which were obtained by the electron impact method.

Human Metabolome Database. The Human Metabolome Database (HMDB) is an online database of small molecule metabolites found in the human body. This freely available electronic database contains detailed information about 42,000 small molecule metabolites. The project has been developed by the Canadian Institutes of Health Research, Alberta Innovates—Health Solutions, and by The Metabolomics Innovation Centre (Human Metabolome Database, 2016). This database was designed to give helpful information about (1) chemical data, (2) clinical data, and (3) molecular biology/biochemistry data. In the field of mass spectrometry, to identify or characterize human metabolites, HMDB allows to query NMR spectroscopy, GC MS spectrometry and LC/MS spectrometry data. In fact, HMDB includes nearly 10,000 NMR, GC MS, and LC/MS spectra. Additionally, there are other databases related to HMDB including Urine Metabolome, DrugBank, Toxin and Toxin Target Database (T3DB), Small Molecule Pathway Database (SMPDB), and Food Database (FooDB). The HMDB provides information about 3100 small molecule metabolites found in human urine (Urine Metabolome Database, 2016). Up to date, the DrugBank database (Drugbank, 2016) helps to get information about more than 8200 drug and drug metabolites. T3DB (Toxin and Toxin Target Database, 2016) contains information on 3673 toxins such as pollutants, pesticides, drugs, and food toxins. SMPDB (Small Molecule Pathway Database, 2016) has more than 618 pathway diagrams. FooDB is the most recent database in HMDB site (Food Database, 2016). This database contains information on 28,000 food components and food additives. Among others, FooDB allows to obtain data on (1) MS and MS/MS spectra, (2) nomenclature, and (3) concentrations in various foods.

Fiehn Library. This database has been developed by West Coast Metabolomics Center (California, USA) (Kind et al., 2009, p. 10,038). The library has been developed using several mass spectrometers including TOF or single quadrupole using electron impact ionization. Currently, this database contains over 1000 identified metabolites. MS spectra along with retention time allow to obtain a comprehensive metabolic profiling.

mz Cloud. mz cloud is a free database, which contains information on 6000 compounds and more than 1,518,041 high-resolution MS/MS spectra (m/z Cloud Database, 2016). This spectra information is shown to the user with spectral trees. Thus, the users could find the information in a more intuitive way. Most data displayed have been collected with Thermo mass spectrometers. MS/MS and multistage MS^n spectra were acquired at different CEs, precursor *m/z*, and isolation widths using CID and HCD. mz Cloud provides information about (1) the chemical structure, (2) computationally and manually annotated fragments (peaks), (3) identified adducts and multiply charged ions, and (4) molecular formulas, predicted precursor structure. Additionally, mz cloud will allow registered users to create individual, public, or private spectral libraries increasing the laboratory workflow.

A summary of mz cloud interface work is illustrated in Fig. 2.16, on the search of MS-spectra information of dopamine. First of all, the user should indicate the compound name (Fig. 2.16A). In this section, additional complementary information for instance related to the mass spectrometer and ionization mode used is available. Then, the user can select the different MS^n spectra available in the spectral tree (Fig. 2.16B). Finally, parameter such as CID or HCD applied in the selected MS^n spectrum could be checked.

The following databases are not readily available. However, because of their wide use and acceptance by scientific community it is worthwhile mentioning them. These databases are the Wiley Registry MS/MS and the NIST MS/MS database.

The Wiley Registry MS/MS. This MS/MS spectral library includes nearly 775,500 MS/MS spectra and 741,000 chemical structures. Thus, sensitive, specific, and robust identification of small molecules such as illicit drugs, pharmaceutical compounds, pesticides, and other small bioorganic molecules could be carried out (Oberacher, 2013, p. 312). These MS/MS spectra were obtained with different high-resolution mass spectrometers (Q-TOF) employing positive or negative ESI at least at 10 different CEs (Wiley, 2016). Its full potential has been checked in several cross-validation studies using sample spectra extracted from tandem mass spectral libraries from literature and acquired on various types of tandem mass spectrometers (Oberacher, Whitley, & Berger, 2013, p. 487).

The NIST MS/MS database. This database was developed by National Institute of Standards and Technology (NIST, Maryland, USA). NIST MS/MS database, which is widely used for scientific community, is supplied by John Wiley and Sons (NIST Standard Reference Database, 2016). The aim of this library is to provide reference mass spectral data for the identification of compounds through the fragmentation of their ions generated by ESI. MS spectra were obtained using several types of mass spectrometers, including low- and high-resolution mass spectrometers, including ion trap, QqQ, and Q-TOF. ESI electronic impact ionization has been employed. NIST MS/MS database currently provides more than 193,119 and 41,165 MS/MS spectra of small molecules and biologically active peptides, respectively (NIST Standard Reference Database, 2016).

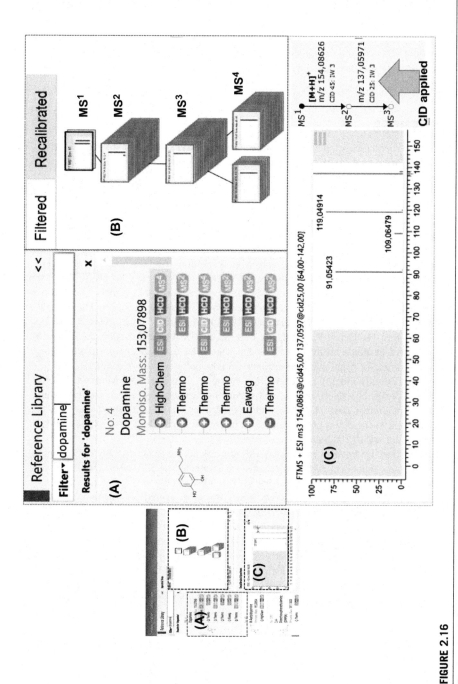

FIGURE 2.16

Basic search for a suspect compound using *m/z* cloud database: (A) selection of the suspect compound; (B) MS^n spectra available in the spectral tree; (C) MS or MS/MS spectrum selected.

ACKNOWLEDGMENTS

The authors acknowledge funding from the Spanish Ministerio de Economía y Competitividad (MINECO) (CTQ-2015-71321-P and JCI-2014-19573) and Junta de Andalucía (P12-FQM-2242). The use of brand, trade, or firm names in this publication is for identification purpose only and does not constitute endorsement by the authors.

REFERENCES

Abranko, L., García-Reyes, J. F., & Molina-Díaz, A. (2011). In-source fragmentation and accurate mass analysis of multiclass flavonoid conjugates by electrospray ionization time-of-flight mass spectrometry. *Journal of Mass Spectrometry, 46*(5), 478−488.

Andrews, G. L., Simons, B. L., Young, J. B., Hawkridge, A. M., & Muddiman, D. C. (2011). Performance characteristics of a new hybrid quadrupole time-of-flight tandem mass spectrometer. *Analytical Chemistry, 83*(13), 5442−5446.

Arnhard, K., Gottschall, A., & Pitterl, F. (2015). Applying "Sequential Windowed Acquisition of All Theoretical Fragment Ion Mass Spectra", (SWATH) for systematic toxicological analysis with liquid chromatography-high-resolution tandem mass spectrometry. *Analytical and Bioanalytical Chemistry, 407*(2), 405−414.

Bateman, K. P., Kellmann, M., Muenster, H., Papp, R., & Taylor, L. (2009). Quantitative-qualitative data acquisition using a benchtop Orbitrap mass spectrometer. *Journal of the American Society for Mass Spectrometry, 20*(8), 1441−1450.

Belmonte Vallés, N., Uclés, S., Besil, N., Mezcua, M., & Fernández-Alba, A. R. (2015). Analysis of pesticide residues in fruits and vegetables using gas chromatography-high resolution time-of-flight mass spectrometry. *Analytical Methods, 7*(5), 2162−2171.

Beucher, L., Dervilly-Pinel, G., Cesbron, N., Penot, M., Gicquiau, A., Monteau, F., et al. (2016). Specific characterization of non-steroidal selective androgen receptor modulators using supercritical fluid chromatography coupled to ion-mobility mass spectrometry: Application to the detection of enobosarm in bovine urine. *Drug Testing and Analysis.* http://dx.doi.org/10.1002/dta.1951, (in press).

Beucher, L., Dervilly-Pinel, G., Prevost, S., Monteau, F., & Le Bizec, B. (2015). Determination of a large set of beta-adrenergic agonists in animal matrices based on ion mobility and mass separations. *Analytical Chemistry, 87*(18), 9234−9242.

Bloomfield, N. & Le Blanc, Y. (2009). Mass defect triggered information dependent acquisition. United States Patent US7589318 B2.

Brown, L. J., Smith, R. W., Toutoungi, D. E., Reynolds, J. C., Brstow, A. W. T., Ray, A., et al. (2012). Enhanced analyte detection using in-source fragmentation of field asymmetric waveform ion mobility spectrometry-selected ions in combination with time-of-flight mass spectrometry. *Analytical Chemistry, 84*(9), 4095−4103.

Brown, L. J., Toutoungi, D. E., Davenport, N. A., Reynolds, J. C., Kaur-Atwal, G., Boyle, P., et al. (2010). Miniaturized ultra high field asymmetric waveform ion mobility spectrometry combined with mass spectrometry for peptide analysis. *Analytical Chemistry, 82*(23), 9827−9834.

Cameron, A. E., & Eggers, D. F., Jr. (1948). An ion "velocitron". *Review of Scientific Instruments, 19*, 605−647.

Chemspider Database. (2016). (Online) Available from http://www.chemspider.com/.

Cho, K., Mahieu, N., Ivanisevic, J., Uritboonthai, W., Chen, Y. J., Siuzdak, G., et al. (2014). isoMETLIN: A database for isotope-based metabolomics. *Analytical Chemistry, 86*(19), 9358–9361.

Cho, Y., Ahmed, A., Islam, A., & Kim, S. (2015). Developments in FT-ICR MS instrumentation, ionization techniques, and data interpretation methods for petroleomics. *Mass Spectrometry Reviews, 34*(2), 248–263.

Comisarow, M. B., & Marshall, A. G. (1974). Fourier transform ion cyclotron resonance spectroscopy. *Chemical Physics Letters, 25*(2), 282–283.

Comisarow, M. B., & Marshall, A. G. (1996). The early development of Fourier transform ion cyclotron resonance (FT-ICR) spectroscopy. *Journal of Mass Spectrometry, 31*(6), 581–585.

Dawson, J. H. J., & Guilhaus, M. (1989). Orthogonal-acceleration time-of-flight mass spectrometer. *Rapid Communications in Mass Spectrometry, 3*(5), 155–159.

Denisov, E., Damoc, E., Lange, O., & Makarov, A. (2012). Orbitrap mass spectrometry with resolving powers above 1,000,000. *International Journal of Mass Spectrometry, 325–327*, 80–85.

Doerr, A. (2014). DIA mass spectrometry. *Nature Methods, 12*, 35.

Drugbank. (2016). (Online) Available from http://www.drugbank.ca/.

Eckers, C., Wolff, J. C., Haskins, N. J., Sage, A. B., Giles, K., & Bateman, R. (2000). Accurate-mass liquid chromatography/mass spectrometry on orthogonal acceleration time-of-flight mass analyzers using switching between separate sample and reference sprays. 1. Proof of concept. *Analytical Chemistry, 72*(16), 3683–3688.

Eliuk, A., & Makarov, S. (2015). Evolution of Orbitrap mass spectrometry instrumentation. *Annual Review of Analytical Chemistry, 8*, 61–80.

Eljarrat, E., & Barceló, D. (2002). Congener-specific determination of dioxins and related compounds by gas chromatography coupled to LRMS, HRMS, MS/MS and TOFMS. *Journal of Mass Spectrometry., 37*(11), 1105–1117.

Far, J., Delvaux, C., Kune, C., Eppe, G., & De Pauw, E. (2014). The use of ion mobility mass spectrometry for isomer composition determination extracted from Se-rich yeast. *Analytical Chemistry, 86*(22), 11246–11254.

Fenn, J. B., Mann, M., Meng, C. K., Wong, S. F., & Whitehouse, C. M. (1989). Electrospray ionization for mass spectrometry of large biomolecules. *Science, 246*(4926), 64–71.

Ferrer, I., & Thurman, E. M. (2005). Measuring the mass of an electron by LC/TOF-MS: A study of twin ions. *Analytical Chemistry, 77*(10), 3394–3400.

Fjeldsted, J. (2003). *Time-of-flight mass spectrometry. Technical overview.* Available from https://www.agilent.com/cs/library/technicaloverviews/Public/5989-0373EN%2011-Dec-2003.pdf.

Fjeldsted, J. C. (2016). Advances in time-of-flight mass spectrometry. In S. Pérez, P. Eichhorn, & D. Barceló (Eds.), *Comprehensive analytical chemistryApplications of time-of-flight and Orbitrap mass spectrometry in environmental, food, doping and forensic analysis.* Amsterdam: Elsevier.

Gallien, S., Duriez, E., Demeure, K., & Domon, B. (2012). Selectivity of LC-MS/MS analysis: Implication for proteomics experiments. *Journal of Proteomics, 81*, 148–158.

Geiger, T., Cox, J., & Mann, M. (2010). Proteomics on an Orbitrap benchtop mass spectrometer using all-ion fragmentation. *Molecular & Cellular Proteomics, 9*(10), 2252–2261.

Gilbert-López, B., Geltenpoth, H., Meyer, C., Michels, A., Hayen, H., Molina-Díaz, A., et al. (2013). Performance of dielectric barrier discharge ionization mass spectrometry for pesticide testing: A comparison with atmospheric pressure chemical ionization and electrospray ionization. *Rapid Communications in Mass Spectrometry, 27*(3), 419–429.

Giles, K., Pringle, S., Worthington, K. R., Little, D., Wildgoose, J. L., & Bateman, R. H. (2004). Applications of a travelling wave-based radio-frequency-only stacked ring ion guide. *Rapid Communications in Mass Spectrometry, 18*(20), 2401—2414.

Goscinny, S., Joly, L., De Pauw, E., Hanot, V., & Eppe, G. (2015). Travelling-wave ion mobility time-of-flight mass spectrometry as an alternative strategy for screening of multi-class pesticides in fruits and vegetables. *Journal of Chromatography A, 1405*, 85—93.

Guevremont, R. (2004). High-field asymmetric waveform ion mobility spectrometry: A new tool for mass spectrometry. *Journal of Chromatography A., 1058*, 3—19.

Hernández, F., Sancho, J. V., Ibañez, M., Abad, E., Portolés, T., & Mattioli, L. (2012). Current use of high-resolution mass spectrometry in the environmental sciences. *Analytical and Bioanalytical Chemistry, 403*(5), 1251—1264.

Horai, H., Arita, M., Kanaya, S., Nihei, Y., Ikeda, T., Suwa, K., et al. (2010). MassBank: a public repository for sharing mass spectral data for life sciences. *Journal of Mass Spectrometry, 45*(7), 703—714.

Hu, Q., Noll, R., Li, H., Makarov, A., & Hardman, M. (2005). The Orbitrap: A new mass spectrometer. *Journal of Mass Spectrometry, 40*(4), 430—443.

Human Metabolome Database. (2016). (Online) Available from http://www.hmdb.ca/.

isoMETLIN Database. (2016). (Online) Available from https://isometlin.scripps.edu.

Kanu, A. B., Dwivedi, P., Tam, M., Matz, L., & Hill, H. H. (2008). Ion mobility-mass spectrometry. *Journal of Mass Spectrometry, 43*(1), 1—22.

Kelly, R. T., Tolmachev, A. V., Page, J. S., Tang, K., & Smith, R. D. (2010). The ion funnel: Theory, implementations and applications. *Mass Spectrometry Reviews, 29*(2), 294—312.

Kind, T., Wohlgemuth, G., Lee, D. Y., Lu, Y., Palazoglu, M., Shahbaz, S., et al. (2009). FiehnLib-mass spectral and retention index libraries for metabolomics based on quadrupole and time-of-flight gas chromatography/mass spectrometry. *Analytical Chemistry, 81*(21), 10038—10048.

Kingdon, K. H. (1923). A method for the neutralization of electron space charge by positive ionization at very low gas pressures. *Physical Review, 21*, 408—418.

Kolakowski, B. M., & Mester, Z. (2007). Review of applications of high-field asymmetric waveform ion mobility spectrometry (FAIMS) and differential mobility spectrometry (DMS). *Analyst, 132*, 842—864.

Kurulugama, R. T., Darland, E., Kuhlmann, F., Stafford, G., & Fjeldsted, J. (2015). Evaluation of drift gas selection in complex sample analyses using a high performance drift tube ion mobility-QTOF mass spectrometer. *Analyst, 140*, 6834—6844.

Kurulugama, R. T., Imatani, K., & Taylor, L. (2013). *The agilent ion mobility Q-TOF mass spectrometer system. Technical overview*. Available from http://www.agilent.com/cs/library/technicaloverviews/public/5991-3244EN.pdf.

Ladumor, M. K., Tiwari, S., Patil, A., Bhavsar, K., Jhajra, S., Prasad, B., et al. (2016). High-resolution mass spectrometry in metabolite identification. In S. Pérez, P. Eichhorn, & D. Barceló (Eds.), *Applications of time-of-flight and Orbitrap mass spectrometry in environmental, food, doping and forensic analysis*. Amsterdam: Elsevier.

Lange, O., Damoc, E., Wieghaus, A., & Makarov, A. (2014). Enhanced fourier transform for Orbitrap mass spectrometry. *International Journal of Mass Spectrometry, 369*, 16—22.

Lapthorn, C., Pullen, F., & Chowdhry, B. Z. (2013). Ion Mobility spectrometry-mass spectrometry (IMS-MS) of small molecules: Separating and assigning structure to ions. *Mass Spectrometry Reviews, 32*(81), 43—71.

Law, K. P., & Lim, Y. P. (2013). Recent advances in mass spectrometry: data independent analysis and hyper reaction monitoring. *Expert Review of Proteomics, 10*(6), 551—566.

Li, X. Q., Zhang, Q. H., Ma, K., Li, H. M., & Guo, Z. (2015). Identification and determination of 34 water-soluble synthetic dyes in foodstuff by high performance liquid chromatography-diode array detection-ion trap time-of-flight tandem mass spectrometry. *Food Chemistry, 182*, 316–326.

Lin, L., Lin, H., Zhang, M., Dong, X., Yin, X., Qu, C., et al. (2015). Types, principle, and characteristics of tandem high-resolution mass spectrometry and its applications. *RSC Advances, 5*(130), 107623–107636.

m/z Cloud Database. (2016). (Online) Available from https://www.mzcloud.org/.

Ma, L., Wen, B., Ruan, Q., & Zhu, M. (2008). Rapid screening of glutathione-trapped reactive metabolites by linear ion trap mass spectrometry with isotope pattern-dependent scanning and postacquisition data mining. *Chemical Research in Toxicology, 21*(7), 1477–1483.

Ma, S., & Chowdhury, S. K. (2013). Data acquisition and data mining techniques for metabolite identification using LC coupled to high-resolution MS. *Bioanalysis, 5*(10), 1285–1297.

Makarov, A. (1999) Mass spectrometer. United States Patent US 5886346.

Makarov, A. (2000). Electrostatic axially harmonic orbital trapping: A high performance technique of mass analysis. *Analytical Chemistry, 72*(6), 1156–1162.

Makarov, A., Denisov, E., Kholomeev, A., Balschun, W., Lange, O., Strupat, K., et al. (2006). Performance evaluation of a hybrid linear ion trap/Orbitrap mass spectrometer. *Analytical Chemistry, 78*(7), 2113–2120.

Makarov, A., & Hardman, M.E. (2006) Mass spectrometry method and apparatus. United States Patent US 6998609 B2.

Mamyrin, B.A. (1966). (Doctoral dissertation). Leningrad: Physico-technical Institute, USSR Academy of sciences.

Mann, M., Hendrickson, R. C., & Pandey, A. (2001). Analysis of proteins and proteomes by mass spectrometry. *Annual Review of Biochemistry, 70*, 437–473.

Marshall, A. G., & Chen, T. (2015). 40 years of Fourier transform ion cyclotron resonance mass spectrometry. *International Journal of Mass Spectrometry, 377*, 410–420.

Marshall, A. G., & Hendrickson, C. L. (2008). High-resolution mass spectrometers. *Annual Review of Analytical Chemistry, 1*, 579–599.

Marshall, A. G., Hendrickson, C. L., & Jackson, G. S. (1998). Fourier transform ion cyclotron resonance mass spectrometry: A primer. *Mass Spectrometry Reviews, 17*(1), 1–35.

Martins, C. P. B., Bromirski, M., Prieto Conaway, M. C., & Makarov, A. A. (2016). Orbitrap mass spectrometry: Evolution and applicability. In S. Pérez, P. Eichhorn, & D. Barceló (Eds.), *Applications of time-of-flight and Orbitrap mass spectrometry in environmental, food, doping and forensic analysis*. Amsterdam: Elsevier.

Massbank Database. (2016). (Online) Available from http://www.massbank.jp.

Metlin Database. (2016). (Online) Available from https://metlin.scripps.edu/index.php.

Michalski, A., Damoc, E., Lange, O., Denisov, E., Nolting, D., Müller, M., et al. (2012). Ultra-high resolution linear ion trap Orbitrap mass spectrometer (Orbitrap Elite) facilitates top down LC-MS/MS and versatile peptide fragmentation modes. *Molecular & Cellular Proteomics*. http://dx.doi.org/10.1074/mcp.O111.013698.

Mol, H. G. J., Tienstra, M., & Zomer, P. (2016). Evaluation of gas chromatography − electron ionization-full-scan high resolution Orbitrap mass spectrometry for pesticide residue analysis. *Analytica Chimica Acta, 935*, 161–172.

Morris, H. R., Paxton, T., Dell, A., Langhorne, J., Berg, M., Bordoli, R. S., et al. (1996). High-sensitivity collisionally-activated decomposition tandem mass spectrometry on a novel quadrupole/orthogonal-acceleration time-of-flight mass spectrometer. *Rapid Communications in Mass Spectrometry, 10*(8), 889–896.

Murray, K. K., Boyd, R. K., Eberlin, M. N., Langley, G. J., Li, L., & Naito, Y. (2013). Definitions of terms relating to mass spectrometry (IUPAC Recommendations 2013). *Pure and Applied Chemistry, 85*(7), 1515–1609.

Nicolardi, S., Bogdnov, B., Deelder, A. M., Palmblad, M., & Van Der Burgt, Y. E. M. (2015). Developments in FTICR-MS and its potential for body fluid signatures. *International Journal of Molecular Sciences, 16*(11), 27133–27144.

Nier, K. A., Yergey, A. L., & Gale, P. J. (Eds.). (2015), *Historical Perspectives, Part A: The development of mass spectrometry: Vol. 9. The encyclopedia of mass spectrometry.* Amsterdam: Elsevier.

Niessen, W. M. A., & Falck, D. (2015). Introduction to mass spectrometry, a tutorial. In J. Kool, & W. M. A. Niessen (Eds.), *Analyzing biomolecular interactions by mass spectrometry.* Weinheim: Wiley-VCH Verlag GmbH & Co. KGaA.

Nikolaev, E. N. (2015). Some notes about FT ICR mass spectrometry. *International Journal of Mass Spectrometry, 377*, 421–431.

Nist Standard Reference Database 1a V14. (2016). (Online) Available from https://www.nist.gov/srd/nist-standard-reference-database-1a-v14#.

Oberacher, H. (2013). Applying tandem mass spectral libraries for solving the critical assessment of small molecule identification (CASMI) LC/MS challenge 2012. *Metabolites, 3*(2), 312–324.

Oberacher, H., Whitley, G., & Berger, B. (2013). Evaluation of the sensitivity of the 'Wiley registry of tandem mass spectral data, MSforID' with MS/MS data of the 'NIST/NIH/EPA mass spectral library'. *Journal of Mass Spectrometry, 48*(4), 487–496.

Olsen, J. V., Macek, B., Lange, O., Makarov, A., Horning, S., & Mann, M. (2007). High-energy C-trap dissociation for peptide modification analysis. *Nature Methods, 4*, 709–712.

Olsen, J. V., Schwartz, J. C., Griper-Raming, J., Nielsen, M. L., Damoc, E., Denisov, E., et al. (2009). A dual in pressure linear ion trap Orbitrap instrument with very high sequencing speed. *Molecular & Cellular Proteomics, 8*(12), 2759–2769.

Perry, R., Cooks, R. G., & Noll, R. (2008). Orbitrap mass spectrometry: Instrumentation, ion motion and applications. *Mass Spectrometry Reviews, 27*(6), 661–699.

Peterson, A. C., Balloon, A. J., Westphall, M. S., & Coon, J. J. (2014). Development of a GC/quadrupole-Orbitrap mass spectrometer, Part II: New approaches for discovery metabolomics. *Analytical Chemistry, 85*(20), 10044–10051.

Peterson, A. C., Mcalister, G. C., Quarmby, S. T., Griep-Raming, J., & Coon, J. J. (2010). Development and characterization of a GC-enabled QLT-Orbitrap for high-resolution and high-mass accuracy GC/MS. *Analytical Chemistry, 82*(20), 8618–8628.

Pringle, S. D., Giles, K., Wildgoose, J. L., Williams, J. P., Slade, S. E., Thalassinos, K., et al. (2007). An investigation of the mobility separation of some peptide and protein ions using a new hybrid quadrupole/travelling wave IMS/oa-TOF instrument. *International Journal of Mass Spectrometry, 261*(1), 1–12.

Rauniyar, N. (2015). Parallel reaction monitoring: a targeted experiment performed using high resolution and high mass accuracy mass spectrometry. *International Journal of Molecular Sciences, 16*(12), 28566–28581.

Roemmelt, A. T., Steuer, A. E., Poetzsch, M., & Kraemer, T. (2014). Liquid chromatography, in combination with a quadrupole time-of- flight instrument (LC QTOF), with sequential window acquisition of all theoretical fragment-ion spectra (SWATH) acquisition: systematic studies on its use for screenings in clinical and forensic toxicology and comparison with information-dependent acquisition (IDA). *Analytical Chemistry, 86*(23), 11742−11749.

Rose, R., Damoc, E., Denisov, E., Makarov, A., & Heck, A. J. (2012). High-sensitivity Orbitrap mass analysis of intact macromolecular assemblies. *Nature Methods, 9*, 1084−1086.

Scheltema, R. A., Hauschild, J. P., Lange, O., Hornburg, D., Denisov, E., Damoc, E., et al. (2014). The Q exactive HF, a benchtop mass spectrometer with a prefilter, high performance quadrupole and an ultra-high field Orbitrap analyzer. *Molecular & Cellular Proteomics, 13*(12), 3698−3708.

Schmidt, A., Gehlenborg, N., & Bodenmiller, B. (2008). An integrated, directed mass spectrometric approach for in-depth characterization of complex peptide mixtures. *Molecular & Cellular Proteomics, 7*(11), 2138−2150.

Schwemer, T., Rüger, C. P., Sklorz, M., & Zimmermann, R. (2015). Gas chromatography coupled to atmospheric pressure chemical ionization FT-ICR mass spectrometry for improvement of data reliability. *Analytical Chemistry, 87*(24), 11957−11961.

Senko, M. W., Remes, P. M., Canterbury, J. D., Mathur, R., Song, Q., Eliuk, S. M., et al. (2013). Novel parallelized quadrupole/linear ion trap/Orbitrap tribrid mass spectrometer improving proteome coverage and peptide identification rates. *Analytical Chemistry, 85*(24), 11710−11714.

Shimadzu Technical Report. (2009). *Accurate mass measurement of high concentration samples by LCMS-IT-TOF.* Retrieved from http://www.ssi.shimadzu.com/products/literature/lcms/TechReport%20Vol27_LCMS_TEM_AP.pdf.

Silcock, P., Cojocariu, C., Roberts, D., Quarmby, S., Guckenberger, B., Cole, J., et al. (2015). A fully integrated GC Orbitrap system opens a new chapter in GC-MS. *Proceedings of 63rd American Society for Mass Spectrometry.*

Smith, C. A., Maille, G. O., Want, E. J., Qin, C., Trauger, S. A., Brandon, T. R., et al. (2005). METLIN: A metabolite mass spectral database. *Therapeutic Drug Monitoring, 27*(6), 747−751.

Smith, R. W., Toutoungi, D. E., Reynolds, J. C., Bristow, A. W. T., Ray, A., Sage, A., et al. (2013). Enhanced performance in the determination of ibuprofen 1-β-o-acyl glucuronide in urine by combining high field asymmetric waveform ion mobility spectrometry with liquid chromatography-time-of-flight mass spectrometry. *Journal of Chromatography A, 1278*, 76−81.

Stephan, S., Hippler, J., Köhler, T., Deeb, A. A., Schmidt, T. C., & Schmitz, O. J. (2016). Contaminant screening of wastewater with HPLC-IM-qTOF-MS and LC+LC-IM-qTOF-MS using a CCS database. *Analytical and Bioanalytical Chemistry, 408*(24), 6545−6555.

Stephens, W. E. (1946). Proceedings of the American Physical Society. *Physical Reviews, 69*, 691.

Stupat, K., Kovtoun, V., Biu, H., Viner, R., Stafford, G., & Horning, S. J. (2009). MALDI produced ions inspected with a linear ion trap-Orbitrap hybrid mass analyzer. *Journal of the American Society for Mass Spectrometry, 20*(8), 1451−1463.

Tautenhahn, R., Cho, K., Uritboonthai, W., Zhu, Z., Patti, G. J., & Siuzdak, G. (2012). An accelerated workflow for untargeted metabolomics using the METLIN database. *Nature Biotechnology, 30*, 826−828.

Thalassinos, K., Slade, S. E., Jennings, K. R., Scrivens, J. H., Wildgroose, J., Hoyes, J., et al. (2004). Ion mobility mass spectrometry of proteins in a modified commercial mass spectrometer. *International Journal of Mass Spectrometry, 236*, 55−63.

Thermo Scientific Technical Report. (2016). *Thermo scientific Q-Exactive GC Orbitrap GC-MS/MS system (PS10458-EN 0316S).* http://planetorbitrap.com/data/fe/file/PS-10458-GC-MS-Q-Exactive-Orbitrap-PS10458-EN.pdf.

Toxin And Toxin Target Database. (2016). (Online) Available from http://www.t3db.ca/.

Tranchida, P. Q., Franchida, F. A., Dugo, P., & Mondello, L. (2016). Comprehensive two-dimensional gas chromatography-mass spectrometry: Recent evolution and current trends. *Mass Spectrometry Reviews, 35*(4), 524−534.

The Food Database. (2016). (Online) Available from http://foodb.ca/.

The Small Molecule Pathway Database. (2016). (Online) Available from http://smpdb.ca/.

The Spectral Data Base For Organic Compounds. (2016). (Online) Available from http://sdbs.db.aist.go.jp/sdbs/cgi-bin/cre_index.cgi.

The Urine Metabolome Database. (2016). (Online) Available from http://www.urinemetabolome.ca.

Van Babel, B., & Abad, E. (2008). Long-term worldwide QA/QC of dioxins and dioxin-like PCBs in environmental samples. *Analytical Chemistry, 80*(11), 3956−3964.

Van Babel, B., Geng, D., Cherta, L., Nacher-Mestre, J., Portolés, T., Ábalos, M., et al. (2015). Atmospheric pressure chemical ionization tandem mass spectrometry (APGC/MS/MS) an alternative to high-resolution mass spectrometry (HRGC-HRMS) for the determination of dioxins. *Analytical Chemistry, 87*(17), 9047−9053.

Van Mourik, L. M., Leonards, P. E. G., Gaus, C., & De Boer, J. (2015). Recent developments in capabilities for analysing chlorinated paraffins in environmental matrices: A review. *Chemosphere, 136*, 259−272.

Wiley, W. C. (1956). Bendix time-of-flight mass spectrometer. *Science, 124*(3226), 817−820.

Wiley. (2016). (Online) Available from http://www.wiley.com/WileyCDA/WileyTitle/productCd-1118037448.html.

Wiley, W. C., & McLaren, L. H. (1955). Time-of-flight mass spectrometer with improved resolution. *Review of Scientific Instruments, 26*, 1150−1157.

Wolff, J. C., Eckers, C., Sage, A. B., Giles, K., & Bateman, R. (2001). Accurate mass liquid chromatography/mass spectrometry on Quadrupole orthogonal acceleration time-of-flight mass analyzers using switching between separate sample and reference sprays. 2. Applications using the dual electrospray ion source. *Analytical Chemistry, 73*(11), 2605−2612.

Yamashita, M., & Fenn, J. B. (1984). Electrospray ion source. Another variation on the free-jet theme. *Journal of Physical Chemistry, 88*(20), 4451−4459.

Zhan, Z., Xing, J., Sun, Z., Ting, E. Z. W., & Chew, Y. L. (2015). A comparative study of targeted screening method by LC-MS/MS and un-targeted screening method by LC-TOF in residual pesticides analysis. *Presented at American Society for Mass Spectrometry.* WP054.

Zhang, F., Yu, C., Wang, W., Fan, R., Zhang, Z., & Guo, Y. (2012). Rapid simultaneous screening and identification of multiple pesticide residues in vegetables. *Analytica Chimica Acta, 757*, 39−47.

Zhang, H., Zhang, D., Rayb, K., & Zhu, M. (2009). Mass defect filter technique and its applications to drug metabolite identification by high-resolution mass spectrometry. *Journal of Mass Spectrometry, 44*(7), 999−1016.

Zimmermann, R., Groger, T., Schäffer, M., Weggler, B., Wendt, J., Sklorz, M., et al. (2014). One dimensional and comprehensive two-dimensional gas chromatography coupled to a multi-reflection, ultra high resolution time-of flight mass spectrometer for the characterization of complex mixtures. In *Presented at International Symposium on Hyphenated Techniques in Chromatography and Hyphenated Chromatographic Analyzers.*

Analytical Strategies Used in HRMS

3

Ana Agüera, Ana Belén Martínez-Piernas, Marina Celia Campos-Mañas

CIESOL, Joint Centre of the University of Almería — CIEMAT, Almería, Spain

3.1 INTRODUCTION

Today, it is well recognized that liquid chromatography coupled to tandem mass spectrometry (LC-MS/MS) is the most widely applied instrumental technique for pesticide residue analysis in food. Various reasons such as the wider range of analytes that are amenable by this technique, including polar, nonvolatile, and thermolabile pesticides, and the excellent quantitative performance justify this fact. Triple quadrupole (QqQ) and quadrupole linear ion trap (QqLIT) analyzers represent the best option for quantitative analysis without any doubt. When operating in multiple-reaction monitoring (MRM) mode, accurate determination, identification, and confirmation of multiple analytes (typically more than 150) are obtained with a high reproducibility, sensitivity, and selectivity. Presence of the (de)protonated molecules or adduct ions and two characteristic product ions (MRM transitions) is recognized as useful criteria for reliable confirmation of the pesticides in the samples. Quantitative performance is also benefited from the broad linear dynamic range of the detector, typically up to four to five orders of magnitude. Consequently, these instruments are indispensable tools in official laboratories devoted for routine analysis of defined pesticide lists, where quantification is the main goal to comply with regulatory controls. However, each time the need to identify compounds out of "frequent" target lists is more evident. Therefore, qualitative screening methods able to monitor a large number of compounds not included in the scope of the laboratories have been implemented. This methodology offers laboratories a cost-effective means to extend their analytical scope to analytes that can be only eventually present in the samples. It is in this area where high-resolution mass spectrometry (HRMS), using time-of-flight (TOF) and Orbitrap mass analyzers, and more frequently their couplings Q-TOF and Q-Orbitrap, is gaining popularity because it provides strong evidence of the identity of the suspected analytes. Instrumental improvements experimented in the last ten years in terms of mass-resolving power (25,000—140,000 FWHM), mass accuracy (<2 ppm), sensitivity (and detectability), and dynamic range have contributed to this fact. In addition, powerful software for data acquisition and evaluation has opened a wide range of possibilities for the identification of nontarget species and unknown compounds.

Applications in High Resolution Mass Spectrometry. http://dx.doi.org/10.1016/B978-0-12-809464-8.00003-8

By taking advantage of these properties, various analytical strategies can be used in HRMS, depending on the objective of the analysis. Some of them will be reviewed in this chapter.

3.2 ADVANTAGES OF HIGH-RESOLUTION MASS SPECTROMETRY IN PESTICIDE ANALYSIS

HRMS instruments present several advantages with respect to conventional QqQ systems in pesticide residue analysis. They show unique capabilities for unequivocal identification mainly based on the accurate mass measurements. In addition, they present increased selectivity, which is probably the most important characteristic when trace amounts of hundreds of analytes have to be determined in very complex samples, containing thousands of matrix components frequently present at significantly higher concentrations; such is the case of food analysis. Finally, operating in full-scan mode permits the detection of an unlimited number of compounds with a minimum effort in the optimization of analytical parameters and offers the possibility of retrospective analysis of data, which are not possible using QqQ. However, to date they have exhibited lower sensitivity and narrower dynamic range than modern QqQ instruments, which have limited their use in quantitative applications. Nevertheless, recent studies have demonstrated their performance in quantitative analysis of pesticide residues in food (Gómez-Ramos, Rajski, Heinzen, & Fernández-Alba, 2015). The development of powerful data processing software packages is contributing substantially to the implementation of these instruments and it is in this field where more development is expected. Powerful software tools are needed to explore the high amount of full-scan MS and MS/MS data generated during the analysis of such complex food matrices in an automatic routine.

3.2.1 SELECTIVITY IN HIGH-RESOLUTION MASS SPECTROMETRY: ACCURATE MASS AND RESOLUTION IN QUALITATIVE ANALYSIS

The accurate mass of a target ion is valuable information for its unequivocal identification. Modern instruments are able to provide very low mass errors, which allows a reliable confirmation of the analytes in the samples. The DG SANTE (SANTE/11945/2015) recognizes a tolerance in mass accuracy of 5 ppm as an adequate identification criterion, but frequently, most of the modern HRMS instruments are able to provide mass accuracies lower than 2 ppm.

However, mass accuracy can be affected by different factors. Presence of coeluting isobaric matrix interferences can alter accurate mass measurements, resulting in false negative and positive results. Because of the large amount of matrix components present in food samples, these coelutions are relatively frequent, especially when the number of pesticides included in the analytical method is high (Gómez-Ramos, Ferrer, Malato, Agüera, & Fernández-Alba, 2013). Chromatographic gradients are inefficient to resolve this problem in all cases, and higher mass resolutions, providing enhanced selectivity, are required to discriminate such isobaric species. In most of the cases, resolution >20,000 (at full width half height, FWHM) and mass accuracy <5 ppm, which is reached for most of TOF instruments, are sufficient to separate pesticides of interest from matrix interferences. However, higher resolution is possible using Orbitrap mass analyzers, which can reach values up to 140.000 FWHM.

High resolution is not only useful to discriminate analyte ions from matrix interferences, reducing false negative report, but also necessary to distinguish pesticide ions with the same nominal mass, when differences in retention time (RT) of both species are not significant (Pérez-Ortega et al., 2016). Therefore, it is easy to think that a higher resolution provides better results. However, this is not so true, especially when analytes are present at low concentrations because very often an increase in resolution results in a diminution of the sensitivity. In TOF instruments, it is possible to operate in "low-resolution" and "high-resolution" modes. High-resolution mode is recommended for qualitative analysis, while it is operating in low-resolution mode where best results in terms of sensitivity and dynamic range are obtained. Thus, selection between both operation modes depends on the objective of the analysis. Orbitrap mass analyzers allow selection of different resolution values, typically corresponding to 17,500; 35,000; 70,000; 100,000; or up to 140,000 FWHM in modern instruments. Again, an increase in resolution does not always contribute to get better results and some aspects have to be considered. Thus, it must be taken into account that resolution values in Orbitrap systems refer to an ion at m/z 200 because the resolution at which an ion is measured is inversely proportional to the square root of its m/z. Orbitrap instruments produce the highest resolution for low m/z ions. This behavior is opposite to the typical TOF performance, which provides the highest resolution for relatively high m/z ion masses.

Users have to verify whether sufficient resolutions are obtained for the m/z range of their particular application. Most of the (Q)TOF-MS(/MS) and Orbitrap MS(/MS) applications reported for the analysis of pesticides in fruits and vegetables use resolutions ranging from 5000 to 20,000 FWHM for (Q)TOF systems and typically higher than 30,000–50,000 FWHM for Orbitrap instruments.

Once a resolution value is selected, the selectivity strongly depends on the width of the extracted mass window for the selected ion. Reducing the mass window width can significantly improve the selectivity, whenever the mass accuracy and precision provided by the instrument is taken into account. Otherwise, extremely narrow mass widths can lead to badly defined chromatographic peaks, where some data points are absent. On the other hand, very restrictive mass windows can be inadequate when isobaric coeluting interferences are present, affecting the accurate mass of the target ion. In this case, the analyte can be undetected because its "experimental" accurate mass falls outside the selected window. As an example, a mass extraction window of 2.5 ppm was proposed by Jia, Chu, Ling, Huang, and Chang (2014) in a high-throughput screening of more than 330 pesticide and veterinary drug residues in baby food by Q-Orbitrap mass spectrometry. This optimized value provided a high selectivity and reduced probability of false positives.

False negatives can also be reported when suspect screening workflows are applied. In this analytical strategy, RT of the target analytes is not included in the searching list, and the application of automatic search algorithms by the instrument software can lead to some false negatives when isobaric interferences are present.

3.2.2 IMPROVING SELECTIVITY BY TANDEM MASS SPECTROMETRY INFORMATION

Information provided by full-scan HRMS data is essential for improving identification capabilities in pesticide analysis, but in many cases it is insufficient. One of the challenges encountered is the number of false positives and negatives detected during data processing when only the expected RT and the exact mass of the (de)protonated or adduct ions are considered for the identification (Mol et al., 2010). Even when careful optimization of the operational parameters affecting automated analyte detection is performed, the report of false positives is common. Differentiation of structural isomers can also result in an unsolvable problem. They cannot be distinguished in full scan because they have exactly the same mass and elemental composition. Isotopic pattern is also identical and only differences in MS/MS fragmentation can be useful for a correct identification. Consequently, the inclusion of fragments in MS and product ions in MS/MS as identification criteria has been evaluated and reported as a relevant strategy for reducing the number of false positives during screening. Mol, Zomer, and De Koning (2012) evaluated the use of additional adducts, M+1 and M+2 isotopes, and fragment ions as diagnostic ions and their suitability as identification criterion. Adducts did not represent a good choice because of their scarce incidence and high variability. Isotopes were highly selective and reproducible but lacked sufficient sensitivity. Finally, the use of fragment ions resulted in a favorable choice from a sensitivity point of view, although in some cases showed less selective, and the average variability of the ion ratio was higher. Identification requirements of DG SANTE (SANTE/11945/2015) for HRMS establish that at least two ions have to be recorded: one molecular ion, (de)protonated molecule or adduct ion with mass accuracy of 5 ppm, plus one MS/MS product ion. Furthermore, the ratio of the peak area of the two masses must be within ±30% (relative) of the average ratio for the compound in the measured standards from the same run.

Several strategies are available for providing useful diagnostic fragment ions. When a single-stage HRMS instrument is used, improvement in identification capabilities can be obtained by increasing the ionization voltage in the source, the so-called "in-source" fragmentation. Fragment ions are generated between the atmospheric pressure source and the high-vacuum region of the mass analyzer. Ions generated in the source are accelerated by the application of voltages, prompting collisions with surrounding species, which can produce sufficient energy to yield (diagnostic) fragment ions. The improvement in selectivity obtained by this strategy is usually accompanied by a reduction in sensitivity. Pérez-Ortega et al. (2016) have evaluated in depth this fragmentation process using a single TOF-MS system, but the results obtained in terms of information provided are not so satisfactory as using other alternative fragmentation techniques.

Modern tandem mass spectrometers, incorporating a quadrupole mass filter, enable the simultaneous acquisition of full-scan MS and MS/MS on all peaks observed in a given sample with improved selectivity. In this case, ions selected in the quadrupole filter are fragmented in a dedicated collision cell (collision-induced dissociation, CID). HRMS spectra can be acquired using different available acquisition techniques, which can roughly be divided into two categories: data-dependent acquisition (DDA), which is the older and more commonly applied method, and data-independent acquisition (DIA) strategies that have been developed recently (Fig. 3.1).

In targeted analysis, the DDA mode is recommended. While operating under DDA acquisition mode, only selected ions within a narrow m/z window, typically 1−3 Da wide, are transferred to the collision cell to generate product ions. Under these conditions, the MS^2 spectra are highly selective and minimal interferences are present. A list of target parent ions and RT window for each one are predefined. Data acquisition is carried out in full-scan mode, and only when any of the parent ions from the target list are detected at the correct RT, the instrument switches to MS/MS mode. Once the product ion spectra are recorded, the system returns to the full-scan mode. Gómez-Ramos et al. (2015) evaluated the identification capability of this strategy using LC/Q-Orbitrap MS. One hundred and thirty nine pesticide residues were analyzed in various fruit and vegetable samples. Two different resolutions were selected, 70,000 in full-scan and 17,500 in MS/MS. The use of a

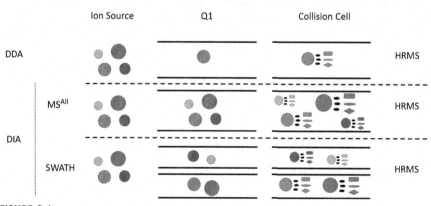

FIGURE 3.1

Scheme of strategies available for providing tandem mass spectrometry fragmentation in high-resolution mass spectrometry (HRMS). *DDA*, data-dependent acquisition; *DIA*, data-independent acquisition.

Adapted from Zhu, X., Chen, Y., & Subramanian, R. (2014). Comparison of information-dependent acquisition, SWATH, and MS^All techniques in metabolite identification study employing ultrahigh-performance liquid chromatography−quadrupole time of-flight mass spectrometry. Analytical Chemistry, 86(2), 1202−1209.

lower resolution in MS/MS was advantageous because it consumed less analysis time without any damage in the selectivity, thus assuring more points per peak in full-scan mode. In addition, an increase to 35,000 affected neither the number of compounds confirmed nor the mass accuracy of the ions. In the same cases, mass errors of MS/MS fragments were higher than 5 ppm, and thus, the mass tolerance for fragments was set at 10 ppm to avoid false negatives. Under these conditions, 100 real samples were analyzed and the results were compared with those performed by LC-QqQ-MS/MS. Results demonstrated that MS/MS spectra were useful in the identification of false positives coming from isobaric matrix compounds, providing similar identification potential as two transitions and ion ratio criterion. However, some limitations have been observed as the loss of MS/MS information for the less abundant ions because acquisition of MS/MS is triggered by the detection of ions above a certain threshold. In addition, the maximum number of precursors selected per cycle and active exclusion for compounds overlapping or with close RT can affect acquisition of MS/MS spectra. Therefore, additional injections in targeted MS/MS mode can be required. In targeted MS/MS, scans for specific precursor ions are always carried out across the whole RT window. This results in a high consumption of the scan time that can affect the number of data points registered in the full-scan chromatogram and it is not applicable to large-scale screening methods.

In contrast to DDA, another strategy known as "all ion mode," MS^{All} or MS^E, has been applied to generate useful MS/MS fragmentation in pesticide residue analysis (Pérez-Ortega et al., 2016). This technique performs a DIA, recording all the product ions generated in the collision cell without previous precursor ion isolation in Q1. This operation mode lacks the specificity of IDA but has benefits of acquiring all the information all the time without time windows. Thus, this operation mode is not limited by the scan speed of the instrument, compared to DDA methods where only selected ions are chosen for fragmentation. To get adequate fragmentation spectra for each precursor ion, different collision energies can be applied in a single injection. The strength of MS^{All} method is a high MS/MS acquisition hit rate. A higher number of MS/MS are acquired compared with DDA (Zhu, Chen, & Subramanian, 2014). Given the number of potentially coeluting analytes/matrix components in food samples, the use of information-independent acquisition mode seems to be a good choice because several MS/MS features can be collected from coeluting species without the loss of sensitivity. The MS^{All} technique makes use of deconvolution algorithms to produce cleaner MS/MS spectra and benefits from sharp peaks provided by ultrahigh-performance liquid chromatography because good chromatographic separation is required to acquire a high-quality spectrum. This acquisition mode has been proven ideal for qualitative purposes and allows for retrospective analysis using the accurate mass full acquisition (Kinyua et al., 2015).

To overcome the scarce selectivity and improve the quality of DIA MS/MS spectra, a new strategy has been proposed. It is known as variable data—independent acquisition (vDIA) or SWATH (Sequential Windowed Acquisition of All Theoretical Fragment Ion Mass Spectra) and consists in dissociate compounds in selected isolation width windows (e.g., 21 Da) rather than the entire mass range to limit the number of potential interfering compounds. Thus, narrow mass ranges are isolated in Q1 that are each independently and consecutively analyzed. The use of narrow isolation windows reduces the matrix interferences and effectively increase the dynamic range of DIA because highly abundant species only affect a narrow portion of the m/z space. The number and width of the segments can be selected by the users. Therefore, they can be adjusted according to the type and number of ions present in the samples. Narrower segments are created at the low m/z ranges, where higher number of ions are expected, and they can be wider at higher m/z. Despite the increase in selectivity and sensitivity, it must be considered that under this strategy a duty cycle includes a single MS scan followed by a series of product ion scans. These longer cycles can conduct to a diminution in the points per chromatographic peak recorded.

DIA strategies have found broad applicability in proteomics. Small molecule applications including pesticide analysis in food have primarily been realized with broadband DIA (Mol, Zomer et al., 2012; Pérez-Ortega et al., 2016). Because of its recent development, scarce applications of segmented fragmentations are still reported. Zomer and Mol (2015) described the validation of a method using vDIA in a Q-Orbitrap system. As a compromise between the number of fragmentation events, resolving power, and number of data points across a chromatographic peak, the proposed method used one full-scan and five consecutive fragmentation events. To keep the overall cycle time below 1 s, a resolving power of 70,000 was selected for the full-scan event and a lower resolving power of 35,000 was chosen for the vDIA scans. The precursor ion range (m/z 100−1000) was divided into four isolation windows of 100 Da equally divided over m/z 100−500 and a fifth event covering m/z 500−1000. The lower resolving power for the vDIA scans did not compromise the mass accuracy of the measurement of the fragment ions. Fig. 3.2 shows the benefit of acquiring fragment ions by vDIA compared to MS^{All} acquisition. Enhanced selectivity was obtained when using vDIA, as shown for the detection of carbaryl in orange, probably by the fact that matrix interferences, which also produce a fragment at m/z 145.0648, have a precursor ion outside the vDIA window of 195−305, and it is not fragmented in the vDIA scan used for detection of the carbaryl fragment (Zomer & Mol, 2015).

Besides improved selectivity, sensitivity also was improved because the use of a narrow m/z range of the precursor ions in each vDIA scan allows for an increase in the proportion of ions from the compound of interest, which are accumulated in the C-trap, thereby increasing the sensitivity for that compound. Although still scarcely studied, it is expected that future applications and improvements in deconvolution software contribute to the implementation of these DIA approaches.

3.2.3 QUANTITATIVE PERFORMANCE

Quantitative performance of an instrument is usually referred to the sensitivity, precision, and linearity aspects. Traditionally, (Q)TOF-MS has offered worse quantitative capabilities than QqQ mainly because they offer a narrow dynamic range and

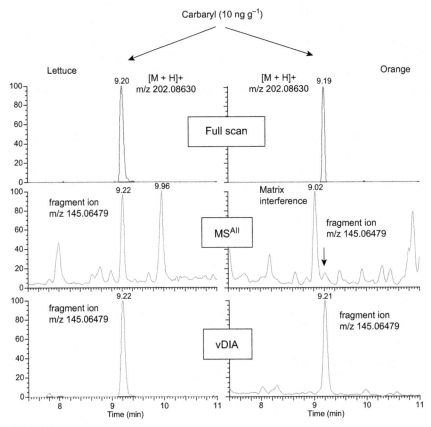

FIGURE 3.2

Extracted ion chromatograms (exact mass ±5 ppm) for carbaryl at 10 ng/g in lettuce and orange: (top) [M+H]$^+$ m/z 202.08630 from the full-scan event without fragmentation; (middle) fragment ion m/z 145.06479 from the fragmentation event acquired by MS^{All}; and (bottom) the same fragment ion but acquired by variable data−independent acquisition (vDIA).

Adapted from Zomer, P., & Mol, H.G.J. (2015). Simultaneous quantitative determination, identification and qualitative screening of pesticides in fruits and vegetables using LC-Q-Orbitrap™-MS. Food Additives & Contaminants: Part A, 32(10), 1628–1636.

poor sensitivity. Therefore, workflows for pesticide analysis in vegetable samples have been proposed including screening and identification by LC-HRMS, based on RT and accurate mass, and confirmation/quantitation by LC-QqQ-MS, based on two MRM transitions (García-Reyes, Hernando, Ferrer, Molina-Díaz, & Fernández-Alba, 2007). However, even though some authors still consider QqQ and QqLIT as the best alternatives for target analysis of pesticide residues (Villaverde et al., 2016), recent applications and comparative studies to assess the effectiveness of HRMS cast doubt on such claims (Gómez-Ramos et al., 2013, 2015; Grimalt, Sancho, Pozo, & Hernández, 2010; Kaufmann et al., 2012; Lacina, Urbanova, Poustka, & Hajslova, 2010; Rajski, Gómez-Ramos, & Fernandez-Alba, 2014; Wang, Chow, Chang, & Wong, 2014; Zomer & Mol, 2015). Instrumental developments in HRMS are contributing to expand quantitative applications.

In TOF instruments, saturation of the detector was pointed out as the main cause of the limited linearity. New-generation instruments have incorporated enhanced mathematical algorithms to overcome this limitation. A high-speed analog-to-digital converter to record the ion arrival time, compared to the previous time-to-digital converter approach, is able to measure the abundance of a given ion when it increases, which allows increasing dynamic range to four orders of magnitude. Thus, this weakness has almost disappeared in the modern instruments.

By its peculiar mode of operation, Orbitrap mass spectrometer shows better dynamic range than TOF-MS. Orbitrap instruments incorporate the C-trap device located between the transfer section and the Orbitrap analyzer. The C-trap accumulates the ions entering from the transfer section, and after filling with a sufficient number of ions, they are injected into the Orbitrap. An automatic gain control (AGC) functionality is incorporated to control the maximum number of ions (typically 3,000,000 ions) that can be accumulated to prevent detrimental effects caused by overfilling. The collection of ions is stopped by the closure of an electronic aperture (gate lens) as soon as the maximum number of ions (target capacity) is reached. Sampling only a variable part of the incoming ion beam would be expected to affect quantification. However, this does not happen, because the reported ion abundance is automatically corrected by a factor corresponding to the shortened ion injection time.

Despite the benefits of AGC technology, limitation of the number of ions entering the Orbitrap also can affect the detection capability of trace compounds, when they are present in complex matrices, because a large amount of matrix components present in the extracts consume the vast majority of the available ion capacity of the C-trap. Shortening the ion injection time reduces not only the number of collected matrix ions, but also the absolute number of analyte ions, thus reducing the detectability of low-abundance analyte ions. To overcome this limitation, the use of the multiplexing feature of Q-Orbitrap instruments has been recently proposed (Kaufmann & Walker, 2016). Under this strategy, quadrupole is used as a broad-band mass filter permitting the sequential collection of selected mass ranges. The consecutively collected ions are stored in the C-trap and injected as a single ion cloud into the Orbitrap analyzer. Mass ranges can be customized by variating the isolation time of low-abundance versus high-abundance mass range segments to provide the attenuation of those with intense matrix signals and the prolongation of ion injection times for those presenting low abundance. This approach has provided promising results (Kaufmann & Walker, 2016), revealing a higher sensitivity and an extended intrascan dynamic range for low-abundance analytes in the presence of intensive matrix interferences. Thus, future applications are expected in this field.

Another advantage of Orbitrap MS is the higher mass resolution. Resolution becomes a major issue when complex samples are analyzed. Insufficient resolution affects not only the identification of the analytes but also the quantitative performance because of the inability to use the required narrow mass-extraction windows, which has a significant impact on selectivity and peak shape. The selection of the appropriate resolving power depends on the ratio of the analyte concentration relative to coeluting matrix interferences. Consequently, the optimum setting depends on the type of sample, the analyte concentration, and the sample preparation before LC-MS analysis (Kellmann, Muenster, Zomer, & Mol, 2009). A resolving power of $\geq$50,000 could be required to avoid peak distortion when using narrow mass-extraction windows, needed for appropriate selectivity at low levels in complex matrices. Zomer and Mol (2015) demonstrated the suitability of an LC-Q-Orbitrap MS for quantitative analysis of pesticides in fruit and vegetable samples using a resolving power of 70,000 FWHM. The same resolution was also applied by Gómez-Ramos et al. (2015) who compared the identification and quantitation capabilities of an LC/Q-Orbitrap MS and an LC-QqQ MS/MS system in the analysis of 139 pesticides in 100 real fruit and vegetable samples. Results reveal that the two instruments have similar capabilities for quantification.

The use of higher resolving power can result in an enhanced selectivity and accuracy of mass assignment. However, some limitations arise from the fact that in Orbitrap mass resolution is inversely related to the scanning speed. This means that higher resolution requires higher acquisition times and consequently a lower number of spectra per second are recorded, thus affecting the reproducibility of a precise chromatographic peak shape, mainly when narrow peak widths are detected (few seconds). A compromise between mass resolution and chromatographic profile has to be reached in quantitative analysis to obtain a minimum of 10 points required for an adequate definition of a chromatographic peak.

Taking into account the latest studies, it is expected that HRMS become more and more considered as key instruments in quantitative analyses. Their implementation in routine analysis has been strengthened by the development of new data processing software, which incorporates targeted quantitative analysis capabilities in their workflows, allowing the processing of batches containing numerous samples.

3.3 DATA ANALYSIS WORKFLOWS IN HIGH-RESOLUTION MASS SPECTROMETRY

Typically, full scan-MS and MS/MS chromatograms obtained during the analysis of pesticides in food are very rich in information and contain thousands of ions from both analytes present in the samples and matrix interferences. Thus, powerful software tools are required to automatically explore the high-resolution and accurate mass data generated to get reliable results, especially when very complex matrices are analyzed. The instrumentation vendors have developed their own software packages, which include options to operate under different screening strategies.

Three main data processing strategies are currently recognized, which include target screening (when reference standards are available), suspect screening (intended to the identification of "suspected substances" selected based on prior information but for which no reference standards are available), and finally nontarget screening (when neither prior information nor reference standards are available) (Krauss, Singer, & Hollender, 2010, p. 943) (Fig. 3.3). The two first options are targeted strategies because they involve previous analytes selection and the use of customized databases including information of the exact mass, adducts formation (typically $[M+H]^+$, $[M-H]^-$, $[M+Na]^+$, $[M+NH_4]^+$), RT (in target analysis), and fragments of MS/MS spectra. This scheme, which is widely accepted in environmental analysis, is somewhat confusing in the field of pesticide analysis.

It is clear that target analysis refers to the identification and quantification of a set of previously selected analytes. Target compounds are optimized to get the best response in the LC-MS system and a thorough validation study is performed to meet the requirements of DG SANTE. In this case, target analysis ignores anything except the target analytes present in the samples. Under this concept, target analysis of pesticides has been traditionally undertaken by QqQ-MS. In contrast, there is some confusion about what is meant by nontarget analysis. In some cases, it means

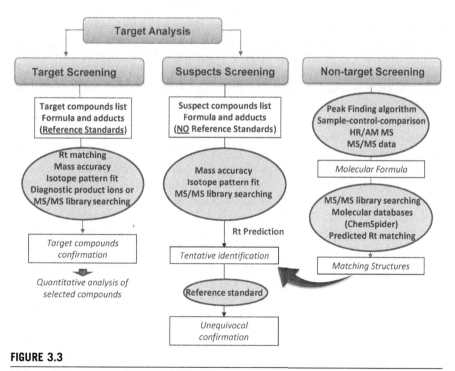

FIGURE 3.3

Main workflows used in high-resolution mass spectrometry (HRMS) methods.

screening against a large database, without any preconceptions as to what will be found, or it might mean retrospective analysis of a data set contrast for compounds not specifically anticipated. For other nontargeted analyses, it may be understood as the detection of complete "unknowns" in food. Obviating these differences, this section will refer to the workflows used for qualitative screening of pesticides with or without inclusion of RT information. True unknown studies are less frequent and in most cases are intended to pesticide metabolites identification.

Generally speaking, raw data processing for identification of analytes of interest in complex samples involves some basic steps, which include peak picking, mathematical mass spectral deconvolution, and determination of molecular ions by adduct detection. Automated component detection provides rapid and effective peak detection in a preselected mass range (generally from m/z 100–1000). Raw data are reviewed to determine which masses are due to chromatographic events and which are due to background. The key is obtaining clean and background-free mass spectra, which permit that very low-level components can be successfully detected in complex matrices by substantially reducing data complexity. Background masses are excluded from the processed file leaving only chromatographically significant information for further processing and review. A relative abundance higher or equal to 0.010% and an absolute abundance higher than or equal to the height of 1000 counts are usually selected as compounds filters, although these values must be selected carefully to avoid losses of compounds present at very low concentrations. Quality of the analysis greatly depends on the efficiency of this process. The resolution is also selected at low/high values in TOF instruments or at a discrete value in Orbitrap analyzers, depending on the type of analysis and according to recommendations described earlier. Data acquisition usually includes simultaneous full-scan MS and MS/MS fragmentation events. Thus, type and conditions of fragmentation are also selected.

Once peaks corresponding to all the potential compounds of interest are extracted, different data-processing approaches are generally available. Databases in table format are created including a list of masses or formulas for the targeted compounds. It is remarkable that these lists can be easily modified based on the compounds of interest of the respective laboratories. In addition to compound name, formula, isotope, adduct, and RT, other columns can be included and edited based on the methods and needs. The "extraction mass," used for the compounds search can be entered manually or be calculated automatically based on the formula, isotope, and adduct provided. The (de)protonated molecule used to be the selected ion, except for some analytes, which display adducts (sodium, ammonium, or methanol), or yield fragment ions (in-source CID) with higher relative abundance. Once a compounds table is generated, it can be saved for future processing. Several compound lists can be created and alternatively applied to samples processing depending on the objectives of the analysis. When RT is available, a true target analysis is performed and selected masses are searched in the corresponding RT window. If RT is not included (suspect screening), target masses are explored along the whole chromatogram. Because reference standards are not required, the number of substances that can be screened qualitatively is theoretically unlimited. However, as many compounds are included in the database, a higher probability of false positives detection is obtained. Positive findings are reported when both accurate mass and MS/MS spectrum match with those of the suspect pesticide. Information about elemental composition of the candidates is very useful, and the presence of characteristic isotopic clusters (i.e., due to Cl atoms) facilitates the identification of compounds containing these elements. Availability of commercial spectral libraries is mandatory to discriminate between isobaric compounds presenting different structures based on their typical fragmentation patterns. Even when the observation of these parameters provides a high level of confidence in the identification of the suspect analytes, final confirmation is only obtained by the acquisition and analysis of the corresponding reference standard. As an example, Moschet, Piazzoli, Singer, and Hollender (2013) evaluated the efficiency of a suspect screening strategy using LC coupled to Q-Orbitrap MS without the prior purchase of reference standards for assessing the exposure of 185 rarely investigated pesticides and their transformation products (TPs) in 76 surface water samples. Only the exact mass was used as a priori information. Thresholds for blank subtraction, peak area, peak shape, signal-to-noise, and isotopic pattern were applied to automatically filter the initially picked peaks. From roughly 60,000 peaks that were detected initially in all samples, 90% was automatically reduced by the application of the optimized strategy and 16 substances were confirmed with reference standards.

Despite the proved success of this suspect screening strategy, manual check of positive findings is required and it represents a time-consuming task, especially for complex matrices where the probability of false positive detection increases. This supposes a handicap to efficient laboratory workflow, and consequently, most of the screening methods reported in literature include the RT as search parameter (targeted screening), especially if they are intended to regulatory analysis applications, where all identifications require comparison to a reference standard.

The objective of these targeted screening methods is to provide yes/no responses for a selected number of analytes. Once a database has been established, software parameters for analytes detection need to be optimized to obtain a fit-for-purpose balance between false positives and false negatives reported by the software. Identification is generally based on RT, accurate mass of the quasi-molecular ion, isotopic pattern, and MS/MS fragmentation. Tolerances for these parameters usually include ±5 ppm error in accurate mass of the quasi-molecular ion, a RT window between ±0.05 and 0.2 min, a 10%−20% of variation in the isotopic profile, and 70%−80% match between mass spectra of the target analytes in the sample and a mass library when available.

The DG SANTE (SANTE/11945/2015) recognizes a tolerance of 5 ppm in the accurate mass as an adequate identification criterion, but frequently, most of the modern HRMS instruments are able to provide mass accuracies lower than 2 ppm when analyzed for pesticides in fruit and vegetable samples. As an example, Gilbert-López et al. (2010) reported mass errors below 2 ppm in more than 90% of the 35 pesticides determined using LC-TOF-MS in fruit-based soft drinks.

However, the selection of lower tolerance can diminish the number of false positives but at the expense of a higher report of false negative results. As rejection of false negatives can be performed based on fulfilling other criteria or ultimately by confirmation with analytical standards, values below 5 ppm are not usually employed.

The RT window tolerance has been investigated by various authors (Mezcua, Malato, García-Reyes, Molina-Díaz, & Fernández-Alba, 2009). In general, RTs in today's chromatographic systems are sufficiently stable to justify an absolute tolerance of ±0.1 min (Mol et al., 2015). However, a higher tolerance can benefit to early eluting compounds, which present a higher variability.

Latest generation of HR mass spectrometers are highly reliable in accurately measuring relative isotope abundances (RIA) of small organic molecules. Thus, new instruments incorporate algorithms into their compound identification software that routinely take into account isotope ratios for generating elemental compositions, thereby providing an additional criterion for confirming the proposed formulas. The correct assignation of the molecular formula in target analysis provides a higher grade of certainty in the identification, but it is especially useful in nontarget analysis of unknown identification. Tolerances between 10 and 20% of the theoretical values are usual. Absolute ion abundance is the main factor influencing the RIA measurement accuracy. Xu et al. (2010) defined intensity thresholds above which errors were consistently 20% below their theoretical value.

Finally, characteristic MS/MS fragmentation of target analytes can also be used to enhance identification capability of screening methods. The use of MS/MS spectral libraries results in a relevant tool. However, automated identification of small molecules, searching in MS/MS libraries, can fail because of their limited coverage and the heterogeneity of LC-MS/MS spectra between different instruments and operating conditions, which makes them difficult to compare and hinder its transferability. To improve MS/MS spectral match some libraries include MS/MS spectra obtained under different conditions (various collision energies, different matrices) for the same compound. Some instrument vendors include in their software packages MS/MS libraries (e.g., the AB SCIEX iMethod Meta MS/MS spectral library, which contains MS/MS spectra for over 2500 compounds) to facilitate the automatic search of MS/MS spectra of positive findings. A good alternative is the use of homemade spectral libraries, where MS/MS spectra of pesticides are recorded under the same operational conditions than in the samples. An example is reported by Wang et al. (2015) for the qualitative screening and identification of 317 pesticides in vegetables and fruits using a homemade MS/MS spectral library. Four independent spectra of each pesticide were collected at four different collision energies. A match score ≥70 was selected as the identification criterion. Automatic data processing yielded false negative results in real samples showing spectral library retrieval scores lower than 70. This fact was explained by the low concentration of the pesticides in the samples together with the complexity of the matrix. A manual background subtraction and detailed inspection of the spectra were required to fulfill the preset criterion.

Continuing with the workflow, once the criteria are selected, extracted ion chromatograms are generated based on RT matching combined with accurate mass measurements of the targeted species. MS/MS information is further automatically evaluated if the detected signal exceeds the user-defined intensity threshold or S/N. Positive hits are listed with indication of the degree of compliance with the parameters selected. Commonly, automatic detection has to be followed by a quick manual verification of the hits reported by the software, to reduce false detects (Malato, Lozano, Mezcua, Agüera, & Fernandez-Alba, 2011). An example of the results displayed by the instrument software after the application of the described workflow is shown in Fig. 3.4.

3.3.1 QUALITATIVE SCREENING METHOD VALIDATION

An ideal qualitative screening method should be able to detect a large number of target analytes at concentrations higher than their reporting limit (0.01 mg/kg) in a wide range of matrices. To facilitate its implementation in routine analysis, the method has to be automated, rapid, and easy to use, with minimal intervention from the analyst (Mol, Reynolds, Fussell, & Štajnbaher, 2012). Consequently, the operating parameters and MS-identification criteria must be carefully selected and optimized to minimize the number of false positives and/or negatives reported.

Qualitative screening method validation is usually a very time-consuming task, because of the great number of analyses required to demonstrate rates of false positives and negatives for given analytes at the desired concentration range in diverse samples. AOAC International standards for Performance-Tested Methods certification of screening methods (e.g., test kits) entailed the analysis of ≥ 30 different samples of each matrix at each analyte concentration of interest (which must include blanks). In Europe, the EU guidance document (SANTE/11945/2015) prescribes that each compound should be tested using a set of at least 20 samples that include different matrices (each at least in duplicate) covering the intended use of the method, at the anticipated screening detection limit (SDL).

The validation of a qualitative screening method is usually focused on pesticides detectability. The confidence of detection of an analyte at a certain concentration level should be established and an SDL is calculated. SANTE Guidelines (SANTE/11945/2015) define SDL as the lowest level at which an analyte has been detected (not necessarily meeting the MS-identification criteria) in at least 95% of the samples (i.e., an acceptable false-negative rate of 5%). Providing that compounds tentatively detected by these methods have to be confirmed by an appropriate confirmatory method, there is no need for a strict criterion for the number of false positive detects. Detectability of analytes in the samples diminishes, and consequently the probability of false negative results increases, in the presence of complex matrices, especially when pesticides are present at low concentration levels.

Mol, Zomer et al. (2012) evaluated the analytical capabilities of a screening method for pesticides in vegetables and fruits using a single stage Orbitrap-MS. The study involved 21 different vegetable and fruit commodities, a screening

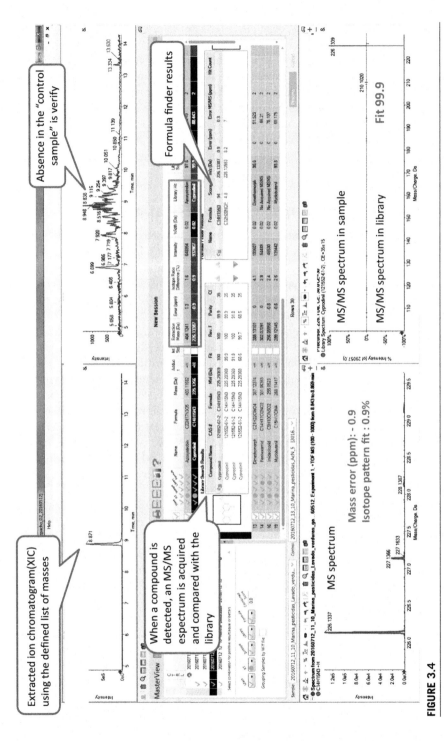

FIGURE 3.4

An example of the results displayed by the instrument software after the application of a targeted screening workflow.

database of 556 pesticides for the evaluation of false positives, and a test set of 130 pesticides spiked to the commodities at 0.01, 0.05, and 0.20 mg/kg for the evaluation of false negatives. The number of false positives out of 11,676 pesticide/commodity combinations targeted was 36 (0.3%). The percentage of false negatives, assessed for 2730 pesticide/commodity combinations, was 13%, 3%, and 1% at the 0.01-, 0.05-, and 0.20-mg/kg level, respectively (slightly higher with fully automated detection). Following the SANCO/12495/2011 protocol for validation of screening methods, the SDL was determined for 130 pesticides and found to be 0.01, 0.05, and ≥ 0.20 mg/kg for 86, 30, and 14 pesticides, respectively. This was sufficient for testing MRL compliance of the majority of the pesticide/commodity combinations tested.

A similar study was performed by the same authors (Mol et al., 2015) but, in this case, using a hybrid Q-Orbitrap system. An overall detection rate of 92% was achieved at the 10-ng/g level, increasing to 98% at 200 ng/g. Automatic detection by the software produced an acceptable number of false negatives, and a low number of initial false detects, which on manual verification, were reduced to zero. This manual verification took on average less than 2 min per sample, which was considered acceptable for routine application.

3.3.2 NONTARGET ANALYSIS

The use of true nontarget approaches in pesticide residue analysis in food is scarce. Nontarget screening methods are intended to identify unexpected or unknown compounds, such as metabolites, present in food matrices. The application of these nontarget strategies, common in other research areas such as metabolomics or environmental analysis, entails great complexity, when applied to food matrices because of their inherent complexity. Data analysis workflow combines nontarget peak finding, sample—control comparison, empirical formula finding, MS/MS library searching, and search in compound databases combined with MS/MS interpretation (Knolhoff & Croley, 2016).

Because of the unknown nature of the contaminants, no compound lists are available. Powerful nontarget peak finding algorithms are required to find chromatographic features, which could correspond with true compounds. Because of the complexity of the food samples, a large amount of peaks can be generated and efficient filters are required to reduce the number of masses to be investigated. Blank or control samples are analyzed to differentiate matrix and sample specific signals from true contaminations. Thus, only peaks significantly more abundant in the sample (typically 10 times) than in the blanks are considered. Compounds known to be inherent to the samples can also be removed by using "exclude list" containing masses that are systematically not considered in the analysis.

The application of statistical methods, such as principal component analysis (PCA) or linear discrimination analysis, can be useful for discriminating peaks of interest from the endogenous matrix components. Cotton et al. (2014) demonstrated the applicability of these strategies in the detection of a banned pesticide in honey.

Data mining for chlorine-containing molecular species was applied using partial least squares discriminant analysis and PCA.

Once a relevant peak is detected, a molecular formula is assigned based on the accurate mass, isotopic pattern, and MS/MS data. Usually, a list of candidate formulas is ranked as a function of the accurate mass error obtained by comparing the experimental and the theoretical mass. The molecular formulas proposed can be searched in commercial databases trying to assign a structure. Some molecular databases also include MS/MS spectra for data comparisons. There are a number of molecular databases available, including ChemSpider (http://www.chemspider.com), PubChem (Bolton, Wang, Thiessen, & Bryant, 2008, pp. 217–240), Chemical Abstracts Service[SM] (CAS) database searched through SciFinder (https://scifinder.cas.org), and Metlin (a metabolite database) (Smith et al., 2005). Some instruments incorporate hyperlinks in their software to easily start a ChemSpider search and MS/MS interpretation.

The theoretical MS/MS pattern is automatically compared with the experimental MS/MS spectrum and ranked by a confidence score. ChemSpider ranks the structures based on the number of references, making it easy to identify the most likely structure. The predicted structure of a selected product ion is highlighted in the structure display to allow results evaluation. An example is shown in Fig. 3.5 for the identification of diuron. The molecular formula ($C_9H_{10}Cl_2N_2O$) was searched against ChemSpider, and diuron was proposed as the most probable structure, occupying the first position in the ranking. This structure was compared against the measured MS/MS spectrum. In addition, experimental MS/MS pattern could be easily explained by the theoretical fragmentation pattern using the proposed structures.

This procedure is very useful in the identification of unknowns; however, it is a quite frequent finding that a large number of compounds match a given molecular formula and manual search of a large amount of compounds in different molecular databases results in a tedious and time-consuming task. Thus, the development of automated workflows for searching multiple molecular formulas in multiple databases should be very useful for facilitating this task, although the manual inspection of the results is unavoidable. When no positive findings are found using the molecular formula, structural elucidation of unknowns is based on the interpretation of their MS/MS spectra. This entails a higher grade of difficulty, especially when no information about the origin of the nontarget species is available. Unknown compounds can be predicted if their MS/MS spectra have similar fragmentation patterns to related compounds that have been previously characterized; such is the case of TPs identification (Agüera, Gómez-Ramos, & Fernandez-Alba, 2012, pp. 61–110).

The use of MS/MS fragmentation prediction tools is also helpful. In the majority of cases, these tools require a proposed molecular structure for the corresponding MS/MS spectral prediction. Then the theoretical spectra are used to discard or confirm the molecule by comparison with the observed MS/MS spectrum. Some of these tools include ACD/MS Fragmenter, MolFind, MetFrag, MIDAS (Metabolite Identification via Database Searching), and SmartMass (Menikarachchi et al., 2012; Wolf, Schmidt, Muller-Hannemann, & Neumann, 2010). Some instrument

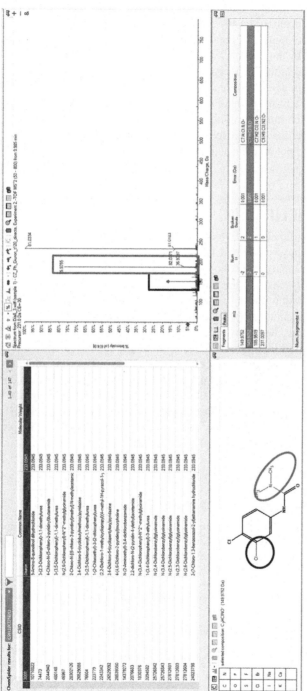

FIGURE 3.5

Identification of diuron by search against ChemSpider database. Comparison of the theoretical tandem mass spectrometry (MS/MS) pattern with the experimental MS/MS spectrum and assignation of MS/MS fragments.

vendors also include these analytical tools in their workflows, such as Molecular Structure Correlator (Agilent) and Mass Frontier (Thermo).

The development of nontargeted screening strategies is challenging and requires great effort by the analysts, which is not always successful. Thus, new developments are expected in this field. Knolhoff and Croley (2016) in a recent review point out data analysis as the bottleneck of nontargeted screening analyses. Improvements through the advancement of data processing tools, which ensure sufficient data quality through chromatographic separations and adequate ion abundance, are required. Furthermore, the development of comprehensive molecular databases will aid in high-throughput analysis and in the automation of the procedure.

3.4 CONCLUSIONS

Recent instrumental developments and the proposal of efficient and flexible workflows have converted HRMS as a true alternative to conventional QqQ-MS in pesticide analysis in food. The increasing applications in this field in the last ten years demonstrate the interest of laboratories devoted to this type of analysis in incorporating these techniques in routine analysis. The potential of high-resolution accurate mass in the selectivity and confirmation capability of the analysis represents the main strength of these techniques. New generations of TOF- and Orbitrap-based instruments have dramatically increased both resolution and mass accuracy, thus allowing very efficient qualitative screening methods, with reduced number of false positive and negative detects. Quantification capabilities have also been improved with enhanced linear ranges and reproducibility, and initial but solid steps are being done in this field. With all this, it can be said that HRMS represents an alternative of future for routine laboratories involved in pesticide residue analysis.

ACKNOWLEDGMENTS

Authors acknowledge the Ministry of Economy and Competitiveness (Spanish Government) (Project ref. CTQ2013-46398-R), the Regional Government of Andalusia (Project ref. RNM-1739), and the European Regional Development Fund (ERDF) for support.

REFERENCES

Agüera, A., Gómez-Ramos, M. M., & Fernandez-Alba, A. R. (2012). *Chemical evaluation of water treatment processes by LC−(Q)TOF-MS: Identification of transformation products in TOF-MS within food and environmental analysis.* Oxford, UK: Elsevier.

Bolton, E., Wang, Y., Thiessen, P. A., & Bryant, S. H. (2008). *Annual reports in computational chemistry.* Oxford, UK: Elsevier.

Cotton, J., Leroux, F., Broudin, S., Marie, M., Corman, B., Tabet, J. C., et al. (2014). High-resolution mass spectrometry associated with data mining tools for the detection of pollutants and chemical characterization of honey samples. *Journal of Agricultural and Food Chemistry, 62,* 11335−11345.

García-Reyes, J. F., Hernando, M. D., Ferrer, C., Molina-Díaz, A., & Fernández-Alba, A. R. (2007). Large scale pesticide multiresidue methods in food combining liquid chromatography − time-of-flight mass spectrometry and tandem mass spectrometry. *Analytical Chemistry, 79*(19), 7308−7323.

Gilbert-López, B., García-Reyes, J. F., Mezcua, M., Ramos-Martos, N., Fernández-Alba, A. R., & Molina-Díaz, A. (2010). Multi-residue determination of pesticides in fruit-based soft drinks by fast liquid chromatography time-of-flight mass spectrometry. *Talanta, 81*(4−5), 1310−1321.

Gómez-Ramos, M. M., Ferrer, C., Malato, O., Agüera, A., & Fernández-Alba, A. (2013). Liquid chromatography high resolution mass spectrometry for pesticide residue analysis in fruit and vegetables: Screening and quantitative studies. *Journal of Chromatography A, 1287,* 24−37.

Gómez-Ramos, M. M., Rajski, L., Heinzen, H., & Fernández-Alba, A. R. (2015). Liquid chromatography Orbitrap mass spectrometry with simultaneous full scan and tandem MS/MS for highly selective pesticide residue analysis. *Analytical and Bioanalytical Chemistry, 407*(21), 6317−6326.

Grimalt, S., Sancho, J. V., Pozo, O. J., & Hernández, F. (2010). Quantification, confirmation and screening capability of UHPLC coupled to triple quadrupole and hybrid quadrupole time-of-flight mass spectrometry in pesticide residue analysis. *Journal of Mass Spectrometry, 45*(4), 421−436.

Jia, W., Chu, X., Ling, Y., Huang, J., & Chang, J. (2014). High-throughput screening of pesticide and veterinary drug residues in baby food by liquid chromatography coupled to quadrupole Orbitrap mass spectrometry. *Journal of Chromatography A, 1347,* 122−128.

Kaufmann, A., Dvorak, V., Crüzer, C., Butcher, P., Maden, K., Walker, S., et al. (2012). Study of high-resolution mass spectrometry technology as a replacement for tandem mass spectrometry in the field of quantitative pesticide residue analysis. *Journal of AOAC International, 95*(2), 528−548.

Kaufmann, A., & Walker, S. (2016). Extension of the Q Orbitrap intrascan dynamic range by using a dedicated customized scan. *Rapid Communications in Mass Spectrometry, 30*(8), 1087−1095.

Kellmann, M., Muenster, H., Zomer, P., & Mol, H. (2009). Full scan MS in comprehensive qualitative and quantitative residue analysis in food and feed matrices: How much resolving power is required? *Journal of the American Society for Mass Spectrometry, 20*(8), 1464−1476.

Kinyua, J., Negreira, N., Ibáñez, M., Bijlsma, L., Hernández, F., Covaci, A., et al. (2015). A data-independent acquisition workflow for qualitative screening of new psychoactive substances in biological samples. *Analytical and Bioanalytical Chemistry, 407*(29), 8773−8785.

Knolhoff, A. M., & Croley, T. R. (2016). Non-targeted screening approaches for contaminants and adulterants in food using liquid chromatography hyphenated to high resolution mass spectrometry. *Journal of Chromatography A, 1428,* 86−96.

Krauss, M., Singer, H., & Hollender, J. (2010). LC-high resolution MS in environmental analysis: From target screening to the identification of unknowns. *Analytical and Bioanalytical Chemistry, 397*(3), 943−951.

Lacina, O., Urbanova, J., Poustka, J., & Hajslova, J. (2010). Identification/quantification of multiple pesticide residues in food plants by ultra-high-performance liquid chromatography-time-of-flight mass spectrometry. *Journal of Chromatography A, 1217*(5), 648–659.

Malato, O., Lozano, A., Mezcua, M., Agüera, A., & Fernandez-Alba, A. R. (2011). Benefits and pitfalls of the application of screening methods for the analysis of pesticide residues in fruits and vegetables. *Journal of Chromatography A, 1218*(42), 7615–7626.

Menikarachchi, L. C., Cawley, S., Hill, D. W., Hall, L. M., Hall, L., Lai, S., et al. (2012). Mol-Find: A software package enabling HPLC/MS-based identification of unknown chemical structures. *Analytical Chemistry, 84*, 9388–9394.

Mezcua, M., Malato, O., García-Reyes, J. F., Molina-Díaz, A., & Fernández-Alba, A. R. (2009). Accurate-mass databases for comprehensive screening of pesticide residues in food by fast liquid chromatography time-of-flight mass spectrometry. *Analytical Chemistry, 81*(3), 913–929.

Mol, H., Lommen, A., Zomer, P., Van Der Kamp, H., Van Der Lee, M., & Gerssen, A. (2010). Data handling and validation in automated detection of food toxicants using full scan GC-MS and LC-MS. *LC GC Europe, 23*(4), 200–210.

Mol, H. G. J., Reynolds, S. L., Fussell, R. J., & Štajnbaher, D. (2012). Guidelines for the validation of qualitative multi-residue methods used to detect pesticides in food. *Drug Testing and Analysis, 4*(Suppl. S1), 10–16.

Mol, H. G. J., Zomer, P., & De Koning, M. (2012). Qualitative aspects and validation of a screening method for pesticides in vegetables and fruits based on liquid chromatography coupled to full scan high resolution (Orbitrap) mass spectrometry. *Analytical and Bioanalytical Chemistry, 403*(10), 2891–2908.

Mol, H. G. J., Zomer, P., García López, M., Fussell, R. J., Scholten, J., De Kok, A., et al. (2015). Identification in residue analysis based on liquid chromatography with tandem mass spectrometry: Experimental evidence to update performance criteria. *Analytica Chimica Acta, 873*, 1–13.

Moschet, C., Piazzoli, A., Singer, H., & Hollender, J. (2013). Alleviating the reference standard dilemma using a systematic exact mass suspect screening approach with liquid chromatography-high resolution mass spectrometry. *Analytical Chemistry, 85*(21), 10312–10320.

Pérez-Ortega, P., Lara-Ortega, F. J., García-Reyes, J. F., Gilbert-López, B., Trojanowicz, M., & Molina-Díaz, A. (2016). A feasibility study of UHPLC-HRMS accurate-mass screening methods for multiclass testing of organic contaminants in food. *Talanta, 160*, 704–712.

Rajski, L., Gómez-Ramos, M. M., & Fernandez-Alba, A. R. (2014). Large pesticide multiresidue screening method by liquid chromatography-Orbitrap mass spectrometry in full scan mode applied to fruit and vegetables. *Journal of Chromatography A, 1360*, 119–127.

SANCO/12495/2011. *Method validation and quality control procedures for pesticides residues analysis in food and feed*. Available from http://www.crl-pesticides.eu/library/docs/fv/SANCO12495-2011.pdf.

SANTE/11945/2015. *Guidance document on analytical quality control and method validation procedures for pesticides residues analysis in food and feed*. Available from http://ec.europa.eu/food/plant/docs/plant_pesticides_mrl_guidelines_wrkdoc_11945_en.pdf.

Smith, C. A., O'maille, G., Want, E. J., Qin, C., Trauger, S. A., Brandon, T. R., et al. (2005). METLIN: A metabolite mass spectral database. *Therapeutic Drug Monitoring, 27*, 747–751.

Villaverde, J. J., Sevilla-Morán, B., López-Goti, C., Alonso-Prados, J. L., & Sandín-España, P. (2016). Trends in analysis of pesticide residues to fulfil the European Regulation (EC) No. 1107/2009. *Trends in Analytical Chemistry, 80*, 568−580.

Wang, Z., Chang, Q., Kang, J., Cao, Y., Ge, N., Fan, C., et al. (2015). Screening and identification strategy for 317 pesticides in fruits and vegetables by liquid chromatography-quadrupole time-of-flight high resolution mass spectrometry. *Analytical Methods, 7*(15), 6385−6402.

Wang, J., Chow, W., Chang, J., & Wong, J. W. (2014). Ultrahigh-performance liquid chromatography electrospray ionization Q-Orbitrap mass spectrometry for the analysis of 451 pesticide residues in fruits and vegetables: method development and validation. *Journal of Agricultural and Food Chemistry, 62*(42), 10375−10391.

Wolf, S., Schmidt, S., Muller-Hannemann, M., & Neumann, S. (2010). In silico fragmentation for computer assisted identification of metabolite mass spectra. *BMC Bioinformatics, 11*, 148.

Xu, Y., Heilier, J.-F., Madalinski, G., Genin, E., Ezan, E., Tabet, J.-C., et al. (2010). Evaluation of accurate mass and relative isotopic abundance measurements in the LTQ-Orbitrap mass spectrometer for further metabolomics database building. *Analytical Chemistry, 82*(13), 5490−5501.

Zhu, X., Chen, Y., & Subramanian, R. (2014). Comparison of information-dependent acquisition, SWATH, and MS[All] techniques in metabolite identification study employing ultrahigh-performance liquid chromatography−quadrupole time of-flight mass spectrometry. *Analytical Chemistry, 86*(2), 1202−1209.

Zomer, P., & Mol, H. G. J. (2015). Simultaneous quantitative determination, identification and qualitative screening of pesticides in fruits and vegetables using LC-Q-Orbitrap™-MS. *Food Additives & Contaminants: Part A, 32*(10), 1628−1636.

FURTHER READING

Arnhard, K., Gottschall, A., Pitterl, F., & Oberacher, H. (2015). Applying 'sequential windowed acquisition of all theoretical fragment ion mass spectra' (SWATH) for systematic toxicological analysis with liquid chromatography-high-resolution tandem mass spectrometry. *Analytical and Bioanalytical Chemistry, 407*(2), 405−414.

Commission Decision of 12 August 2002 implementing council directive 96/23/EC concerning the performance of analytical methods and the interpretation of results (2002/657/ EC). Available from http://eur-lex.europa.eu/legal-content/EN/TXT/PDF/?uri=CELEX: 32002D0657&from=ES.

Grund, B., Marvin, L., & Rochat, B. (2016). Quantitative performance of a quadrupole-orbitrap-MS in targeted LC-MS determinations of small molecules. *Journal of Pharmaceutical and Biomedical Analysis, 124*, 48−56.

Hufsky, F., Scheubert, K., & Böcker, S. (2014). Computational mass spectrometry for small-molecule fragmentation. *TrAC: Trends in Analytical Chemistry, 53*, 41−48.

Kaufmann, A. (2012). High mass resolution versus MS/MS. In A. R. Fernández-Alba (Ed.), *TOF-MS within food and environmental analysis: Comprehensive analytical chemistry* (Vol. 58, pp. 169−215). Amsterdam, The Netherlands: Elsevier (Chapter 4).

Kaufmann, A., Butcher, P., Maden, K., Walker, S., & Widmer, M. (2011). Quantitative and confirmative performance of liquid chromatography coupled to high-resolution mass spectrometry compared to tandem mass spectrometry. *Rapid Communications in Mass Spectrometry, 25*(7), 979–992.

Solís, R. R., Rivas, F. J., Martínez-Piernas, A., & Agüera, A. (2016). Ozonation, photocatalysis and photocatalytic ozonation of diuron: Intermediates identification. *Chemical Engineering Journal, 292*, 72–81.

Zomer, P., Gerssen, A., Lommen, A., Mol, H. G. J., Van Der Lee, M., & Van Der Kamp, H. (2010). Data handling and validation in automated detection of food toxicants using full scan GC-MS and LC-MS. *LC GC Europe, 23*(4), 200–210.

Current Legislation on Pesticides

4

Helen Botitsi[1], Despina Tsipi[1], Anastasios Economou[2]

General Chemical State Laboratory, Athens, Greece[1]; National and Kapodistrian University of Athens, Athens, Greece[2]

4.1 INTRODUCTION

The International Union of Pure and Applied Chemistry (IUPAC) defines a pesticide as any substance or mixture of substances intended for preventing, destroying, or controlling any pest (Holland, 1996). The use of pesticides is one of the most important ways of protecting plants and plant products against harmful organisms, including weeds, and improving agricultural production. Nowadays, more than 1700 pesticides belonging to more than 100 chemical classes are in use worldwide for food production. Information on synthetic and commercially available pesticides can be found at "The Pesticide Manual" which is available online (Turner, 2015) and the electronic database (Compendium of Pesticide Common Names, 2016). Due to possible risks imposed to human health and the environment, pesticides are the most thoroughly tested chemicals in the world. Internationally, the legislative framework is becoming more and more demanding and strict for the authorization of pesticides, the establishment of the maximum residue limits (MRLs), and the monitoring of pesticide residues in food of plant and animal origin.

4.2 PESTICIDES

4.2.1 IDENTITY AND PHYSICOCHEMICAL PROPERTIES

The systematic names of chemicals are derived from the IUPAC and the Chemical Abstracts Service (CAS). In addition to a systematic name, CAS assigns a registry number to each chemical. The Technical Committee 81 of the International Organization for Standardization (ISO) has devised a system for naming pesticides that includes ISO 257:2004 (ISO, 2004), ISO 765:2016 (ISO, 2016a), ISO 1750:1981 (ISO, 1981), and amendments.

Pesticides currently used worldwide, belonging to different chemical classes have different physicochemical properties (IUPAC, 2016; PAN NA, 2016; Stephenson, Ferris, Holland, & Nordberg, 2006).

Evaluation of pesticides begins with clear identification of their physical and chemical properties, which are usually measured under well-defined experimental

Applications in High Resolution Mass Spectrometry. http://dx.doi.org/10.1016/B978-0-12-809464-8.00004-X

conditions, according to well-established protocols recognized by national and international agencies [e.g., the US Environmental Protection Agency (EPA) guidelines, the Organization for Economic Co-operation and Development (OECD), and European Union (EU) protocols]. The main physicochemical properties of pesticides—water solubility, vapor pressure, volatility, stability in water, photodegradation, water—octanol partition coefficient, acid—base properties—define their relationship with different domains such as the pesticide—environment interactions, mode of application of pesticides, and analytical determination (Barceló & Hennion, 1997).

4.2.2 PESTICIDES CLASSIFICATION

Pesticides are classified either by their chemical class (e.g., organochlorines, organophosphates) or based on their target action (e.g., acaricides, herbicides, insecticides) or by their biochemical mode of action (MoA). The classification of pesticides based on their action on target organisms is presented in Table 4.1. Compounds of different chemical groups such as carbamates, organochlorines, organophosphorus, organotins, phthalimides, pyrethroids, pyrazoles, strobilurins, thiazolidines, thiocarbamates, and thioureas are used as acaricides (to kill mites and ticks). Many of them are also classified as insecticides, intended to kill insects, or to disrupt their growth.

Fungicides are used to kill fungi or to inhibit their development in plants, stored products, or soil. Chemical groups include amides, benzimidazoles, carbamates, dicarboximides, dithiocarbamates, imidazoles, morpholines, pyridines, pyrimidines, quinolines, strobilurins, thiazoles, triazoles, and ureas. Phenoxy acids, triazines, pyrazoles, thioureas, and ureas are some of the chemical classes of herbicides used to kill plants, or to inhibit their growth or development. Avermectin compounds (abamectin), carbamates, (e.g., carbofuran, aldicarb), and some organophosphorus compounds (e.g., fenamiphos, cadusafos, chlorpyrifos) can be used as nematicides, i.e., to kill nematodes in plants or soil. The class of plant growth regulators that can alter the expected growth, flowering, or reproduction rate of plants, includes antiauxins (e.g., clofibric acid), auxins (e.g., 2,4-D, 2,4-DB, dichlorprop, 2,4,5-T), cytokines, defoliants (e.g., ethephon, endothall, tribufos), gametocides (maleic hydrazide), and gibberellins.

Table 4.1 Classification of Pesticides Based on Their Action on Target Organisms

Acaricides	Algicides	Antifeedants
Avicides	Bactericides	Bird repellents
Fungicides	Herbicide safeners	Herbicides
Insecticides	Insect attractants	Insect repellents
Mammal repellents	Mating disrupters	Nematicides
Plant activators	Plant growth regulators	Rodenticides

The chemical groups involved in the different pesticide classes are not equivalent in terms of number of compounds, e.g., the organophosphorus may contain about 90 different compounds currently used, while the neonicotinoids contain very few compounds; both groups are classified as insecticides (Casida & Durkin, 2013a, 2013b; Tomizawa & Casida, 2005). Individual compounds or even chemical groups of active ingredients can occur in more than one class of pesticides; for example, organophosphorus, organochlorines, and pyrethroid groups are used as insecticides, acaricides, and/or nematicides.

Pesticides are bioactive compounds, intended to disrupt a primary target site in the pest. Enzymes, receptors, or channel sites at which specific binding initiates the physiological change can act as target sites of pesticides. For a bioactive molecule used as pesticide, a defined MoA describes the specific biochemical interaction to which its bioactivity—and its toxicity—is mainly attributed. About 100 different biochemical targets MoAs in pest insects, weeds, and fungi have been identified for the major groups of insecticides, herbicides, and fungicides (Casida, 2009). Metabolomic studies have greatly contributed to the discovery of the MoA of pesticides, the assessment of their toxicological risk, and the discovery of new bioactive compounds (Aliferis & Jabaji, 2011; Aliferis & Tokousbalides, 2011). The majority of the commercially available insecticides target the functionality of the nervous system of insects (Casida, 2009; IRAC, 2014). Herbicides disrupt mainly the plants photosynthesis and pigment synthesis, whereas fungicides exert inhibitory action on several vital biochemical systems of microorganisms (Casida, 2009).

A major limiting factor in the continuous use of pesticides is the emergence of resistance developed by pests (Casida, 2009; Casida & Durkin, 2013b). The importance of pesticide management has led to the establishment of the Resistance Action Committees, i.e., the Herbicide Resistance Action Committee (http://hracglobal.com/), the Fungicide Resistance Action Committee (http://www.frac.info/), and the Insecticide Resistance Action Committee (http://www.irac-online.org/) to define resistance groups.

4.2.3 PESTICIDE METABOLITES AND TRANSFORMATION PRODUCTS

In plants, animals, and the environment, pesticides can be transformed into a large number of degradation products, commonly defined as transformation products (TPs), through biological, chemical, and physical processes. Metabolism studies are necessary to understand the fate of pesticides, identify the metabolites, and provide data for human dietary risk assessment.

The qualitative and quantitative nature of pesticide residues in plants and livestock is dependent on the processes of absorption (the movement of the pesticide across membranes), distribution (transport within the biological system), metabolism (biological or chemical transformation of pesticides), and elimination (Skidmore & Ambrus, 2004). A great number of complex biotransformation pathways may occur within biological systems during the metabolism of pesticides

(Dorough, 1980). Pesticide metabolism may initially involve oxidation, reduction, and hydrolytic reactions such as aliphatic or alicyclic hydroxylation, aromatic hydroxylation etc. and conjugation reactions mainly with glutathione, sugars, and amino acids (Roberts & Hutson, 1999; Skidmore & Ambrus, 2004).

Several pesticide metabolites and TPs have been identified in food and environmental matrices using mass spectrometric techniques with low- and high-resolution mass analyzers. In many studies new "unknown" TPs have been identified; several reviews are focused on the inherent advantages of mass analyzers (TOF-MS, Orbitrap and their hybrid platforms, QqTOF-MS, Q-Orbitrap, LTQ-Orbitrap) for pesticide metabolites and TP identification and structure elucidation (Del Mar Gómez-Ramos, Ferrer, Malato, Agüera, & Fernández-Alba, 2013; Farré, Picó, & and Barceló, 2014; Fernández-Alba and García-Reyes, 2008; García-Reyes, Hernano, Molina-Díaz, & Fernández-Alba, 2007; Hernández, Portolés, Pitarch, & López, 2011; Kaufmann, 2012; Martínez Vidal, Plaza-Bolaños, Romero-González, & Garrido Frenich, 2009; Soler & Picó, 2007).

4.3 LEGISLATION

Human exposure to pesticides at low concentration levels through the food chain as a result of their extensive use in agricultural practice, may pose a risk for public health. Therefore, strict legislation and other measures concerning the protection of public health have been promoted by national and international authorities. Setting rules for the approval of active substances used as plant protection products, the establishment of MRLs in food commodities, and the careful monitoring of pesticide residues in foodstuffs and environment are the main objectives of legislation (Masiá, Morales Suarez-Varela, Llopis-Gonzalez, & Picó, 2016). Table 4.2 summarizes the legislation status of pesticides around the world.

4.3.1 PESTICIDES AUTHORIZATION

Pesticide products are subject to extensive biological, chemical, and toxicological tests before authorization. The test requirements have been specified by the national agencies, and they are generally carried out according to internationally tested guidelines harmonized by the OECD and developed by experts of OECD countries and the Joint FAO/WHO Meeting on Pesticide Residues (FAO, 2016) and regularly updated based on evolving scientific knowledge (OECD, 2009).

In the EU, the application procedure, evaluation of the active substances, and their approval are covered by Regulation (EC) 1107/2009 of the European Parliament and the Council defining the legal framework concerning the placing of plant protection products on the market (EC, 2009b). Following the Regulation (EC) 1107/2009, more recent Commission Regulations specify the data required for active substances and plant protection products to be approved and placed on the market (EU, 2013a, 2013b). For a new active substance (pesticide), the minimum

Table 4.2 Legislation on Pesticides

Country/ Organization	Legislation
European Union	EU pesticides database (http://ec.europa.eu/food/plant/pesticides/eu-pesticides-database)
	Method validation and quality control procedures for pesticide residues analysis in food and feed (SANTE 11945/2015)
	Pesticide residue analytical methods (SANCO/825/00 rev. 8.1)
United States	e-Code of federal regulations (eCFR) (http://www.ecfr.gov/cgi-bin/text-idx?SID
	Official methods of AOAC International http://www.aoac.org/pubs/pubjaoac.html
	Pesticide analytical manual, Food and Drug administration http://www.cfsan.fda.gov/~frf/pami1.html
China	China pesticides database http://202.127.42.84/tbt-sps/indexEnglish.do National food safety standard—MRLs for pesticides in foods (GB2763-2014)
Japan	MRLs database http://www.m5ws001.squarestart.ne.jp/foundation/search.html Analytical methods for residual compositional substances of agricultural chemicals, feed additives, and veterinary drugs in food (Syoku-an No.0124001)
Australia	FSANZ food standards code http://www.foodstandards.gov.au/code/Pages/default.aspx Standard 1.4.2—MRLs
	Guidelines on analytical method's performance http://apvma.gov.au.node/1031
Codex Alimentarius	Codex MRLS http://www.fao.org/fao-who-codexalimentarius/standards/ pesticide-mrls
	CAC/GL 40-1993: Guidelines on good laboratory practice in pesticide residue analysis http://www.fao.org/fao-who-codexalimentarius/standards/list-of-standards/en/

submitted data should include the following information: identity of the active substance; physical and chemical properties of the active substance; analytical methods; toxicological and metabolism studies; residues in or on treated products, food and feed; and fate and behavior in the environment and ecotoxicological studies.

In the United Stated, the EPA registers (i.e., approves) the use of pesticides and establishes tolerances (the maximum amounts of residues that are permitted in or on a food (EPA, 2016a). EPA, before the registration of a new pesticide or the authorization of a new use for a registered pesticide, must ensure that the pesticide, applied according to label directions, can be used with a reasonable certainty of no harm to human health and without posing unreasonable risks to the environment. To make such determinations, EPA requires more than 100 different scientific studies and tests from applicants.

4.3.2 MAXIMUM RESIDUE LIMITS

"Residue" means one or more substances present in or on plants or plant products, edible animal products, drinking water, or elsewhere in the environment resulting from the use of a pesticide, including their metabolites, breakdown, or reaction products, considered being of toxicological significance. "Metabolite" means any metabolite or a degradation product of a pesticide, formed either in organisms or in the environment. A metabolite is deemed relevant if there is a reason to assume that it has intrinsic properties comparable to the parent substance in terms of its biological target activity, that it poses a higher or comparable risk to organisms than the parent substance, or that it has certain toxicological properties that are considered unacceptable (EC, 2009a). Two "residue definitions" for each pesticide may be needed for regulatory purposes: one for the monitoring of residues and MRL enforcement purposes and the other for performing risk assessment purposes taking into account toxicologically relevant compounds.

To ensure the food safety for consumers and to facilitate the international trade, international bodies have established *MRLs*, for pesticide residues in food commodities. According to the definition of IUPAC, the MRL is the maximum concentration of a residue that is legally permitted or recognized as acceptable in, or on, a food agricultural commodity or animal feedstuff as set by Codex or the EU or a national regulatory authority (Holland, 1996). The term "tolerance" used in the United States is, in most instances, synonymous with MRL, normally expressed as mg/kg fresh weight. The pesticide MRLs for foodstuff and feed were first established in 1961 by the Codex Alimentarius Commission (CAC, www.codexalimentarius.net) of the Food and Agricultural Organization (FAO) of the World Health Organization (WHO), which is responsible for setting MRLs (Codex Alimentarius, 2016) and *extraneous maximum residue limits* (EMRLs) at a global level to assure food safety for human consumption and fair international trade. The Codex Committee on Pesticide Residues (CCPR) has developed a classification of food and feed, adopted by the CAC, which comprises three classes of primary food and feed commodities and two classes of processed food commodities, including 19 types of commodities (FAO/WHO/CAC, 1993, pp. 1–183). To date, the CAC includes MRLs for 218 compounds in 309 food commodities. A free-searchable database allows access to the Codex MRLs and EMRLs for Pesticides adopted by the CAC (http://www.fao.org/fao-who-codexalimentarius/standards/pestres/search/en/). These MRLs are recognized by the World Trade Organizations' Agreement on Sanitary and Phytosanitary Measures (WTO, 2013) as food safety standards.

In principle MRLs are set on the basis of the following (FAO, 2016):

1. Supervised agricultural residue trials
2. Using appropriate consumer intake models

3. Data from toxicological tests on the pesticide that allow for the fixing of an "acceptable daily intake" (ADI)
4. If the estimated daily consumer intake for all commodities calculated under (2) is lower than the ADI calculated under (3), then the residue level under (1) is set as the MRL.

It is important to note that the MRLs are not maximum toxicological limits. They are based on good agricultural practice, and foods derived from commodities that comply with the respective MRLs are intended to be toxicologically acceptable.

For pesticides that are not still in use or that have been banned (e.g., aldrin, chlordane, DDT, dieldrin, endrin, heptachlor), an *EMRL* has been established. This is the maximum concentration of a pesticide residue arising from environmental sources including the pesticides which have been nationally banned but, because of its persistent properties, the residues may still exist in food commodities.

In the EU, the Commission is fixing MRLs for pesticides currently or formerly used in or outside the EU based on recommendations by the European Food Safety Authority (EFSA). The Regulation of European Commission 396/2005/EC (EC, 2005), which entered into force on September 01, 2008, has fixed MRLs for ~500 pesticide active substances in 378 commodities.

The Annexes (I to VII) of Regulation (EC) 396/2005 specify the MRLs and the products to which they apply. Annex I is the list of products to which the MRLs apply. Annex I has been established by Commission Regulation (EC) 178/2006 (EC, 2006). It contains more than 300 products of plant and animal origin. Annex II is the list of EU definitive MRLs and it consolidates the existing EU legislation before September 1, 2008 (EC, 2008). It specifies MRLs for ~250 pesticides. Annex III is the list of the so-called EU temporary MRLs. It is the result of the harmonization process, as it lists pesticides for which, before September 1, 2008, MRLs were only set at national level. It specifies MRLs for more than 450 pesticides. Annex IV is the list of currently 52 pesticides for which no MRLs are needed because of their low risk. Annex V will contain the list of pesticides for which a default limit other than 0.01 mg/kg will apply while Annex VI will contain the list of conversion factors of MRLs for processed commodities. Annex VII contains a list of pesticides used as fumigants for which the Member States are allowed to apply special derogations before the products are placed on the market. Subsequent commission regulations have amended the initial Regulation (EC) 396/2005 in terms of commodities to which MRLs apply and modifications of MRLs for different pesticides or inclusion of new pesticides based on recommendations by EFSA. To consolidate the newly introduced legislation, the EU pesticides database provides a search tool to find out which active substances are approved in Europe together with a reference to the EU legislation (EU, 2016a, 2016b).

EU rules cover pesticides currently or formerly used in agriculture in or outside the EU (around 1100). If a pesticide is not specifically enlisted, a general default MRL of 0.01 mg/kg applies. In general, MRLs in the European Food Regulation (EC) 396/2005 are in the range 0.01−10 mg/kg. In particular, for vegetables, fruits, and cereals intended for the production of baby foods, an MRL of 0.01 mg/kg is applicable to most pesticides. MRLs below 0.01 mg/kg have been established for a few other toxic pesticides, while the use of some very toxic pesticides is prohibited according to Commission Directive 2003/13/EC (Commission Directive, 2003) and Commission Directive, 2006/125/EC (Commission Directive, 2006). In the United States, EPA is responsible for regulating the pesticides used to protect crops and for setting limits on the amount of pesticides that may remain in or on foods marketed in the United States. These limits on pesticides left on foods are called "tolerances" in the United States (EPA, 2016b). In setting the tolerance, EPA must make a safety finding that the pesticide can be used with "reasonable certainty of no harm." To make this finding, EPA considers (1) the toxicity of the pesticide and its breakdown products; (2) how much of the pesticide is applied and how often; and (3) how much of the pesticide (i.e., the residue) remains in or on food by the time it is marketed and prepared.

It should be emphasized that pesticide tolerances are still set to equal or slightly exceed the maximum residues found in the manufacturers' field trials. As such, the tolerances still represent enforcement tools and should not be confused with safety standards even though the EPA does consider possible health risks before establishing tolerances.

In 2008, the OECD, aiming at the harmonization of the calculation of MRLs across its member countries, commissioned an expert group to propose a new MRL calculation procedure. The OECD MRL calculator became available in March 2010 (OECD, 2011).

4.3.3 MONITORING PROGRAMS

In the EU, the Member States, according to the Regulation (EC) 396/2005, sample and analyze different combinations of product/pesticide residue annually within an EU multiannual coordinated control program (EU-MACCP) to ensure compliance with maximum residue levels of pesticides and to assess the consumer exposure to pesticide residues (EU, 2016a, 2016b). The objectives of the program are (1) regulatory monitoring to enforce tolerances and (2) total diet study to better estimate actual exposure of the population to pesticide residues in food (the food samples are representative of the nationwide diet). Additionally, each Member State shall design and perform its national monitoring control program annually.

EFSA has the responsibility of the monitoring program. The Member States, after implementation of the monitoring programs, of pesticide residue levels in food commodities, EU-MACCP, and National, submit their results to EFSA and the European Commission. According to Article 32 of Regulation (EC) 396/2005, EFSA is responsible for drawing up an annual report on pesticide residues on the basis of the results provided by the reporting countries. On the basis of the results of EU monitoring programs, EFSA may derive some recommendations aimed at improving the enforcement of the European pesticide residue legislation.

Moreover, the Rapid Alert System for Food and Feed (RASFF) has been established to provide the control authorities with an effective tool for exchange of information on measures taken to ensure food safety by establishing a network for the notification of a direct or indirect risk to human health deriving from food or feed (EC, 2002b).

The organization of the official controls within the EU is based on the general rules established by Regulation (EC) 882/2004 (EC, 2004), where the general requirements for methods of sampling and analysis and laboratories are laid down in Articles 11 and 12. According to Article 12 of this regulation, the competent authority of the Member States shall designate laboratories that may carry out the analysis of samples taken during official controls. However, they may only designate laboratories that operate and are assessed and accredited in accordance with the following European standards: (1) ISO/IEC 17025:2005 (ISO/IEC, 2005), (2) ISO/IEC 17011:2004 (ISO/IEC, 2004), and (3) taking into account criteria for different testing methods laid down in Community feed and food law.

To contribute to a high quality and uniformity of analytical results of EU official control laboratories, an analytical network of European Reference Laboratories (EURLs, www.eurl-pesticides.eu.), National Reference Laboratories (NRLs), and Official National Laboratories (OFLs) is designated (EURL−NRL−OFL network) (EC, 2004). Currently, there are four EURLs established in the EU for pesticide residues: fruits and vegetables; cereals and feed; food of animal origin; and single-residue methods.

Similarly, in the United States, Federal, State, and local network of laboratories called the Food Emergency Response Network, which is responsible for the detection and identification of both chemical and biological threats to the food supply, has been established following the Bioterrorism Act of 2002 (FDA, 2002), by the Centre for Disease Control, the US Department of Agriculture (USDA), and the Food and Drug Administration (FDA). The FDA performs its regulatory monitoring program to enforce pesticide tolerances (FDA, 2015) and acquires data on particular commodity and pesticide combinations by carrying out market basket surveys under its Total Diet Study Program (TDS) (FDA, 2016). The FDA is charged with enforcing tolerances in both imported food and domestic food shipped into interstate commerce [except for meat, poultry, and certain egg products, which are monitored by the USDA (FDA, 2015)]. Unlike the FDA's regulatory monitoring program, the TDS is not designed to enforce pesticide tolerances but rather to provide an estimate of dietary pesticide residue exposure to the general US population and to specific US population subgroups. As such, it determines pesticide residues not in the raw commodity, but in food that are prepared table-ready for consumption. In addition, the FDA works with EPA to set "action levels" and enforcement guidelines for residues of pesticides, such as DDT, that may remain in the environment after their use is discontinued. The guidelines are set at levels that protect public health.

The USDA conducts two pesticide residue monitoring programs: the USDA's National Residue Program (NRP) (USDA, 2016a) and the Pesticide Data Program (PDP) (USDA, 2016b). The NRP includes samples of meat, poultry, and raw eggs that are analyzed for pesticide residues as well as for animal drugs and environmental contaminants. The NRP is a collaborative interagency program established to protect the public from exposure to harmful levels of chemical residues in meat, poultry, and egg products produced or imported into the United States. The USDA Food Safety and Inspection Service (FSIS), the EPA, and the Department of Health and Human Services of the FDA are the federal agencies primarily involved in managing this program. FSIS publishes its quarterly reports to summarize the chemical residue results for the US NRP for meat, poultry, and processed egg products (USDA, 2016a); the report is intended to supplement the annual US residue report known as the Red Book, which FSIS continues to publish.

The USDA's PDP carried out by the USDA's Agricultural Marketing Service (AMS) and directed at raw agricultural products and various processed foods has an important role in the implementation of the Food Quality Protection Act (USDA, 2016b). This law directs the Secretary of Agriculture to collect pesticide residue data on commodities most frequently consumed by infants and children. The PDP is the primary source of residue data used by the EPA to assess dietary exposure during the review of the safety of existing pesticide tolerances. In contrast with FDA's regulatory monitoring programs that are designed primarily to enforce pesticide tolerances, PDP's sampling protocols are designed to more accurately reflect residue levels that reach consumers and, as such, provide a more reliable tool for assessing human dietary risk from exposure to pesticides.

4.4 ANALYTICAL QUALITY CONTROL—METHOD VALIDATION

Nowadays, there are several internationally recognized organizations offering guidelines on method validation and related topics. Some examples are the OECD Guidance Document for Single Laboratory Validation of Quantitative Analytical Method (OECD, 2014); the Association of Official Analytical Chemists (AOAC, 2002a); the American Society for Testing and Material (ASTM, 2011); the European Committee for Normalization (CEN, 2008); the Cooperation on International Traceability in Analytical Chemistry (CITAC, 2002); the European Analytical Chemistry group (EURACHEM) (Eurachem Guide: The Fitness for Purpose of Analytical Methods— A Laboratory Guide to Method Validation and Related Topics) (Magnusson & Örnemark, 2014); and the IUPAC (Thompson, Ellison, & Wood, 2002).

In the ISO/IEC 17025:2005 standard, method validation is defined as "the confirmation by examination and the provision of objective evidence that the particular requirements for a specific intended use are fulfilled" (ISO/IEC, 2005). In the Eurachem Guide for Method Validation the following definition is elaborated: "Method Validation is the process of defining an analytical requirement and confirming that the method under consideration has capabilities consistent with that the application requires. Inherent in this is the need to evaluate the method's performance and very important the judgment of method suitability" (Magnusson & Örnemark, 2014). The judgment of method suitability is very important: in the past, method validation tended to concentrate only on evaluating the performance characteristics and not on the documentation and demonstration of whether the method is fit for the particular analytical purpose (Thompson et al., 2002).

In the International Vocabulary of Metrology (VIM) (JCGM, 2012), "validation" is defined as verification, where the specified requirements are adequate for an intended use (VIM:§2.45), while "verification" is defined as the provision of objective evidence that a given item fulfills specified requirements (VIM:§2.44). According to the standard ISO/IEC 17025:2005, the laboratory shall validate nonstandard methods, in-house methods, standard methods used outside their intended scope, amplifications, and modifications of standard methods while standard methods, e.g., ISO, ASTM methods, can be verified and not fully validated before being introduced in the laboratory (ISO/IEC, 2005).

There are different approaches for method validation with obviously different requirements (Taverniers, De Loose, & Van Bockstaele, 2004a). There is the interlaboratory approach where collaborative studies are performed based on certain protocols and standards (AOAC, 2002b; Horwitz, 1995). These standards require a minimum number of laboratories and test materials to be included in the collaborative trial to fully validate the analytical method (Horwitz, 1988, 1995). Standard methods adopted by internationally recognized standardization bodies, such as ISO, CEN, AOAC, are internationally validated by collaborative studies to check method accuracy and robustness. In other cases, the single laboratory approach is another alternative for method validation (AOAC, 2002a; Thompson et al., 2002).

Requirements for how to carry out method validation are usually specified in guidelines within a particular sector as in the case of pesticide residues (CAC, 1993a; SANTE/11945/2015) or veterinary drug residues (Commission Decision 2002/657/EC) (EC, 2002a). Wherever specified requirements exist, it is recommended to be fulfilled to ensure that the particular validation terminology and statistics used are interpreted in a manner consistent within the relevant sector (Stöckl, D'Hondt, & Thienpont, 2009).

The method performance parameters to be studied and evaluated during method validation include selectivity/specificity, accuracy (trueness and precision), linearity, range, detection limit (the lowest amount of analyte to be detected), limit of quantification (the lowest amount of analyte that can be measured), ruggedness, and robustness. Definitions of these parameters from the Eurachem Guide (Magnusson & Örnemark, 2014), the VIM (Barwick & Prichard, 2011), and IUPAC (Thompson et al., 2002) will be briefly discussed (Masson, 2007; Vanatta & Coleman, 2007).

Selectivity in analytical chemistry relates to the extent to which the method can be used to determine particular analytes in mixtures or matrices without interferences from other components of similar behavior. While IUPAC recommends the term "selectivity," some others use the term "specificity"; specificity is considered to be the ultimate in selectivity meaning that no interferences are supposed to occur (Magnusson & Örnemark, 2014; Vessman et al., 2001).

Working range is defined as the interval over which the method provides results with an acceptable uncertainty. During method validation, it is necessary to confirm that the method can be used over this interval; method linearity and calibration procedure should be considered.

The *analytical sensitivity* is defined as the change in instrument response, which corresponds to a change in the measured quantity, i.e., the slope of the response curve. Many methods include a sample preparation step to extract the target analyte before measurement by the instrument. In these cases, there are two working ranges: the method working range given in the scope of the method and the instrument working range. Both of them should be assessed in the course of the validation. The method working range should be established for each matrix covered in the method scope due to different interferences and related matrix effects.

The *detection limit*, or *limit of detection*, *LOD*, is the lowest quantity of a substance that can be distinguished from the absence of that substance (a *blank value*) within a stated confidence limit (generally 1%) (Evard, Kruve, & Leito, 2016a, 2016b; Magnusson & Örnemark, 2014). The detection limit is usually estimated from the mean of the blank, the standard deviation of the blank (S_0), and some confidence factor.

$$LOD = \text{Mean of blank} + 3*S_0 \qquad (4.1)$$

S_0 is based on results near zero concentration adjusted for any averaging or blank correction used in practice. The samples used for the estimation of the LOD should be either "blank" samples that do not contain the analyte or test samples with concentrations of analyte close to or below the expected LOD (Shrivastava & Gupta, 2011). These samples can be prepared by spiking a blank sample. When blank samples are not available, reagent blanks can be used. Reagent blanks not subjected to the whole measurement procedure will give the instrument LOD that should be clearly distinguished by the method detection limit. To obtain a method detection limit, this must be based on the analysis of samples that have been taken through the whole analytical procedure (and their results have been calculated with the same equation applied for test samples). The method detection limit is the useful parameter for method validation (Magnusson & Örnemark, 2014).

The *limit of quantitation*, *LOQ*, is the lowest level of a substance that can be determined with acceptable performance. The LOQ is estimated from the mean of the blank, S_0, and some confidence factor (Shrivastava & Gupta, 2011).

$$LOQ = \text{Mean of blank} + 10*S_0 \qquad (4.2)$$

The standard deviation of results at concentrations near zero (S_0), used for the estimation of LOD and LOQ, can be obtained under repeatability conditions. In cases where blank values differ from day-to-day, intermediate precision conditions are preferred; if quality control (QC) samples at low concentration levels are available, their standard deviation can be used for the estimation of LOD and LOQ (Evard et al., 2016a, 2016b; Kruve et al., 2015a).

Measurement *accuracy* expresses the closeness of a single result (X_i) to a reference value (μ); a measurement result of a test sample can be expressed as a combination of the mean value (X_m), the systematic error (δ), and the random error (σ_i). Method validation seeks to investigate the accuracy of results by assessing both systematic and random effects on single results. Accuracy is dependent on trueness and precision as stated in the ISO 5725−1:1994 (ISO, 1994a).

$$X_i = X_m \pm \delta \pm \sigma_i \tag{4.3}$$

Trueness is the expression of how close to a reference value is the mean of an infinite number of results (Barwick & Prichard, 2011; ISO, 1994a). Because it is not possible to take an infinite number of measurements, trueness cannot be measured but can be expressed quantitatively in terms of "bias." The determination of bias is based on the comparison of the mean of the results (X_m) obtained with the method in validation process with a suitable reference number (X_{ref}) (Hibbert, 2007). For the determination of bias, three approaches are available: (1) the analysis of reference material; (2) recovery experiments using spiked samples; and (3) comparison with results obtained with another method. In the absence of suitable reference materials, recovery experiments can be used to estimate bias by spiking matrix blanks or test samples with the target analyte over a range of concentrations (ISO, 1994d; Thompson et al., 2000; Thorburn Burns, Danzer, & Townshend, 2002).

Precision (ISO, 1994a) is a measure of closeness of results, usually expressed by statistical parameters that describe the spread of results, i.e., the standard deviation calculated from results obtained by replicate measurements under specified conditions. These conditions determine the type of precision obtained (Barwick & Prichard, 2011; Magnusson & Örnemark, 2014). Measurement repeatability and measurement reproducibility represent the two extreme measures of precision that can be obtained (ISO, 1994b, 1998). Repeatability is a measure of the variability of results when a measurement is performed by a single analyst using the same equipment over a short timescale. This gives the smallest variation of results. Reproducibility is a measure of the variability of results between laboratories and it represents the largest variation in results. Between these two extreme cases, intermediate precision gives an estimation of the variation when measurements are made in a single laboratory but under conditions more variable than the repeatability conditions (e.g., many analysts over a longer timescale) (ISO, 1994c). This precision can be representative of all sources of variation that will occur in a single laboratory under routine conditions (Meyer, 2007). All these different degrees of precision are illustrated in Fig. 4.1.

Precision is dependent on analyte concentration and is usually determined at a number of concentrations across the working range of the method. This could include a particular concentration of interest such as a regulatory limit plus concentrations at the limits of the measuring range. When the standard deviation is proportional to the analyte concentration, precision may be expressed as a relative standard deviation (RSD). Materials used for the evaluation of precision should be representative of test samples in terms of matrix and analyte concentration, homogeneity, and stability. The replicates should be independent, meaning that the whole process should be repeated for each replicate. The minimum number of replicates is usually between 5 or 6 and 15 (Barwick & Prichard, 2011; Kruve et al., 2015b; Magnusson & Örnemark, 2014).

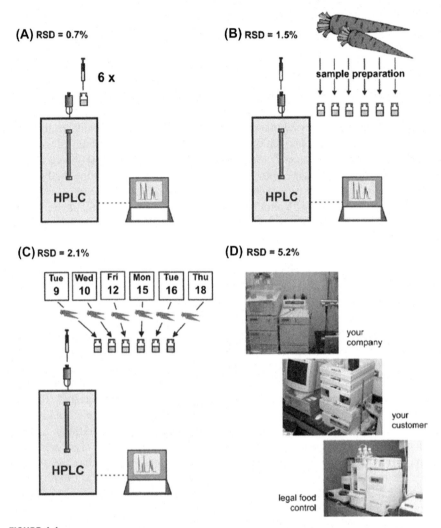

(A) RSD = 0.7%

6 x

HPLC

(B) RSD = 1.5%

sample preparation

HPLC

(C) RSD = 2.1%

Tue	Wed	Fri	Mon	Tue	Thu
9	10	12	15	16	18

HPLC

(D) RSD = 5.2%

your company

your customer

legal food control

FIGURE 4.1

The sixfold injection of the same sample (A) will yield a lower repeatability (measured as standard deviation) than the sixfold preparation of the same sample (B). If the six preparations are performed on different days (C) the standard deviation will be even higher and the number is called "intermediate precision" (or "day-to-day precision" in this case). If the same sample is investigated in different laboratories (D) the resulting standard deviation is highest and is called "reproducibility."

Reproduced with permission from Meyer, V.R. (2007). Measurement uncertainty.
Journal of Chromatography A, 1158, 15–24.

The method **total error**, described by the method accuracy, can be quantitatively expressed with the uncertainty estimate combining the method systematic error, i.e., the trueness expressed by the bias, and the random error, i.e., the method precision and related standard deviation. A comprehensive diagram of the links between the terms used to describe the quality of results obtained with a method is illustrated in Fig. 4.2 (Magnusson & Örnemark, 2014).

Method validation gives information on the method capabilities and limitations that may occur in the routine use, so it helps to design a proper internal QC system (CITAC, 2002; Masson, 2007; Prichard, 1998; Taverniers et al., 2004a; Taverniers, De Loose, & Van Bockstaele, 2004b; Vanatta & Coleman, 2007). Specific controls should be applied to the method to verify that it is performing as expected and that the data produced are still fit for purpose (FFP) each time; this can be achieved by methods of QC. QC can take many forms, both inside the laboratory (internal) and between the laboratory and other laboratories (external) (Prichard, 1998). Internal QC refers to procedures undertaken by the laboratory for the monitoring of operations and measurement results to ensure the validity of results (Nordtest, 2011); this may include replicate analysis of stable test samples, blanks, standard solutions for the calibration, spiked samples, blind samples, and QC samples. The introduction of the quality assurance process in laboratories has considerably increased the use of standard or reference materials as analytical control materials and control charts as QC tool (ISO, 2012, 2013, 2014a). Shewhart charts are applied as a mean control chart for detecting bias in an analytical method, but they are able to detect shifts of major magnitude; the ISO standard 7870-2:2013 (ISO, 2013) gives the rules for the recognition of an out-of-control situation. Exponentially weighted moving average (EWMA) control charts make possible faster detection of small to moderate shifts of the process mean (ISO, 2016b). The joint use of an EWMA control chart and a Shewhart control chart can reassure fast detection of both small and large shifts. The monitoring of the process variability usually requires the use of another

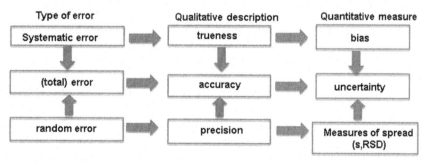

FIGURE 4.2

Illustration of the links between some fundamental concepts used to describe quality of measurement results.

Reproduced from the Eurachem Guide (2014). The Fitness for Purpose of Analytical Methods (2nd ed.).

technique; the R-charts can be used to monitor the analytical precision within each analytical run (ISO, 2014a). In cases where the commonly used Shewhart control chart may either be not applicable or less efficient in detecting the patterns of variation, other more specialized control charts can be used such as the moving average charts (ISO, 2014b) or the cumulative sum control charts (ISO, 2011).

Data from method validation regarding trueness and precision are very useful to set criteria for the internal QC. External QC is usually performed by participation in proficiency testing (PT) schemes, which can be defined as the periodic assessment of the competency of individual laboratories. PT helps to check variation between laboratories (reproducibility) and systematic errors (bias). International harmonized protocols exist for the organization of PT schemes based on the standard ISO/IEC 17043:2010 (ISO/IEC, 2010). Important information for laboratories on this issue is included in the Eurachem Guide: "Selection, use and interpretation of PT schemes by laboratories" (Mann & Brookman, 2011). Besides their role as an external QC tool, PT results or samples could be used as an alternative to fulfill some of the quality assurance requirements such as analytical precision, uncertainty assessment, and internal QC (Detaille & Maetz, 2006).

4.4.1 GUIDELINES FOR PESTICIDE RESIDUE ANALYSIS

In the field of pesticide residues analysis, guidance documents and guidelines have been elaborated by international organizations aiming to support laboratories involved in the monitoring of pesticide residues on foodstuffs such as the OECD Guidance Document on Pesticide Residue Analytical Methods (OECD, 2007).

In the Codex Alimentarius, the CCPR is responsible for the elaboration of methods of sampling and analysis, as well as the recommendation of validation and QC procedures for methods for the determination of pesticide residues in food. The CCPR has elaborated guidelines on performance criteria, specific for analytical methods for the determination of pesticide residues in food and feed (CCPR, 2016) and on good laboratory practice in pesticide residue analysis (CAC, 1993a). Additionally, guidelines on the use of mass spectrometry (MS) for identification, confirmation, and quantitative determination of residues are given in CAC (2005). The guidance documents of CCPR refer to qualitative (screening, identification, confirmation) and quantitative analytical methods, which closely follow the recommendations of the document SANTE/11945/2015 (SANTE, 2015). In the EU, the guidance document "Method Validation & Quality Control Procedures for Pesticide Residues Analysis in Food & Feed," SANTE/11945/2015, implemented by January 01, 2016, is published by the Directorate General for Health and Consumers of the European Commission, DG-SANTE. The document describes the method validation and QC requirements to support the validity of data used for checking compliance with MRLs, enforcement actions or assessment of consumer exposure to pesticides. The document is complementary and integral to the requirements in ISO/IEC 17025:2005. The key objectives, stated in the document, are to (1) provide a harmonized cost-effective quality assurance system in the EU; (2) ensure the quality and comparability of analytical results; (3) ensure that acceptable accuracy is achieved; (4) ensure that false positives or false negatives are not reported, and (5) support compliance with ISO/IEC 17025.

The Document SANTE/11945/2015, also referred as AQC (analytical quality control) document, entails mutually acceptable scientific rules for official pesticide residue analysis within the EU as agreed by all Member States of the EU and constitutes a technical guideline in the sense of article 28 of Regulation 396/2005. It should thus be consulted in audits and accreditations of official pesticide residue laboratories according to ISO/IEC 17025 (SANTE, 2015).

In the US legislation, there are no generic guidelines applicable to Analytical QC and each agency publishes its own requirements. The Pesticide Analytical Manual (FDA/PAM, 2015) published by FDA includes analytical methods used in FDA laboratories for regulatory purposes. PAM Volume I contains multiresidue methods (MRMs) while Volume II contains methods designed for the analysis of commodities for residues of only a single compound.

The FDA's Office of Regulatory Affairs has set guidelines for the validation of chemical methods for the FDA Foods Program to be applied in the FDA laboratories (FDA/FVM, 2015). The FDA has also issued a guidance document on the development, evaluation, and application of MS methods for confirming the identity of pesticide residues (FDA, 2013). Methods of analysis used by the USDA's PDP laboratories are available at the USDA (2016b); each method has been demonstrated to produce robust data through extensive method validation and ongoing quality assurance/QC and PT procedures. The PDP standard operation procedures for sampling, sample process and analysis, method validation, and data handling can be found at USDA SOPs (2016).

4.4.1.1 Method Validation for Pesticide Residues

In the EU Guidance Document No. SANTE 11945/2015, the following definition is attributed to "method validation": the process of characterizing the performance to be expected of a method in terms of its scope, specificity, sensitivity, accuracy, repeatability, and within-laboratory reproducibility. Some information on all characteristics, except within-laboratory reproducibility, should be established before the analysis of samples, whereas data on reproducibility and extensions of scope may be produced from Analytical QC during the analysis of samples. Wherever possible, the assessment of accuracy should involve analysis of certified reference materials and the participation in proficiency tests or other interlaboratory comparisons (SANTE, 2015).

In the guidelines on performance criteria for methods of analysis for the determination of pesticide residues elaborated by the CCPR (2016), it is stated that the process of method validation is intended to demonstrate that a method is *FFP*. According to the CCPR document, the validation should specify the analyte (identity and concentration), account for the matrix effect and provide a statistical characterization of the recovery results, and indicate if the rates of false positives and negatives are minimally accepted. To ensure that validation of the method remains appropriate over time, the method should be continuously assessed using routine PT and appropriate QC samples.

In the EU Guidelines, the process of design and preliminary assessment of the performance characteristics of a method including ruggedness is defined as "method development" (SANTE, 2015). Method development and/or method optimization phase for pesticide residues analysis may include testing the matrix effect, the homogeneity of the comminuted samples, the stability of residues during sample processing, and the stability of analytes in solvents and extracts, while method ruggedness could be addressed by testing the effects of changes in the instrument, reagents, pH, temperature, operators, and other factors. According to Codex Guidelines, the limits for experimental parameters defined during ruggedness tests should be prescribed in the method protocol. Permissible deviations should produce no meaningful change in the results so that the fitness for purpose of the method is not put in danger (CCPR, 2016).

According to the SANTE 11945/2015 document, method validation for pesticide residues is a requirement and must be performed before introducing the method in the laboratory to test samples. Method validation must be supported and extended by method performance verification during routine analysis applying a suitable and effective internal QC scheme. SANTE 11945/2015 document also refers to the ongoing method validation that can be incorporated in the method performance verification procedure applied in the routine use of the method. As stated in the EU guidelines, the purpose of ongoing method validation is to demonstrate method robustness through evaluation of mean recovery and within-laboratory reproducibility and applicability to other commodities from the same commodity groups. Ongoing method validation data can reveal if minor adjustments made to the method over time do affect or do not affect method performance, can be used to determine acceptable limits for recovery during routine analysis, and can provide information for estimation of the within-laboratory uncertainty. This is in agreement with the concept of the Analytical Quality Assurance Cycle (AQAC) that can be applied as a quality tool in all kinds of laboratories (Olivares & Lopes, 2012). The AQAC starts with validation, demonstrating that the method, as performed, is FFP for which it is to be used. After validation, the second step is the evaluation of uncertainty, and in the third step, QC demonstrates that the method provides reliable results acting as ongoing validation scheme producing new data to adjust the uncertainty (Olivares & Lopes, 2012).

According to the Analytical QC document (SANTE, 2015), representative matrices may be used to validate MRM and single-residue methods. For at least, one representative commodity from each commodity group must be validated, depending on the intended scope of the method. When the method is applied to a wider variety of matrices, complementary validation data should be acquired from ongoing QC during routine analyses. In the Codex Guidelines on Good Practice in Pesticide Residue Analysis, it is stated that the laboratory must prefer methods having multiresidue and/or multimatrix applicability and that the use of representative matrices is an important tool in validating methods (CAC, 1993a).

Where experience shows similar performance of extraction and cleanup between broadly similar commodities/sample matrices, a representative commodity may be selected to represent each commodity group having common properties and used for method validation. In the Annex A of SANTE 11945/2015 document, there is a table containing all the known food commodities divided into commodity groups. Food commodities are classified into 10 commodity groups, 6 of plant origin and 4 of animal origin, based on their main characteristics, meaning the water content, the acidity, the oil content, etc. Each commodity group contains typical commodity categories and typical representative commodities, which can help the laboratory to choose the more appropriate matrices to perform method validation. In this recent revision of the SANTE 11945/2015 document, a similar classification scheme for feed categories has been also introduced. A similar table of representative commodities and species for commodity groups used for validation is in the CODEX GLP Guidelines (CAC, 1993a). The Codex guideline CAC/GL:41-1993 (CAC, 1993b) describes the portion of the raw agricultural commodity to which the MRL applies and which must be used to prepare the analytical sample for the determination of pesticide residues. Likewise, the Commission Regulation (EU) 752/2014 (EU, 2014) establishes the food and feed products to which MRLs of the Regulation (EC) 396/2005 apply. Similar procedures are proposed by the two legislative documents. For example, in the case of olives the whole commodity after removal of stones should be used but the residues should be calculated and expressed on the whole fruit.

An important factor that must be taken into consideration when a method validation plan is designed is the residue definition of targeted pesticides. As previously discussed, the residue definition of many pesticides may include not only the parent compound but also a number of metabolites and TPs that exhibit similar toxicity. In these cases, the validation should include all the compounds if this is feasible, depending on the analytical capability and on the availability of analytical standards for all the compounds. For example, in the case of spirotetramat, the MRL includes the parent compound and its four metabolites, BYI08330-enol, BYI08330-ketohydroxy, BYI08330-monohydroxy, and BYI08330 enol-glucoside (structures shown in Fig. 4.3), expressed as spirotetramat for commodities of plant origin while the MRL for commodities of animal origin includes the parent compound and its BYI08330-enol metabolite (EFSA, 2013).

Even more complex residue definitions for some pesticides exist. A sound example is the herbicide tepraloxydim; its MRL is defined as the sum of tepraloxydim and its metabolites that can be hydrolyzed either to the moiety 3-(tetrahydro-pyran-4-yl)-glutaric acid or to the moiety 3-hydroxy-(tetrahydro-pyran-4-yl)-glutaric acid, expressed as tepraloxydim. In the EFSA review of the MRLs for tepraloxydim (EFSA, 2011) about 10 compounds are listed as its metabolites (structures shown in Fig. 4.4).

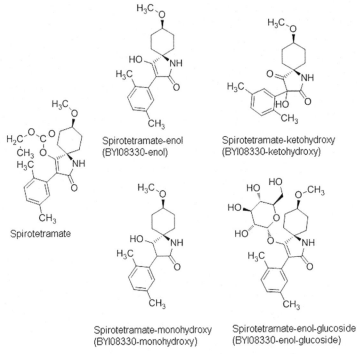

Spirotetramate-enol
(BYI08330-enol)

Spirotetramate-ketohydroxy
(BYI08330-ketohydroxy)

Spirotetramate

Spirotetramate-monohydroxy
(BYI08330-monohydroxy)

Spirotetramate-enol-glucoside
(BYI08330-enol-glucoside)

FIGURE 4.3

Spirotetramat, parent compound and metabolites.

In some cases, the residue definition of a specific compound is different in commodities of plant origin from the respective in foods of animal origin. For example, the MRL of flusilazole in foods of animal origin in addition to the parent compound includes also the metabolite bis(4-fluorophenyl)methylsilanol, the MRL of acetamiprid includes also the N-desmethyl-acetamiprid, the MRL of bixafen includes also bixafen-desmethyl. In other cases, as for the pesticides chlorothalonil, fenpropimorph, spiroxamine, and kresoxim-methyl only their metabolites—2,5,6-trichloro-4-hydroxyphtalonitrile (SDS 3701), fenpropimorph carboxylic acid (BF421-2), spiroxamine carboxylic acid, and BF-490-9, respectively—are included in the residue definition in commodities of animal origin and not the parent compound (EC, 2005).

In the EU guidelines, the initial full method validation shall include all analytes within the scope of the method. The method must be tested to assess sensitivity, matrix effects, the LOQ or reporting limit (RL), selectivity/specificity, trueness in terms of recovery, precision in terms of repeatability and within-laboratory reproducibility, and the method robustness. Different acceptance criteria have also been established to judge the method performance to decide if the method is FFP. Validation parameters and specific acceptance criteria are shown in Table 4.3 (SANTE, 2015).

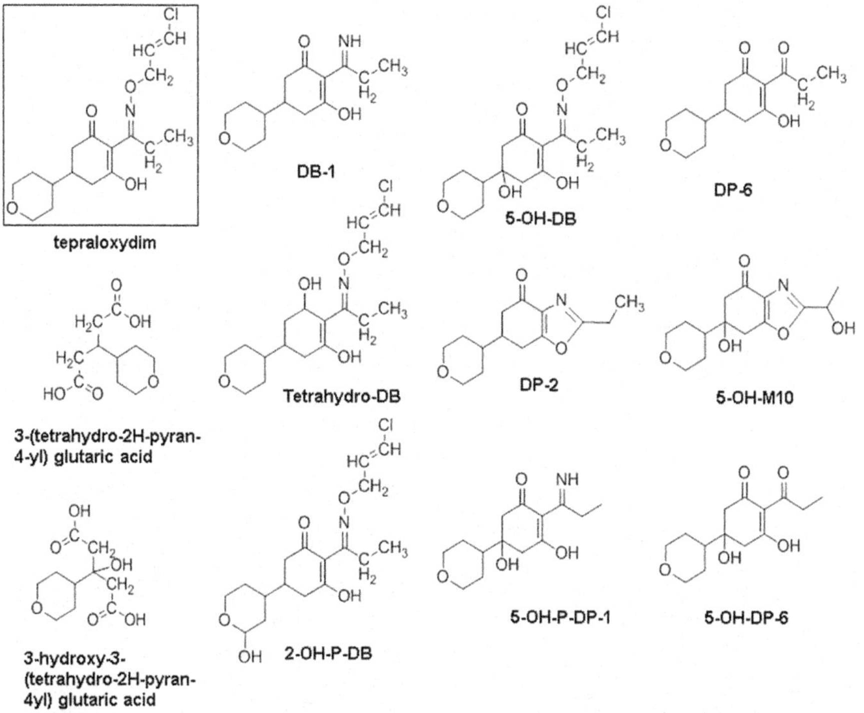

FIGURE 4.4

Tepraloxydim, parent compound and metabolites.

Table 4.3 Validation Parameters and Criteria in SANTE (2015)

Parameter	What/How	Criterion
Sensitivity/linearity	Linearity check calibration curve	Residuals < ±20%
Matrix effect	Matrix-matched and solvent standards	% Matrix effect
LOQ	Lowest spike level meeting the method performance criteria for trueness and precision	≤MRL
Specificity	Response in reagent blank and blank control samples Identification criteria	<30% of RL
Trueness (bias)	Average recovery	70%–120%
Precision (RSD$_r$)	Repeatability—spike levels	<±20%
Precision (RSD$_{wR}$)	Within-laboratory reproducibility	<±20%
Robustness	Ongoing validation/verification	

LOQ, *limit of quantitation;* MRL, *maximum residue limit;* RL, *reporting limit;* RSD, *relative standard deviation.*

For the sensitivity/linearity parameter, linearity should be established from five calibration levels covering the working/concentration range of the method. Visual inspection and/or calculation of residuals are preferred to check the fit of calibration curve; relative individual residuals should be within $\pm20\%$. The use of weighted linear regression ($1/x$) is recommended versus linear regression. Some of the statistical tests available for the evaluation of calibration models are the "Lack-of-fit" test, the "Goodness-of-fit" test, and the Mandel's fitting test (Kruve et al., 2015b; Miller & Miller, 1993; Vanatta & Coleman, 2007).

In the EU guideline, the matrix effect is defined as "the influence of one or more undetected compounds from the sample on the measurement of the analyte concentration or mass." Matrix effects should be assessed at the initial method validation stage along with linearity because linear ranges may vary for different matrices. Solvent calibration curve can be used when the comparison of response from solvent standards and matrix-matched standards is within $\pm20\%$. In practice, matrix-matched calibration (Cuadros-Rodríguez, Bagur-González, Sánchez-Viñas, González-Casado, & Gómez-Sáez, 2007) is adopted for accurate quantification of pesticide residues using GC—MS and LC—MS techniques and this is one officially accepted option in the EU for pesticide residues in food and feed. Several hypotheses on the causes and the underlying mechanisms of matrix effects in LC—MS are thoroughly discussed by Stahnke and Alder (2015) while different approaches to evaluate the ionization suppression phenomenon occurred in LC—MS methods and related effects on method validation are comprehensively reviewed by Kruve et al. (2015b). The efficiency of alternative calibration approaches should be investigated during method development and validation stages (Cuadros-Rodríguez et al., 2007); standard addition and procedural standard calibration can be applied to compensate for matrix effects and low recoveries while the use of internal standard, especially isotopically labeled internal standard (IL-IS), can improve recovery of analytes, increase linearity range (Araujo, 2009; Nilsson & Eklund, 2007), and also correct for various random errors, system response drift, etc. In the EU guideline, it is recommended to apply standard addition for confirmatory quantitative analysis in cases of MRL exceedances and/or if no blank material is available for the preparation of matrix-matched standard solutions. The use of procedural standards can compensate for matrix effects and low extraction recoveries but this is of limited applicability in series of samples of the same type to be processed within the same batch (SANTE, 2015).

In GC—MS, matrix-induced signal enhancement (Erney, Gillespie, Gilvydis, & Poole, 1993; Rahman, Abd El-Aty, & Shim, 2013) is more often encountered compared to matrix-induced suppression (Fajgelj & Ambrus, 2000). An alternative strategy to decrease the matrix-induced enhancement effect in GC—MS analysis is the use of analyte protectants, added both in samples and standards to mask the active sites in the GC system (Anastassiades, Mastovska, & Lehotay, 2003; Mastovska, Lehotay, & Anastassiades, 2005); significant peak quality improvement for analytes can result in the enhancement in detectability and accuracy in quantification as depicted in Fig. 4.5.

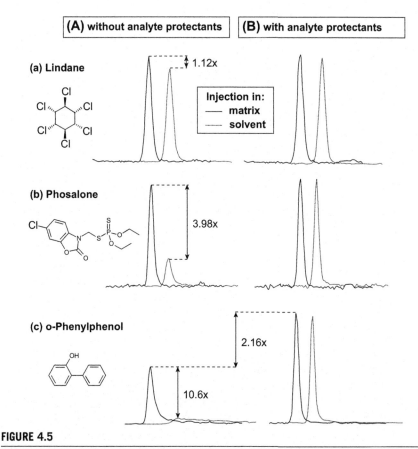

FIGURE 4.5

Response enhancement of pesticides in a mixed fruit extract and in solvent without (A) and with (B) addition of analyte protectants.

Reproduced with permission from Mastovska, K., Lehotay S.J, & Anastassiades, M. (2005). Combination of analyte protectants to overcome matrix effects in routine GC analysis of pesticide residues in food matrices. Analytical Chemistry, 77, 8129–8137.

Another parameter to be tested during method validation is method selectivity/specificity; the response of reagent blank and blank control samples should be lower than 30% of the analyte response corresponding to the RL of the method. Method selectivity can affect the identification of analytes. Requirements and criteria for identification of analytes and confirmation using hyphenated techniques such as gas or liquid chromatography coupled to MS will be discussed separately.

In SANTE 11945/2015, estimation of method trueness and precision can be based on recovery experiments; a minimum of five replicates are required at two different spike levels, the targeted LOQ (limit of quantification) or RL of the method and at least one higher level, 2−10 times the targeted LOQ or the MRL. Acceptable mean recoveries for method validation are within the range 70%−120% with an associated repeatability RSD_r below or equal to 20% for all analytes within the scope of the method. The method LOQ is defined as the lowest spike level of the validation fulfilling these performance criteria. In MRMs, mean recoveries below 70% but demonstrating good precision can be accepted according to the EU guideline. For the estimation of intermediate precision, the within-laboratory reproducibility RSD_{wR} derived from the ongoing method validation and performance verification studies can be used; RSD_{wR} should not exceed 20%. Recoveries and RSDs data from the ongoing method validation and QC studies can be used for the evaluation of method robustness and applicability. Extension of the scope of MRMs can include new matrices and/or new analytes; ongoing method validation can be performed with QC samples used for routine recovery check, prepared in different commodities of the same commodity group and/or spiked with new analytes. In this way, the applicability of the method to different matrices and analytes is tested achieving more realistic estimates of method recoveries, within-laboratory reproducibility, RSD_{wR}, and uncertainty. All the above parameters should be evaluated during the initial and ongoing validation stages of a quantitative analytical method for the determination of pesticide residues in food samples (SANTE, 2015). Similar criteria for the validation for the method individual performance characteristics are included in the Codex guidelines (CCPR, 2016).

The European Commission has supported the G6MA-CT-2000−01012 project on "Metrology of Qualitative Chemical Analysis" (MEQUALAN), focused on the implementation of quality principles in qualitative analysis: traceability, reliability (uncertainty), validation, and internal/external QC for qualitative methods (Rios et al., 2003). The performance of qualitative methods is usually expressed in terms of false response rates—the false-positive and false-negative rates should be determined—and detection capability. Method validation and interlaboratory studies for qualitative methods require the application of a different set of statistical tools and terminology from those used for quantitative method validation (Ellison & Fearn, 2005). Experiments to assess the performance of qualitative methods generally require substantially more replications than validation of quantitative methods.

In the EU guidelines, laboratories are encouraged to use screening methods to extend their scope. Validation requirements for qualitative methods are focused on the establishment of the screening detection limit (SDL), which is the lowest level at which an analyte has been detected in at least 95% of the samples (i.e., an acceptable false-negative rate of 5%). The method selectivity should be also checked using "blank" samples for the possible presence of false detects (SANTE, 2015).

In the FDA Foods Program Guidelines for chemical methods, initial validation of screening methods involving MS analysis may be performed using a limited set of representative analytes and matrices; the performance characteristic that may be evaluated include sensitivity, selectivity, false-positive and false-negative rates, minimum detectable concentration, ruggedness, and confirmation of identity (FDA-FVM, 2015).

4.4.1.2 Quality Assurance

In the EU guideline (SANTE, 2015), the ongoing method performance verification during routine analysis for quantitative methods requires the determination of recoveries of all target analytes to be measured within each batch of analyses. If this is not feasible, the number of analytes may be reduced and a rolling program can be established to cover all analytes, representative, and others, as well as representative commodities, to cover the scope annually. The spiking level may range from $1 \times RL$ to $2-10 \times RL$. The acceptable limits for routine recoveries should be within the range of the mean recovery, $\pm 2 \times RSD$, that have been estimated during initial validation. Alternatively, a default range of 60%−140% can be used for individual recoveries. The construction and use of control charts for recoveries is recommended in the Codex GLP Guidelines as a tool to check for trends in method performance (CAC, 1993a). In the case of screening methods, at least 10 representative analytes that cover all critical points of the method should be spiked to verify the method performance in each batch following a rolling program to cover all analytes from the validated scope (CCPR, 2016; SANTE, 2015).

Another important issue that may affect the quality and reliability of measurement results is the stability of pesticide standard solutions; specified procedures for handling of reference standards should be incorporated in the internal quality control (IQC) system of laboratories. Reference standards of analytes should be of known purity and stored under specified conditions. Preparation and storage conditions of pesticide stock solutions and calibration standard solutions should ensure their stability while suitable statistical tests should be used to check their nominal concentrations (CAC, 1993a; SANTE, 2015).

All accredited laboratories under ISO 17025:2005 are obliged to participate regularly in PT schemes as an external QC procedure. Statistical methods described in ISO 13528:2015 (ISO, 2015) can be applied to demonstrate that the measurement results obtained by laboratories meet specified criteria for acceptable performance. According to EC (2005) all EU laboratories analyzing samples within the framework of official controls on pesticide residues shall participate in the EU proficiency tests (EUPTs) organized annually by the four EURLs. The application of MRMs for pesticide residues analysis involves the reporting of a large number of individual z scores, making the evaluation of the overall performance of the laboratories difficult; useful formulas have been designed to globally evaluate the assessment of the participating laboratories, the sum of weighted z scores, and the sum of squared z scores (Medina-Pastor, Mezcua, Rodríguez-Torreblanca, & Fernández-Alba, 2010), renamed Average of the Squared z Score in the General Protocol for EU Proficiency Tests on Pesticide Residues in Food and Feed. Based on their performance regarding PT scope coverage, reporting false positives and negatives, and the accuracy achieved (z scores), EU laboratories are classified into categories A and B. Laboratories detecting at least 90% of the pesticides present in the PT sample and reporting no false positives are classified as Category A, whereas all the others are included in the Category B (EUPT, 2016). A new three-dimensional (XYZ) classification, with three categories (A, B, or C) within each dimension, has been recently proposed to evaluate performance—underperformance of laboratories participating in the EU Proficiency Test for fruit and vegetables (Valverde, Fernández-Alba, Ferrer, & Aguilera, 2016). Laboratories can be evaluated using the number of detected pesticides (dimension X: scope), the number of acceptable z scores (dimension Y: performance), and the number of false positives (dimension Z: false positives) (Valverde et al., 2016).

The National Metrology Institute of Japan (NMIJ) has organized one PT for the determination of pesticide residues in soybean (Yarita et al., 2015) and one in brown rice (Otake, Yarita, Aoyagi, Numata, & Takatsu, 2014). In both cases, the evaluation of laboratories has been performed using two types of z scores: one was the score based on the consensus values calculated from the analytical results of participants and the other from the values obtained by NMIJ by isotope-dilution MS. Acceptable z scores were achieved by the majority of the participants using both evaluation approaches (Otake et al., 2014; Yarita et al., 2015).

The results of the International Measurement Evaluation Programme 37 study, a PT, which was organized to assess the world-wide performance of food control laboratories (40 EU laboratories and 41 from outside the EU) on the determination of pesticide residues in grapes, showed that the participants performed satisfactorily. However, only 30% of the participants managed to analyze all pesticides satisfactorily, indicating the need for improvement of MRMs (Dehouck et al., 2015).

Each step of the analytical procedure influences the final results in a significant way. There are some critical issues that are not usually incorporated in the method validation part or cannot be expressed and evaluated through a specific method parameter, i.e., the sampling procedure and sample handling. Analysis starts with sampling because it is a very important step to achieve a representative sample for analysis. If this is not ensured, laboratory results will not be realistic even if the best techniques and the most accurate analytical method are applied. The Directive 2002/63/EC (Commission Directive, 2002) and the CAC/GL33-1999 (CAC, 1999) describe the appropriate sampling procedures for food commodities of plant and animal origin for pesticide residues. Samples must be transported under appropriate conditions and delivered to the laboratory in good condition, otherwise should be considered unfit for analysis. Laboratory samples should be stored under well-defined conditions that minimize decay (CAC, 1999; SANTE, 2015). Conversion of a laboratory sample into an analytical sample may need sample preparation that includes removal of parts such as soil, stones, bones, withered leaves, etc. Sample preparation may be subject to random and systematic errors; this step is not included usually in the validation studies. All sample-treatment and processing procedures (drying, mixing, etc. before measurements) should be undertaken within the shortest time possible and under selected and defined conditions to minimize sample decay and pesticide losses (SANTE, 2015). It is highly recommended to analysts to follow standard operating procedures with detailed information and carry them out in a consistent manner, to ensure reliable results. Sampling, sample preparation, sample size reduction, and sample processing procedures should be demonstrated to have no significant effect on the concentration of pesticide residues in the analytical sample. When sample preparation—comminution, cutting, and homogenization—at ambient temperature has a significant influence on the degradation of certain pesticide residues, it is recommended that samples are homogenized at low temperatures (e.g., frozen and/or in the presence of "dry ice"). Whenever there is evidence that a procedure may affect the level of residues, precaution measures should be undertaken (Ambrus et al., 2016; Lehotay & Cook, 2015; SANTE, 2015).

4.4.1.3 Uncertainty

In the standard ISO/IEC 17025:2005 (ISO/IEC, 2005), the determination of the measurement uncertainty (MU) associated with analytical results is a general requirement. The definition of the term measurement uncertainty is: "A parameter associated with the result of a measurement that characterizes the dispersion of the values that could reasonably be attributed to the measurand." A number of guidelines from various organizations (Eurachem, Eurolab, Nordtest) can guide the laboratory through the estimation of typical and expanded MU: The Eurachem/CITAC Guides (Bettencourt da Silva & Williams, 2015; Ellison & Williams, 2012), the Nordtest Handbook for calculation of MU in environmental laboratories (Nordtest, 2004), the ISO/IEC Guide 98-3:2008 (ISO/IEC, 2008). Rozet et al. (2011) have critically summarized the various models used to estimate the overall MU, i.e., the combination of the random and systematic influences providing their pros and cons, and reviewed the main areas of applications.

In the EU guideline, MU describes the range around a measurement result within which the true value is expected to lie within a defined probability (confidence level) (SANTE, 2015). In the case of pesticide residue analysis, the large number of pesticide/food combinations makes the "top-down" approaches more suitable for the evaluation of MU compared to the "bottom-up" approach (Medina-Pastor et al., 2011). Among the top-down approaches, there are two main ways in which MU can be estimated: one includes the Horwitz equation or the FFP-RSD and the other is based on experimental data from the QC work of the laboratory: within-laboratory reproducibility, interlaboratory validation, or a combination of results obtained in proficiency tests. Medina-Pastor et al. evaluated these approaches using the data from the European Proficiency Test Database for Fruits and Vegetables and the MRM validation databases obtained from the National Reference and Official Laboratories in Europe. The main conclusion of the comparative study is that a default expanded MU value of 50% could cover the interlaboratory variability between the European laboratories (Medina-Pastor et al., 2011); the use of the 50% expanded MU is recommended by regulatory authorities in cases of MRL exceedances (SANTE, 2015). Examples for the estimation of MU with the two approaches are included in a special Annex of the EU guideline SANTE 11945/2015.

Precision estimates—repeatability/reproducibility data—and laboratory bias derived from PTs are used, but additional uncertainty sources such as sample sampling, sample preparation, sample processing, and subsampling should also be included in the estimation of MU (CAC, 2006; SANTE, 2015). In the majority of cases, the analytical procedure encompasses sampling, so it is necessary to include the uncertainty related to sampling procedure in the uncertainty-budget calculations. Different ways for the estimation of sampling uncertainty are presented in a dedicated Eurachem Guide "Measurement uncertainty arising from sampling: a guide to methods and approaches" (Ramsey & Ellison, 2007). Because of the significance of the sampling procedure, Bodnar et al. suggested the validation procedure for the chosen sampling technique to be included as a part of the detailed description of the analytical procedure, in the final validation report (Bodnar, Namiesnik and Konieczka, 2013).

Ring tests, proficiency tests, IQC, and method validation can provide information only for the uncertainty of the analytical method and cannot improve the uncertainty derived from the other contributors of the uncertainty of the laboratory phase, the subsampling, the sample preparation, and processing. The effect of sample processing on accuracy and uncertainty of the measured residue values with lettuce, tomato, and maize grain samples was investigated by Ambrus et al. (2016). Results indicated that the concentration of analytes in the final extract was influenced by their physical—chemical properties, the nature of the sample material, the temperature of comminution of sample, and the mass of test portion extracted. Authors suggested to include testing the effect of sample processing in the validation protocols and to check regularly the performance of the complete method within internal QC.

Lehotay and Cook (2015) recommended the implementation of validation of sampling and sample processing protocols and ongoing QC in pesticide residue monitoring programs along with the validation QC data of analytical methods to ensure that analytical results are meaningful for the intended purposes of regulatory registrations, enforcement, and risk assessment.

4.5 MASS SPECTROMETRY IN PESTICIDE RESIDUE ANALYSIS

4.5.1 MASS SPECTROMETRY IDENTIFICATION AND CONFIRMATION

Nowadays, MS is a key technique in the field of pesticide analysis because of its inherent features of multianalyte capabilities, sensitivity, scope of hyphenation with chromatography, selectivity, and potential for qualitative chemical analysis. The analytical power of combined gas or liquid chromatography—mass spectrometry (GC—MS, LC—MS) techniques in trace organic analysis has shown that they are predominant tools in the field of pesticide analysis. Method qualitative performance capabilities should be investigated and its suitability for identification and confirmation of pesticides in food samples should be established according to well-defined criteria of related legislative documents during method validation (Bethem et al., 2003; CAC, 2005; Lehotay et al., 2008; Milman, 2015; SANTE, 2015).

LC—MS and GC—MS identification criteria generally comprise:

1. the number of identification points (IPs) or ions for different MS techniques;
2. tolerance ranges about reference values of chromatographic retention times (t_R);
3. maximum permitted tolerances for relative mass-peak intensities of selective ions;
4. requirements to full mass spectra; and
5. requirements for HRMS and value tolerances for mass accuracy (MA) and resolution.

The CCPR acknowledges that they have no universally accepted criteria for identification but lists the recommendations of the document SANTE/11945/2015 with the exception that the retention time should lie within ±0.2 min or 0.2% relative retention time of standard or matrix-matched solutions for both GC and LC, instead of ±0.1 min stated in SANTE/11945/2015. Additionally, guidelines on the use of MS for identification, confirmation, and quantitative determination of residues are given in the document CAC/GL 56-2005 (CAC, 2005).

In the current EU guidance document (SANTE, 2015), pesticide residues identification using MS coupled to a chromatographic separation system is based on retention time, mass/charge ratio (m/z), and relative abundance (intensity) data. The definitions of different MS terms used in the document SANTE 11945/2015 are included in the IUPAC recommendations for MS (Murray et al., 2013).

The minimum acceptable retention time for the analyte(s) under examination should be at least twice the retention time corresponding to the void volume of the column. The retention time of the analyte in the extract should correspond to that of the calibration standard with a tolerance of ±0.1 min, for both GC and LC. Larger retention time deviations are acceptable where both retention time and peak shape of the analyte match with those of a suitable IL-IS or evidence from validation studies is available.

Guidance for identification based on MS spectra is limited to some recommendations, whereas for identification based on selected ions more detailed criteria are provided.

- Reference spectra for the analyte should be generated using the same instruments and conditions used for analysis of the samples. In case of full scan measurement, careful subtraction of background spectra may be required to ensure that the resultant spectrum from the chromatographic peak is representative.
- Identification using selected ions relies on the correct selection of ions. They must be sufficiently selective for the analyte in the matrix being analyzed and the relevant concentration range. Molecular ions, (de)protonated molecules, or adduct ions are highly characteristic for the analyte and should be included in the measurement and identification procedure whenever possible. In general, especially in single-stage MS, high m/z ions are more specific than low m/z ions (e.g., $m/z < 100$). Although characteristic isotopic ions, especially Cl or Br clusters, may be particularly useful, the selected ions should not exclusively originate from the same part of the analyte molecule.

Different types and modes of mass spectrometric detectors provide different degrees of selectivity, which relates to the confidence in identification. The requirements for identification are given in Table 4.4.

The relative intensities or ratios of selective ions, expressed as a ratio relative to the most intense ion used for identification, should not deviate more than 30% (relative) from those of the calibration or matrix-matched standard solutions (Mol et al., 2015). Larger tolerances may lead to a higher percentage of false-positive results. Similarly, if the tolerances are decreased, then the likelihood of false negatives will increase. Automated data interpretation based on the criteria without complementary interpretation by an experienced analyst is not recommended. For a higher degree of confidence in identification, further evidence may be gained from additional mass spectrometric information such as evaluation of full scan spectra, isotope pattern, adduct ions, additional accurate mass fragment ions, additional product ions (in MS/MS), or accurate mass product ions (SANTE, 2015).

In the United States, the USDA and the FDA each have their own requirements for the analysis of pesticide residues. The USDA criteria are similar to the current EU-SANTE guideline, although the tolerances for retention time and ion ratio differ. The PDP-DATA, effective from 01/02/2016 from the USDA Agricultural Marketing Service, Science & Technology Pesticide Data Program (USDA-AMS-PDP), defines standards to be followed by all laboratories conducting residue studies for the PDP program (USDA, 2016c).

For both GC and LC, if an external standard is used, the retention time of the compound of interest in the standard and the t_R of the same compound in the sample shall be within 0.1 min similar to the SANTE criteria. In contrast to EU legislation and guidelines, the use of MS is preferred but is not obligatory for confirmation.

In the GC/MS and LC/MS Confirmation Criteria, a minimum of three structurally significant ions (meeting the 3:1 S/N ratio) are required for confirmation. For GC/MS, if the molecular ion is present and meets the 3:1 S/N ratio, it is preferable to be included as one of the three ions. Use of fragment ions resulting from water loss to meet the three structurally significant ions requirement is discouraged. The confidence limits of the relative abundance of structurally significant ions used for SIM and/or full scan identification shall be $\pm 20\%$ (absolute) when compared to the same relative abundances observed from a standard solution injection made during the same analytical run. MS spectra produced by "soft" ionization techniques (e.g., GC/MS: CI and for LC/MS: APCI, APPI, ESI, etc.) may require additional evidence for confirmation. If the isotope ratio of the ion(s) or the chromatographic profile of isomers of the analyte is highly characteristic, there may be sufficient information for confirmation. Additional evidence may consist of MS/MS data, use of a different ionization technique, use of a different chromatographic separation system, and for LC/MS systems, altering fragmentation by changing ionization conditions.

The purpose of the guidance document ORA-LAB.010 of the FDA is to provide analytical requirements for regulatory action of pesticide residues and applies to Office of Regulatory Affairs (ORA) laboratories performing pesticide analysis (FDA, 2013).

Table 4.4 Identification Requirements for Different Mass Spectrometry (MS) Techniques (SANTE, 2015)

MS Detector/ Characteristics	Typical Systems	Acquisition	Requirements for Identification	
			Minimum Number of Ions	Others
Unit mass resolution	Quadrupole, ion-trap, TOF	Full scan, limited m/z range, SIM	Three ions	S/N ≥ 3[e] Analyte peaks in the extracted ion chromatogram should fully overlap. Ion ratio within ±30% (relative) of average of calibration standards form same sequence.
MS/MS	Triple quadrupole, ion-trap, Q-trap, Q-TOF, Q-Orbitrap	Selected or multiple reaction monitoring (SRM, MRM), mass resolution for precursor-ion selection equal or better than unit mass resolution	Two product ions	
Accurate mass measurement	High-resolution MS: Q-TOF, Q-Orbitrap, FT-ICR-MS, sector MS	Full scan, limited m/z range, SIM, fragmentation with or without precursor-ion selection, or combinations thereof	Two ions with mass accuracy ≤5 ppm[a,b,c]	
		Combined single-stage MS and MS/MS with mass resolution for precursor-ion selection equal or better than unit mass resolution	*Two ions:* 1 molecular, (de) protonated molecule or adduct ion with mass accuracy ≤5 ppm[a,c] *plus* 1 MS/MS product ion[d]	

FT-ICR-MS, *Fourier transform ion cyclotron mass spectrometry; MS/MS, tandem mass spectrometry; m/z, mass/charge ratio; SIM, selected ion monitoring; TOF, time-of-flight.*
[a] *Preferably including the molecular ion, (de)protonated molecule, or adduct ion.*
[b] *Including at least one fragment ion.*
[c] *<1 mD for m/z < 200.*
[d] *No specific requirement for MA.*
[e] *In case noise is absent, a signal should be present in at least five consecutive scans.*

Table 4.5 Assignment of Identification Points (IPs) in ORA-LAB.010 (the IPs Related to MS Detection Are in Bold)

Low-resolution MS ion	**1 point per ion**
Low-resolution MS/MS precursor ion	**1 point per precursor ion**
Low-resolution MS/MS product ion transition	**1 point per ion**
HR MS ion	**2.0 points per ion**
HR MS precursor ion	**2.0 points per ion**
HR MS product ion (transition)	**2.5 points per ion**
Matching chromatographic retention time	1 point per alternative systems
Selective detection with matching retention time	1 point per detector
Quantitative agreement between alternate columns/detectors	1 point per sample
Isomers with matching retention time	1 point

ORA-Lab.010 adopts the system of IPs (in bold Table 4.5), but also assigns IPs to additional sources of information (four last rows in Table 4.5), including the use on selective detectors. For pesticide identification four IPs are required in ORA-LAB.010 (FDA, 2013).

The ion selection criteria are the following:

1. All selected ions must have a minimum S/N ratio of 3:1. The primary quantitation ions must have a minimum S/N ratio of 10:1.
2. No more than two diagnostic ions may be selected from an isotopic cluster.
3. If the molecular ion abundance is at least 10% of the most abundant ion, it should be selected.
4. Ions should have unique mass differences (i.e., avoid differences of 18 amu due to water loss, 17 amu due to ammonium ion, etc.)
5. For LC−MS, only one molecular ion species may be selected (e.g., avoid ions resulting from the loss of the molecular ion and its adduct with ammonium)

The ion ratio criteria are listed in Table 4.6.

Table 4.6 Maximum Permitted Tolerances for Relative Ion Intensities Using a Range of Mass Spectrometry (MS) Techniques of the Guidance Document ORA-LAB.010

Relative Intensity (% of Base Peak)	Tolerance Window	
	GC−MS	LC−MS
>40%	±10% absolute units	±20% relative units
10%−40%	±25% relative units	±25% relative units
≤10%	±50% relative units	±50% relative units

4.5.2 POTENTIAL OF HIGH-RESOLUTION MASS SPECTROMETRY IN PESTICIDE RESIDUE ANALYSIS

Currently, the use of HRMS has become more prominent because of improvements in commercial instrumentation. In particular, TOF and Orbitrap mass analyzers (with a trend toward Q-TOF and Q-Orbitrap) in combination with LC continue to gain popularity because of their high mass-resolving power [25,000–140,000 full width at half maximum (FWHM)], MA (<2 ppm), sensitivity, dynamic range, potential for targeted and nontargeted acquisition, and improved software for data evaluation. In the HRMS techniques the EU guidelines (SANTE, 2015) require MA <±5 ppm, which is the common standard and technically possible with current instrumentation (Table 4.4). In the m/z range 200–400 (the molecular weights of many pesticides), this corresponds to 1–2 mDa. Similarly, 5 ppm corresponds to 0.5 mDa at m/z 100 and 5 mDa at m/z 1000. A more realistic criterion would be MA <±5 ppm but within 1 mDa for m/z <200. Lehotay, Sapozhnikova, and Mol (2015) extensively discuss the current legislative identification criteria applied with respect to HRMS with specific examples. The same work also cites current research work aiming to propose additional detailed criteria, where IPs are either earned for chromatographic peak capacity, number of diagnostic ions, resolving power, and MA or subtracted for potential chemical interferences of the analyte ions.

The ORA of the FDA, in evaluating direct analysis in real time ambient MS combined with quadrupole (Q)-orbital ion-trap (Orbitrap) HRMS for screening pesticides on fruits and vegetables, proposes the following criteria for positive screening (Lehotay et al., 2015):

1. mass error of all precursor and product ions must be ≤3 ppm;
2. response for the precursor ion must be ≥1000; and
3. in addition to the precursor ion, at least two diagnostic ions must be detected.

The Science and Research Steering Committee of the FDA Foods and Veterinary Medicine (FVM) adopted acceptance criteria for confirmation of identity of chemical residues using HR-MS (FDA/OFVM, 2015). Identification criteria (Table 4.7) refer to:

1. Mass extraction window (MEW): This has to be selected by careful consideration of the resolution of the instrument, drift of the mass axis, and the complexity of the matrix.
2. Signal requirement: If there is noise, an S/N threshold ≥3 is recommended.
3. Retention time: The retention time must match comparison standard within one of the following limits: (1) ≤0.2 min; or (2) within ±2.5%, not to exceed 0.5 min; or (3) within experimental error (multiples of standard deviation) established in the validation method, not to exceed 0.5 min.
4. MA: For identification, the measured exact mass of at least two ions should have a MA of ≤5 ppm in the MS mode and ≤10 ppm in the MS/MS mode.
5. Ion ratios: If the measured exact mass from two or more ions match the MA criterion, it is not necessary to calculate and report ion abundance ratios. If,

Table 4.7 Identification Criteria for Confirmation of Identity of Chemical Residues Using Exact Mass Data (FDA/OFVM, 2015)

MS Mode	MS	MS/MS	MS and MS/MS
EIC: Signal requirement (absolute)	A criterion to be set by one of the following methods: **1.** an S/N threshold ≥3 **2.** an intensity ratio relative to the comparison standard equal or above a preset threshold		
EIC: Retention time, relative to comparison standard	A criterion to be set by one of the following methods: **1.** ≤0.2 min, or **2.** within ±2.5%, not to exceed 0.5 min, or **3.** within an established error range, not to exceed 0.5 min		
MS: Number of structurally significant ions	Minimum 2	Minimum 2	Minimum 2 combined
MS: Mass accuracy	≤5 ppm	≤10 ppm	MS: ≤5 ppm; MS/MS: ≤10 ppm

EIC, *extracted ion chromatogram;* MS, *mass spectrometry;* MS/MS, *tandem mass spectrometry.*

however, the measured mass error is greater than the MA criterion, the ion ratio criteria for nominal mass data as described ORA-LAB.010 shall apply. The same document also proposes general recommendations for nontargeted analysis in terms of chromatography, interpretation of the detected ions, formula generation, and use of search software.

Nowadays, simultaneous trace analysis of hundreds of pesticides preferably in a single run is required. The most commonly used QqQ-MS technology presents certain limitations when analyzing a large number of pesticides. HRMS technology can be a reliable complementary alternative allowing the analysis of a wide range of pesticides in food (Del Mar Gómez-Ramos, Ferrer, Malato, Agüera, & Fernández-Alba, 2013). Latest models of high-resolution mass spectrometers (HRMS) are capable of performing sensitive and reliable quantifications of a large variety of analytes in HR-full scan or, if needed, in a more targeted SIM or MS^2 product scan mode or data-independent-acquisitions (Rochat, 2016). Currently, LC−HRMS analysis is performed with quadrupole-Orbitrap-MS (Q-Orbitrap-MS) or quadrupole-time-of-flight-MS (Q-TOF) platforms; their fast scanning speed [compatible with ultrahigh-performance LC (UHPLC)] together with HRMS-based selective detections is a great advantage when operating in the full scan mode that allows compound identification and retrospective data treatment. Regarding the quantitative performance of recent HRMS instruments in comparison to triple quadrupoles in the selected reactions mode (SRM), there is large body of evidence supporting for the quantitative capability of the latest HRMS instruments based on extracted ion chromatograms (XIC) using narrow MEWs (Grund, Marvin, & Rochat, 2016). A significant drawback in quantitative LC−HRMS analysis is the lack of HRMS-specific guidance for validated quantitative analyses. Moreover, false-positive and false-negative HRMS detections are rare, albeit possible, if inadequate parameters are used. Glauser et al. (2016) investigated two key parameters for the validation of LC−HRMS quantitative analyses: the MA and the MEW that is used to construct the XIC. The authors proposed a FFP MEW width determination procedure based on MA parameters for the validation of quantitative LC−HRMS methods preventing false detections.

The application of UHPLC/ESI Q-Orbitrap MS for the determination of 451 pesticide residues in fruits and vegetables was studied by Wang, Chow, Chang, and Wong (2014). Method validation gave recoveries between 81% and 110% and RSDs below or equal to 20% for more than 90% of pesticide/commodity combinations. LC-Q-Orbitrap MS detection utilizing an improved fully nontargeted way of data acquisition with and without fragmentation was studied by Zomer and Mol (2015). Authors have tested the combination of a full-scan acquisition event without fragmentation at resolving power 70,000, followed by five consecutive fragmentation events (variable data independent acquisition; resolving power 35,000) where all ions from the full-scan range are fragmented. This combination improved selectivity and sensitivity as shown by the validation study that included 184 pesticides in 11 different fruit and vegetable matrices. An SDL of 10 ng/g was achieved for 134 compounds and the identification criteria were met for 93% of the detected pesticide/matrix combinations. In another study, LC−Q-Orbitrap MS working in full scan at a resolution of 70,000 simultaneously with MS/MS scan at a resolution of 17,500 was used to analyze for 139 pesticide residues (Del Mar Gómez-Ramos, Rajski, Heinzen, & Fernández-Alba, 2015). Quantitation and detection was carried out using full scan data while MS/MS data were used only for identification. When full scan with single MS/MS scan is compared with full scan with multiple MS/MS scans, the number of automatically reported pesticides was the same, but full scan with single MS/MS scan ensured more points per peak and better peak area reproducibility. The identification and quantitation capabilities of LC−Q-Orbitrap MS were tested through the analysis of 100 real samples compared to an LC−QqQ MS/MS system revealing that the two instruments were consistent with each other. No false positives or false negatives were reported from both systems (Del Mar Gómez-Ramos et al., 2015). The qualitative and quantitative capabilities of the Q-Orbitrap MS enable efficient and reliable pesticide residue analysis; additionally, the high resolution afforded can increase the selectivity and, in full scan mode, permit the retrospective analysis of the data feature that cannot be achieved with QqQ (Del Mar Gómez-Ramos et al., 2015; Zomer & Mol, 2015).

One of the first studies on the performance of screening methods based on accurate mass measurements using an LC−ESI-QTOF-MS system operating in full scan mode and with automatic identification with accurate-mass databases was presented by Malato, Lozano, Mezcua, Agüera, and Fernandez-Alba (2011). The study involved the analysis of 97 pesticides, in five matrices and at three concentration levels. Sensitivity requirements have been identified as the main obstacle affecting the automatic identification of pesticides, especially in complex matrices (Malato et al., 2011). A UPLC−QTOF−MS method for pesticides detection was developed and reported by García López et al. (2014); samples spiked with 199 pesticides at two different concentrations were analyzed using generic acquisition parameters; the UNIFITM software and the peaks detected were evaluated against a library of 504 pesticides. Using MA (± 5 ppm) with retention time (± 0.2 min) and a low response threshold (100 counts) the validated SDLs were 0.01 mg/kg and 0.05 mg/kg for 57% and 79% of the compounds tested, respectively, with an average of 10 false detects per sample analysis.

The suitability of a single-stage Orbitrap platform for routine pesticides analysis in food safety laboratories has been investigated in many studies. A qualitative screening method for pesticides detection in fruit and vegetables was reported by Mol and coworkers (Mol, Zomer, & de Koning, 2012). Researchers used different commodities spiked with a test mixture of 130 pesticides at three different concentration levels for evaluation of false negatives and a screening database of 556 pesticides for evaluation of false positives. LC-Orbitrap-MS operated in full scan acquisition mode using alternating scan events without/with fragmentation, at a resolving power of 50,000. Analyte detection was based on extraction of the exact mass (± 5 ppm) of the major adduct ion at the database retention time (± 30 s) and the presence of a second diagnostic ion. False positive and negative detection rates were low (Mol et al., 2012). Operational parameters such as resolution, software for the automatic detection, mass tolerance, and retention time extraction window, along with the analytical performance, were assessed in an updated UHPLC-Orbitrap-MS platform (Rajski, Del Mar Gómez-Ramos, & Fernández-Alba, 2014). Three different resolution settings were tested using sample extracts spiked with 170 selected pesticides at four concentration levels. The percentages of detected pesticide rates increased at higher resolution values; a resolution of 70,000 and mass tolerance of 5 ppm were the most adequate screening conditions. Rapid screening and accurate mass confirmation of 116 pesticides in oranges and hazelnuts using an automated online sample preparation method with turbulent-flow chromatography technology coupled to an Orbitrap-MS has been demonstrated by Shi et al. (2011). An interlaboratory comparison for the evaluation of single-stage Orbitrap-MS with a specific automated screening procedure without additional manual review was conducted from three laboratories; similar data derived from the three laboratories for the 247 compounds tested and detectable concentrations only slightly above the range shown for triple quadrupole MS (Eitzer, Hammack, & Filigenzi, 2014). Alder, Steinborn, and Bergelt (2011) have comparatively tested the sensitivity of Orbitrap-MS to that of a widely used (low-resolution) tandem mass spectrometer for the analysis of about 500 pesticides; all pesticides were detectable with sufficient sensitivity when Orbitrap was operated in the full-scan acquisition mode without fragmentation at a resolution of 100,000 FWHM. The quantitative performance of LC—single-stage Orbitrap operated at 50 K FWHM resolution, for the determination of 240 pesticides in three different matrices was similar to the established LC—MS/MS method in terms of accuracy, sensitivity, dynamic range, and selectivity (Kaufmann et al., 2012).

Compared to LC, developments in "full-scan" HRMS instruments for GC are still rather limited (Lehotay et al., 2015). Cervera, Portoles, Pitarch, Beltran, and Hernandez (2012) have evaluated the capability of GC–TOF MS for qualitative and quantitative analysis of pesticide residues; up to five ions using narrow mass window (0.02 Da) at the expected retention time were monitored using a target processing method; method validation revealed some limitations. In another study, GC–QTOF MS combined with a retention index/mass spectrum library—containing both first-stage mass spectra and retention indices—has been evaluated for the analysis of suspected-target pesticide residues in complicated matrices (Zhang et al., 2014). Spiked samples with a test mixture of 165 pesticides were analyzed; satisfactory rates of identification and confirmation were achieved (Zhang et al., 2014). GC–QTOF-MS operating in negative chemical ionization mode and combining full scan with MS/MS experiments using accurate mass analysis has been also tested for the automated determination of pesticide residues (Besil, Uclés, Mezcúa, Heinzen, & Fernández-Alba, 2015). Almost 76% of the 70 compounds included could be identified at the lowest level. Recovery studies at three concentration levels yielded acceptable recoveries and RSD values. Screening and quantification of pesticide residues in fruits and vegetables making use of GC–Q-TOF MS with atmospheric pressure chemical ionization has been investigated by Cervera, Portolés, López, Beltrán, and Hernández (2014). A quantitative validation provided acceptable average recoveries and RSDs for many but not all pesticide-matrix combinations. Finally, the recently introduced GC–EI-Q-Orbitrap MS was found to be highly suited for qualitative screening and quantitative pesticide residue analysis (Mol, Tienstra, & Zomer, 2016).

REFERENCES

Alder, L., Steinborn, A., & Bergelt, S. (2011). Suitability of an orbitrap mass spectrometer for the screening of pesticide residues in extracts of fruits and vegetables. *Journal of AOAC International, 94*(6), 1661–1673.

Aliferis, K. A., & Jabaji, S. (2011). Metabolomics—A robust bioanalytical approach for the discovery of the modes-of-action of pesticides: A review. *Pesticide Biochemistry and Physiology, 100*, 105–117.

Aliferis, K. A., & Tokousbalides, M. C. (2011). Metabolomics in pesticide research and development: Review and future perspectives. *Metabolomics, 7*, 35–53.

Ambrus, A., Buczkó, J., Hamow, K.Á., Juhász, V., Solymosné Majzil, E., Szemánné Dobrik, H., et al. (2016). Contribution of sample processing to variability and accuracy of the results of pesticide residue analysis in plant commodities. *Journal of Agricultural and Food Chemistry, 64*(31), 6071–6081.

Anastassiades, M., Mastovska, K., & Lehotay, S. J. (2003). Evaluation of analyte protectants to improve gas chromatographic analysis of pesticides. *Journal of Chromatography A, 1015*, 163–184.

AOAC. (2002a). *AOAC guidelines for single laboratory validation of chemical methods for dietary supplements and botanicals.* Available from http://www.aoac.org/aoac_prod_imis/AOAC_Docs/StandardsDevelopment/SLV_Guidelines_Dietary_Supplements.pdf.

AOAC. (2002b). *AOAC guidelines for collaborative study procedures to validate characteristics of a method of analysis.* Available from http://www.aoac.org/aoac_prod_imis/AOAC_ Docs/StandardsDevelopment/Collaborative_Study_Validation_Guidelines.pdf.

Araujo, P. (2009). Key aspects of analytical method validation and linearity evaluation. *Journal of Chromatography A, 877,* 2224−2234.

ASTM. (2011). *ASTM: E2857-11. Standard guide for validating analytical methods.* Available from https://www.astm.org/Standards/E2857.

Barceló, D., & Hennion, M. C. (1997). Trace determination of pesticides and their degradation products in water. In *Techniques and instrumentation in analytical chemistry.* The Netherlands: Elsevier Science B.V.

Barwick, V. J., & Prichard, E. (2011). *Eurachem guide: Terminology in analytical measurement − introduction to VIM 3.* Available from http://www.eurachem.org.

Besil, N., Uclés, S., Mezcúa, M., Heinzen, H., & Fernández-Alba, A. R. (2015). Negative chemical ionization gas chromatography coupled to hybrid quadrupole time-of-flight mass spectrometry and automated accurate mass data processing for determination of pesticides in fruit and vegetables. *Analytical and Bioanalytical Chemistry, 407*(21), 6327−6343.

Bethem, R., Boison, J., Gale, J., Heller, D., Lehotay, S., Loo, J., et al. (2003). Establishing the fitness for purpose of MS methods. *Journal of the American Society for Mass Spectrometry, 14,* 528−541.

Bettencourt da Silva, R., & Williams, A. (2015). In *Eurachem/CITAC guide: Setting and using target uncertainty in chemical measurement* (1st ed.) Available from https://www. eurachem.org/images/stories/Guides/pdf/STMU_2015_EN.pdf.

Bodnar, M., Namiesnik, J., & Piotr Konieczka, P. (2013). Validation of a sampling procedure. *Trends in Analytical Chemistry, 51,* 117−126.

CAC. (1993a). *CAC/GL 40-1993. Guidelines on good laboratory practice in pesticide residue analysis.* Revision 2003. Amendment 2010. Available from www.fao.org/input/download/ standards/378/cxg_040e.pdf.

CAC. (1993b). *CAC/GL 41-1993. Portion of commodities to which Maximum Residue Limits apply and which is analyzed.* Amendment 2010. Available from http://www.fao.org/input/ download/standards/43/CXG_041e.pdf.

CAC. (1999). *CAC/GL 33-1999: Recommended methods of sampling for the determination of pesticide residues for compliance with MRLs.* Available at http://www.fao.org/fao-who-codexalimentarius/sh-proxy/en/?lnk=1&url=https%253A%252F%252Fworkspace.fao. org%252Fsites%252Fcodex%252FStandards%252FCAC%2BGL%2B33-1999%252FCXG_ 033e.pdf.

CAC. (2005). *CAC/GL 56-2005. CCPR guidelines on the use of mass spectrometry (MS) for identification, confirmation and quantitative determination of residues.* Available at www. fao.org/input/download/standards/10185/cxg_056e.pdf.

CAC. (2006). *CAC/GL 59-2006. Guidelines on estimation of uncertainty of results.* Amendment 2011. Available from http://www.fao.org/input/download/standards/10692/cxg_ 059e.pdf.

Casida, J. E. (2009). Pest toxicology: The primary mechanisms of pesticide action. *Chemical Research in Toxicology, 22,* 609−619.

Casida, J. E., & Durkin, K. A. (2013a). Anticholinesterase insecticide retrospective. *Chemico-Biological Interactions, 203,* 221−225.

Casida, J. E., & Durkin, K. A. (2013b). Neuroactive insecticides: Targets, selectivity, resistance, and secondary effects. *Annual Review of Entomology, 58,* 99−117.

CCPR. (2016). *Report of the 48th session of the codex committee on pesticide residues, Chongqing, China, 25–30 April 2016*. Proposed draft guidelines on performance criteria for methods of analysis for the determination of Pesticide Residues. Available from http://www.fao.org/fao-who-codexalimentarius/sh-proxy/en/?lnk=1&url=https%253A%252F%252Fworkspace.fao.org%252Fsites%252Fcodex%252FMeetings%252FCX-718-48%252FReport%252FREP16_PRe.pdf.

CEN. (2008). *CEN-GUIDE-13: Validation of environmental test methods*. Available from http://boss.cen.eu/ref/CEN_13.pdf.

Cervera, M. I., Portolés, T., López, F. J., Beltrán, J., & Hernández, F. (2014). Screening and quantification of pesticide residues in fruits and vegetables making use of gas chromatography-quadrupole time-of-flight mass spectrometry with atmospheric pressure chemical ionization. *Analytical and Bioanalytical Chemistry, 406*(27), 6843–6855.

Cervera, M. I., Portoles, T., Pitarch, E., Beltran, J., & Hernandez, F. (2012). Application of gas chromatography time-of-flight mass spectrometry for target and non-target analysis of pesticide residues in fruits and vegetables. *Journal of Chromatography A, 1244*, 168–177.

CITAC/EURACHEM. (2002). *CITAC/EURACHEM guide to quality in analytical chemistry. An aid to accreditation*. Available from https://www.eurachem.org/images/stories/Guides/pdf/CITAC_EURACHEM_GUIDE.pdf.

Codex Alimentarius. (2016). *Pesticide residues in food and feed*. Available at http://www.fao.org/fao-who-codexalimentarius/standards/pestres/en/.

Commission Directive. (2002). Commission Directive 2002/63/EC establishing Community methods of sampling for the official control of pesticide residues in and on products of plant and animal origin and repealing Directive 79/700/EEC. *Official Journal of the European Union, L187*, 30–43.

Commission Directive. (2003). Commission Directive 2003/13/EC of 10 February 2003 amending Directive 96/5/EC on processed cereal-based foods and baby foods for infants and young children. *Official Journal of the European Union, L41*, 33–36.

Commission Directive. (2006). Commission Directive 2006/125/EC of 5 December 2006 on processed cereal-based foods and baby foods for infants and young children. *Official Journal of the European Union, L339*, 16–35.

Compendium of pesticide common names.(2016). Available from http://www.alanwood.net/pesticides/.

Cuadros-Rodríguez, L., Bagur-González, M. G., Sánchez-Viñas, M., González-Casado, A., & Gómez-Sáez, A. M. (2007). Principles of analytical calibration/quantification for the separation sciences. *Journal of Chromatography A, 1158*, 33–46.

Dehouck, P., Grimalt, S., DabrioM ,Cordeiro, F., Fiamegos, Y., Robouch, P., Fernández-Alba, A. R., et al. (2015). Proficiency test on the determination of pesticide residues in grapes with multi-residue methods. *Journal of Chromatography A, 1395*, 143–151.

Del Mar Gómez-Ramos, M., Ferrer, C., Malato, O., Agüera, A., & Fernández-Alba, A. R. (2013). Liquid chromatography-high-resolution mass spectrometry for pesticide residue analysis in fruit and vegetables: Screening and quantitative studies. *Journal of Chromatography A, 1287*, 24–37.

Del Mar Gómez-Ramos, M., Rajski, Ł., Heinzen, H., & Fernández-Alba, A. R. (2015). Liquid chromatography Orbitrap mass spectrometry with simultaneous full scan and tandem MS/MS for highly selective pesticide residue analysis. *Analytical and Bioanalytical Chemistry, 407*(21), 6317–6326.

Detaille, R., & Maetz, P. (2006). Practical uses of proficiency testing as valuable tools for validation and performance assessment in environmental analysis. *Accreditation and Quality Assurance, 11*, 408–413.

Dorough, H. W. (1980). Classification of radioactive pesticide residues in food producing animals. *Journal of Environmental Pathology, Toxicology and Oncology, 3*, 11–19.

EC. (2002a). Commission Decision 2002/657/EC of 12 August 2002 implementing Council Directive 96/23/EC concerning the performance of analytical methods and the interpretation of results. *Official Journal of the European Union, L221*, 8–36.

EC. (2002b). Regulation (EC) No 178/2002 of the European Parliament and of the Council of 28 January 2002 laying down the general principles and requirements of food law, establishing the European Food Safety Authority and laying down procedures in matters of food safety. *Official Journal of the European Union, L31*, 1–24.

EC. (2004). Regulation (EC) No 882/2004 of the European Parliament and of the Council on official controls performed to ensure the verification of compliance with feed and food law, animal health and animal welfare rules. *Official Journal of the European Union, L191*, 1–141.

EC. (2005). Regulation (EC) No 396/2005 of the European Parliament and of the Council of 23 February 2005 on maximum residue levels of pesticides in or on food and feed of plant and animal origin and amending Council Directive 91/414/EEC of the European Parliament and Council. *Official Journal of the European Union, L70*, 1–16.

EC. (2006). Commission Regulation (EC) No 178/2006of 1 February 2006 amending Regulation (EC) No 396/2005 of the European Parliament and of the Council to establish Annex I listing the food and feed products to which maximum levels for pesticide residues apply. *Official Journal of the European Union, L29*, 3–25.

EC. (2008). Commission Regulation (EC) No149/2008 of 29 January 2008 amending Regulation (EC) No 396/2005 of the European Parliament and of the Council by establishing Annexes II, III and IV setting maximum residue levels for products covered by Annex I thereto. *Official Journal of the European Union, L58*, 1–398.

EC. (2009a). Directive 2009/128/EC of the European Parliament and of the Council of 21 October 2009 establishing a framework for Community action to achieve the sustainable use of pesticides. *Official Journal of the European Union, L309*, 71–86.

EC. (2009b). Regulation (EC) No 1107/2009 of the European Parliament and of the Council of 21 October 2009 concerning the placing of plant protection products on the market and repealing Council Directives 79/117/EEC and 91/414/EEC. *Official Journal of the European Union, L309*, 1–50.

EFSA. (2011). Review of the existing maximum residue levels (MRLs) for tepraloxydim according to Article 12 of Regulation (EC) No 396/2005. *EFSA Journal, 9*(10), 1–56, 2433.

EFSA. (2013). Conclusion on the peer review of the pesticide risk assessment of the active substance spirotetramat. *EFSA Journal, 11*(6), 1–90, 3243.

Eitzer, B. D., Hammack, W., & Filigenzi, M. (2014). Interlaboratory comparison of a general method to screen foods for pesticides using QuEChERs extraction with high performance liquid chromatography and high resolution mass spectrometry. *Journal of Agricultural and Food Chemistry, 62*(1), 80–87.

Ellison, S. L. R., & Fearn, T. (2005). Characterising the performance of qualitative analytical methods: Statistics and terminology. *Trends in Analytical Chemistry, 24*(6), 468–476.

Ellison, S. L. R., & Williams, A. (2012). *EURACHEM/CITAC guide: Quantifying uncertainty in analytical measurement* (3rd ed.) Available from https://www.eurachem.org/images/stories/Guides/pdf/QUAM2012_P1.pdf.

EPA. (2016a). *EPA pesticides.* http://www.epa.gov/opp00001/regulating/part-180.html.

EPA. (2016b). *EPA tolerances.* Available from https://www.epa.gov/pesticide-tolerances/about-pesticide-tolerances.

Erney, D. R., Gillespie, A. M., Gilvydis, D. M., & Poole, C. F. (1993). Explanation of the matrix-induced chromatographic response enhancement of organophosphorus pesticides during open tubular column gas chromatography with splitless or hot on-column injection and flame photometric detection. *Journal of Chromatography A, 638,* 57—63.

EU. (2013a). Commission Regulation (EU) No 283/2013 of 1 March 2013 setting out the data requirements for active substances, in accordance with Regulation (EC) No 1107/2009 of the European Parliament and of the Council concerning the placing of plant protection products on the market. *Official Journal of the European Union, L93,* 1—84.

EU. (2013b). Commission Regulation (EU) No 284/2013 of 1 March 2013 setting out the data requirements for plant protection products, in accordance with Regulation (EC) No 1107/2009 of the European Parliament and of the Council concerning the placing of plant protection products on the market. *Official Journal of the European Union, L93,* 85—152.

EU. (2014). Commission regulation (EU) No 752/2014 of 24 June 2014 replacing Annex I to regulation (EC) No 396/2005 of the European Parliament and of the Council (Text with EEA relevance). *Official Journal of the European Union, L208,* 1—71.

EU. (2016a). *EU-pesticides database.* Available from http://ec.europa.eu/food/plant/pesticides/eu-pesticides-database/public/.

EU. (2016b). Commission Implementing Regulation (EU) 2016/662 of 29 April 2016 concerning a coordinated multiannual programme of the Union for 2017, 2018 and 2019 to ensure compliance with maximum residue levels of pesticides and to assess the consumer exposure to pesticide residues in and on food of plant and animal origin. *Official Journal of the European Union, L115,* 2—15.

EUPT. (2016). *EUPT general protocol 2016. General protocol for EU proficiency tests on pesticide residues in food and feed — 6th edition.* Available at http://www.eurl-pesticides.eu/docs/public/tmplt_article.asp?CntID=821&LabID=100&Lang=EN.

Evard, H., Kruve, A., & Leito, I. (2016a). Tutorial on estimating the limit of detection using LC-MS analysis, Part I: Theoretical review. *Analytica Chimica Acta.* In Press (Accepted Manuscript).

Evard, H., Kruve, A., & Leito, I. (2016b). Tutorial on estimating the limit of detection using LC-MS analysis, Part II: Practical aspects. *Analytica Chimica Acta.* In Press (Corrected Proof).

Fajgelj, A., & Ambrus, A. (Eds.). (2000). *Principles and practices of method validation.* Cambridge, England: Royal Society of Chemistry.

FAO. (2016). *Manual on the submission and evaluation of pesticide residues for the estimation of maximum residue levels in food and feed.* FAO Plant Production and Protection Paper 225 (3rd ed.) Rome. Available from http://www.fao.org/agriculture/crops/core-themes/theme/pests/jmpr/jmpr-docs/en.

FAO/WHO/CAC. (1993). Codex Alimentarius. Pesticide residues in food. In *Section 2. Codex classification of foods and animal feeds* (2nd ed., Vol. 2) Rome. Available from www.fao.org/input/download/standards/41/CXA_004_1993e.pdf.

Farré, M., Picó, Y., & Barceló, D. (2014). Application of ultra-high pressure liquid chromatography linear ion-trap orbitrap to qualitative and quantitative assessment of pesticide residues. *Journal of Chromatography A, 1328,* 66—79.

FDA. (2002). *Bioterrorism act.* Available from http://www.fda.gov/RegulatoryInformation/Legislation/ucm148797.htm.

FDA. (2013). *ORA Lab 0.10. Guidance for the analysis and documentation to support regulatory action on pesticide residues.*

FDA. (2015). *Pesticides.* Available from http://www.fda.gov/food/foodborneillness contaminants/pesticides/.

FDA. (2016). *Total diet study.* Available from http://www.fda.gov/food/foodscienceresearch/ totaldietstudy/default.htm.

FDA/FVM. (2015). *FDA FVM program. Guidelines for the validation of chemical methods for the FDA FVM program* (2nd ed.) Available from http://www.fda.gov/downloads/ ScienceResearch/FieldScience/UCM273418.pdf.

FDA/OFVM. (2015). *Acceptance criteria for confirmation of identity of chemical residues using exact mass data within the office of foods and veterinary medicine.* Available from http://www.fda.gov/downloads/ScienceResearch/FieldScience/UCM491328.pdf.

FDA/PAM. (2015). *Pesticide analytical manual.* Available from http://www.fda.gov/Food/ FoodScienceResearch/LaboratoryMethods/ucm2006955.htm.

Fernández-Alba, A. R., & García-Reyes, J. F. (2008). Large scale multi-residue methods for pesticides and their degradation products in food by advanced LC-MS. *Trends in Analytical Chemistry, 27,* 973—990.

García López, M., Fussell, R. J., Stead, S. L., Roberts, D., McCullagh, M., & Rao, R. (2014). Evaluation and validation of an accurate mass screening method for the analysis of pesticides in fruits and vegetables using liquid chromatography—quadrupole-time of flight—mass spectrometry with automated detection. *Journal of Chromatography A, 1373,* 40—50.

García-Reyes, J. F., Hernano, M. D., Molina-Díaz, A., & Fernández-Alba, A. R. (2007). Comprehensive screening of target, non-target and unknown pesticides in food by LC-TOF-MS. *Trends in Analytical Chemistry, 26,* 828—841.

Glauser, G., Grund, B., Gassner, A. L., Menin, L., Henry, H., Bromirsk, M., et al. (2016). Validation of the mass-extraction-window for quantitative methods using liquid chromatography high resolution mass spectrometry. *Analytical Chemistry, 88*(6), 3264—3271.

Grund, B., Marvin, L., & Rochat, B. (2016). Quantitative performance of a quadrupole-orbitrap-MS in targeted LC—MS determinations of small molecules. *Journal of Pharmaceutical and Biomedical Analysis, 124,* 48—56.

Hernández, F., Portolés, T., Pitarch, E., & López, F. J. (2011). Gas chromatography coupled to high-resolution time-of-flight mass spectrometry to analyze trace-level organic compounds in the environment, food safety and toxicology. *Trends in Analytical Chemistry, 30,* 388—400.

Hibbert, D. B. (2007). Systematic errors in analytical measurement results. *Journal of Chromatography A, 1158,* 25—32.

Holland, P. T. (1996). Glossary of terms relating to pesticides. IUPAC Recommendations 1996. *Pure and Applied Chemistry, 78,* 2075—2154. Available from https://www.iupac. org/publications/pac/68/5/1167/pdf/index.html.

Horwitz, W. (1988). Harmonized protocol for the design and interpretation of collaborative studies. *Trends in Analytical Chemistry, 7*(4), 118—120.

Horwitz, W. (1995). IUPAC Protocol for the design, conduct and interpretation of method-performance studies. *Pure and Applied Chemistry, 67*(2), 331—343.

IRAC. (2014). *IRAC MoA classification scheme. Version 7.3.* Available from http://www.irac-online.org/content/uploads/MoA-classification_v7.3_27Feb14.pdf.

ISO. (1981). *ISO 1750:1981. Pesticides and other agrochemicals — common names.* Available from http://www.iso.org/iso/catalogue_detail.htm?csnumber=6370.

ISO. (1994a). *ISO 5725-1:1994. Accuracy (trueness and precision) of measurement methods and results — Part 1: General principles and definitions.* Retrieved from http://www.iso. org/iso/iso_catalogue/catalogue_tc/catalogue_detail.htm%3Fcsnumber%3D11833.

ISO. (1994b). *ISO 5725-2:1994. Accuracy (trueness and precision) of measurement methods and results — Part 2: Basic method for the determination of repeatability and reproducibility of a standard measurement method.* Available from http://www.iso.org/iso/catalogue_detail.htm?csnumber=11834.

ISO. (1994c). *ISO 5725-3:1994. Accuracy (trueness and precision) of measurement methods and results — Part 3: Intermediate measures of the precision of a standard measurement method.* Available from http://www.iso.org/iso/catalogue_detail.htm?csnumber=11835.

ISO. (1994d). *ISO 5725-4:1994. Accuracy (trueness and precision) of measurement methods and results — Part 4: Basic methods for the determination of the trueness of a standard measurement method.* Available from http://www.iso.org/iso/catalogue_detail.htm?csnumber=11836.

ISO. (1998). *ISO 5725-5:1998. Accuracy (trueness and precision) of measurement methods and results —Part 5: Alternative methods for the determination of the precision of a standard measurement method.* Available from http://www.iso.org/iso/catalogue_detail.htm?csnumber=1384.

ISO. (2004). *ISO 257:2004. Pesticides and other agrochemicals — principles for the selection of common names.* Available from http://www.iso.org/iso/iso_catalogue/catalogue_tc/catalogue_detail.htm?csnumber=4158.

ISO. (2011). *ISO 7870-4:2011. Cumulative sum control charts.* Available from http://www.iso.org/iso/catalogue_detail.htm?csnumber=40176.

ISO. (2012). *ISO 7870-3:2012. Acceptance control charts.* Available from http://www.iso.org/iso/iso_catalogue/catalogue_tc/catalogue_detail.htm?csnumber=40175.

ISO. (2013). *ISO 7870-2:2013. Shewhart control charts.* Available from http://www.iso.org/iso/catalogue_detail.htm?csnumber=40174.

ISO. (2014a). *ISO 7870-1:2014. Control charts — Part 1: General guidelines.* Available from http://www.iso.org/iso/catalogue_detail.htm?csnumber=62649.

ISO. (2014b). *ISO 7870-5:2014. Specialized control charts.* Available from http://www.iso.org/iso/iso_catalogue/catalogue_tc/catalogue_detail.htm?csnumber=40177.

ISO. (2015). *ISO 13528:2015. Statistical methods for use in proficiency testing by interlaboratory comparison.* Available from http://www.iso.org/iso/catalogue_detail.htm?csnumber=56125.

ISO. (2016a). *ISO 765:2016 Pesticides considered not to require common names.* Available from http://www.iso.org/iso/home/store/catalogue_ics/catalogue_detail_ics.htm?csnumber=57854.

ISO. (2016b). *ISO 7870-6:2016. EWMA control charts.* Available from http://www.iso.org/iso/catalogue_detail.htm?csnumber=40173.

ISO/IEC. (2004). *ISO/IEC 17011:2004 —Conformity assessment — general requirements for accreditation bodies accrediting conformity assessment bodies.* Available from http://www.iso.org/iso/catalogue_detail?csnumber=29332.

ISO/IEC. (2005). *ISO/IEC 17025:2005 — General requirements for the competence of testing and calibration laboratories.* Available from http://www.iso.org/iso/catalogue_detail.htm?csnumber=39883.

ISO/IEC. (2008). *ISO/IEC guide 98-3:2008 uncertainty of measurement — Part 3: Guide to the expression of uncertainty in measurement (GUM: 1995).* Available from http://www.iso.org/iso/catalogue_detail.htm?csnumber=50461.

ISO/IEC. (2010). *ISO/IEC 17043:2010 Conformity assessment — general requirements for proficiency testing.* Available from http://www.iso.org/iso/iso_catalogue/catalogue_tc/catalogue_detail.htm?csnumber=29366.

IUPAC. (2016). *Pesticides properties database*. Available from http://sitem.herts.ac.uk/aeru/iupac.

JCGM. (2012). *The international vocabulary of metrology—basic and general concepts and associated terms (VIM)* (3rd ed.) Available from www.bipm.org/vim.

Kaufmann, A. (2012). The current role of high-resolution mass spectrometry in food analysis. *Analytical and Bioanalytical Chemistry, 403*, 1233−1249.

Kaufmann, A., Dvorak, V., Crüzer, C., Butcher, P., Maden, K., Walker, S., Widmer, M., et al. (2012). Study of high-resolution mass spectrometry technology as a replacement for tandem mass spectrometry in the field of quantitative pesticide residue analysis. *Journal of AOAC International, 95*(2), 528−548.

Kruve, A., Rebane, R., Kipper, K., Oldekop, M. L., Evard, H., Herodes, K., et al. (2015a). Tutorial review on validation of liquid chromatography−mass spectrometry methods: Part I. *Analytica Chimica Acta, 870*, 29−44.

Kruve, A., Rebane, R., Kipper, K., Oldekop, M. L., Evard, H., Herodes, K., et al. (2015b). Tutorial review on validation of liquid chromatography−mass spectrometry methods: Part II. *Analytica Chimica Acta, 870*, 8−28.

Lehotay, S. J., & Cook, J. M. (2015). Sampling and sample processing in pesticide residue analysis. *Journal of Agricultural and Food Chemistry, 63*(18), 4395−4404.

Lehotay, S. J., Mastovska, K., Amirav, A., Fialkov, A. B., Alon, T., Martos, P. A., deKok, A., & Fernández-Alba, A. R. (2008). Identification and confirmation of chemical residues in food by chromatography-mass spectrometry and other techniques. *Trends in Analytical Chemistry, 27*(11), 1070−1090.

Lehotay, S. J., Sapozhnikova, Y., & Mol, H. G. J. (2015). Current issues involving screening and identification of chemical contaminants in foods by mass spectrometry. *Trends in Analytical Chemistry, 69*, 62−75.

Magnusson, B., & Örnemark, U. (2014). In *Eurachem guide: The fitness for purpose of analytical methods − A laboratory guide to method validation and related topics* (2nd ed.) Available from https://www.eurachem.org/images/stories/Guides/pdf/MV_guide_2nd_ed_EN.pdf.

Malato, O., Lozano, A., Mezcua, M., Agüera, A., & Fernandez-Alba, A. R. (2011). Benefits and pitfalls of the application of screening methods for the analysis of pesticide residues in fruits and vegetables. *Journal of Chromatography A, 1218*(42), 7615−7626.

Mann, I., & Brookman, B. (Eds.). (2011). *Eurachem PT guide: Selection, use and interpretation of proficiency testing (PT) schemes* (2nd ed.) Available from https://www.eurachem.org/images/stories/Guides/pdf/Eurachem_PT_Guide_2011.pdf.

Martínez Vidal, J. L., Plaza-Bolaños, P., Romero-González, R., & Garrido Frenich, A. (2009). Determination of pesticide transformation products: a review of extraction and detection methods. *Journal of Chromatography A, 1216*, 6767−6788.

Masiá, A., Morales Suarez-Varela, M., Llopis-Gonzalez, A., & Picó, Y. (2016). Determination of pesticides and veterinary drug residues in food by liquid chromatography-mass spectrometry: A review. *Analytica Chimica Acta, 936*, 40−61.

Masson, P. (2007). Quality control techniques for routine analysis with liquid chromatography in laboratories. *Journal of Chromatography A, 1158*, 168−173.

Mastovska, K., Lehotay, S. J., & Anastassiades, M. (2005). Combination of analyte protectants to overcome matrix effects in routine GC analysis of pesticide residues in food matrices. *Analytical Chemistry, 77*, 8129−8137.

Medina-Pastor, P., Mezcua, M., Rodríguez-Torreblanca, C., & Fernández-Alba, A. R. (2010). Laboratory assessment by combined z score values in proficiency tests: Experience gained through the European Union proficiency tests for pesticide residues in fruits and vegetables. *Analytical and Bioanalytical Chemistry, 397*(7), 3061−3070.

Medina-Pastor, P., Valverde, A., Pihlstrom, T., Masselter, S., Gamon, M., Mezcua, M., et al. (2011). Comparative study of the main top-down approaches for the estimation of measurement uncertainty in multiresidue analysis of pesticides in fruits and vegetables. *Journal of Agricultural and Food Chemistry, 59*(14), 7609−7619.

Meyer, V. R. (2007). Measurement uncertainity. *Journal of Chromatography A, 1158*, 15−24.

Miller, J. C., & Miller, J. N. (1993). *Statistics for analytical chemistry.* Ellis Horwood PTR Prentice Hall, Analytical Chemistry Series.

Milman, B. L. (2015). General principles of identification by mass spectrometry. *Trends in Analytical Chemistry, 69*, 24−33.

Mol, H. G., Tienstra, M., & Zomer, P. (2016). Evaluation of gas chromatography − electron ionization -full scan high resolution Orbitrap mass spectrometry for pesticide residue analysis. *Analytica Chimica Acta, 935*, 161−172.

Mol, H. G. J., Zomer, P., & de Koning, M. (2012). Qualitative aspects and validation of a screening method for pesticides in vegetables and fruits based on liquid chromatography coupled to full scan high resolution (Orbitrap) mass spectrometry. *Analytical and Bioanalytical Chemistry, 403*, 2891−2908.

Mol, H. G. J., Zomer, P., Lopez, M. G., Fussell, R. G., Scholten, J., de Kok, A., et al. (2015). Identification in residue analysis based on liquid chromatography with tandem mass spectrometry: experimental evidence to update performance criteria. *Analytica Chimica Acta, 873*, 1−13.

Murray, K. K., Boyd, R. K., Eberlin, M. N., Langley, G. J., Liang Li, L., & Naito, Y. (2013). Definitions of terms relating to mass spectrometry (IUPAC Recommendations 2013). *Pure and Applied Chemistry, 85*(7), 1515−1609.

Nilsson, L. B., & Eklund, G. (2007). Direct quantification in bioanalytical LC-MS/MS using internal calibration via analyte/stable isotope ratio. *Journal of Pharmaceutical and Biomedical Sciences, 43*(3), 1094−1099.

Nordtest. (2004). *Handbook for calculation of measurement uncertainty in environmental laboratories. Nordtest Report TR 53. Edition 3.1.* Available from http://www.nordtest.info/index.php/technical-reports/item/handbook-for-calculation-of-measurement-uncertainty-in-environmental-laboratories-nt-tr-537-edition-3.html.

Nordtest. (2011). *Nordtest Report TR 569. Internal quality control. Handbook for chemical laboratories.* Available from http://www.nordtest.info/index.php/technical-reports/item/internal-quality-control-handbook-for-chemical-laboratories-trollboken-troll-book-nt-tr-569-english-edition-4.html.

OECD. (2007). *OECD guidance document on pesticide residue analytical methods, ENV/JM/MONO.* Available from http://www.oecd.org/officialdocuments/publicdisplaydocumentpdf/?cote=env/jm/mono(2007)17&doclanguage=en.

OECD. (2009). Test No 509: Crop field trial. In *OECD guidelines for the testing of chemicals.* Paris, France: OECD Publishing. Available from http://www.oecd-ilibrary.org/environment/oecd-guidelines-for-the-testing-of-chemicals-section-5-other-test-guidelines_20745796;jsessionid=5q30jkgeherme.x-oecd-live-03.

OECD. (2011). *OECD maximum residue limit calculator.* Available from http://www.oecd.org/env/ehs/pesticidesbiocides/oecdmaximumresiduelimitcalculator.htm.

OECD. (2014). *OECD Guidance Document for Single Laboratory Validation of Quantitative Analytical Method-Guidance used in support of pre-and post-registration data requirements for plant protection and biocidal products. ENV/JM/MONO. 20 pp. 1−18).* Available from http://www.oecd.org/officialdocuments/publicdisplaydocumentpdf/?cote=ENV/JM/MONO(2014)20&doclanguage=en.

Olivares, I. R. B., & Lopes, F. A. (2012). Essential steps to providing reliable results using the analytical quality assurance cycle. *Trends in Analytical Chemistry, 35*, 109−121.

Otake, T., Yarita, T., Aoyagi, Y., Numata, M., & Takatsu, A. (2014). Evaluation of the performance of 57 Japanese participating laboratories by two types of z-scores in proficiency test for the quantification of pesticide residues in brown rice. *Analytical and Bioanalytical Chemistry, 406*, 7337–7344.

PANNA. (2016). *Pesticide database.* Available from www.pesticideinfo.org.

Prichard, E. (1998). *Quality in the analytical chemistry laboratory.* Chichester, England: John Wiley & Sons Ltd.

Rahman, Md. M., Abd El-Aty, A. M., & Shim, J. H. (2013). Matrix enhancement effect: A blessing or a curse for gas chromatography?—A review. *Analytica Chimica Acta, 801*, 14–21.

Rajski, Ł., Del Mar Gómez-Ramos, M., & Fernández-Alba, A. R. (2014). Large pesticide multiresidue screening method by liquid chromatography-Orbitrap mass spectrometry in full scan mode applied to fruit and vegetables. *Journal of Chromatography A, 1360*, 119–127.

Ramsey, M. H., & Ellison, S. L. R. (2007). In *Eurachem/CITAC guide: Measurement uncertainty arising from sampling: A guide to methods and approaches.* Available from https://eurachem.org/images/stories/Guides/pdf/UfS_2007.pdf.

Rios, A., Barcelo, D., Buydens, L., Cardenas, S., Heydorn, K., Karlberg, B., et al. (2003). Quality assurance of qualitative analysis in the framework of the European project "MEQUALAN". *Accreditation and Quality Assurance, 8*, 68–77.

Roberts, T., & Hutson, D. (1999). In *Pathways of agrochemicals, part two, insecticides and fungicides, the Royal Society of chemistry* (pp. 107–120).

Rochat, B. (2016). From targeted quantification to untargeted metabolomics: Why LC-high-resolution-MS will become a key instrument in clinical labs. *Trends in Analytical Chemistry, 84*, 151–164.

Rozet, E., Rudaz, S., Marini, R. D., Zimons, E., Boulanger, B., & Hubert, P. (2011). Models to estimate overall analytical measurements uncertainty: Assumptions, comparisons and applications. *Analytica Chimica Acta, 702*, 160–171.

SANTE. (2015). *SANTE/11945/2015. Analytical quality control and method validation procedures for pesticide residues analysis in food and feed.* Available from http://ec.europa.eu/food/plant/docs/plant_pesticides_mrl_guidelines_wrkdoc_11945_en.pdf.

Shi, Y., Chang, J. S., Esposito, C. L., Lafontaine, C., Berube, M. J., Fink, J. A., et al. (2011). Rapid screening for pesticides using automated online sample preparation with a high-resolution benchtop Orbitrap mass spectrometer. *Food Additives & Contaminants, 28*(10), 1383–1392.

Shrivastava, A., & Gupta, V. B. (2011). Methods for the determination of limit of detection and limit of quantitation of the analytical methods. *Chronicles of Young Scientists, 2*(1), 21–25.

Skidmore, M. W., & Ambrus, A. (2004). Pesticide metabolism in crops and livestock. Chapter 3. In D. Hamilton, & S. Crossley (Eds.), *Pesticide residues in food and drinking water: Human exposure and risks* (pp. 63–120). John Wiley & Sons.

Soler, C., & Picó, Y. (2007). Recent trends in liquid chromatography-tandem mass spectrometry to determine pesticides and their metabolites in food. *Trends in Analytical Chemistry, 26*, 103–115.

Stahnke, H., & Alder, L. (2015). Matrix effects in liquid chromatography-electrospray ionization − mass spectrometry. In D. Tsipi, H. Botitsi, & A. Economou (Eds.), *Mass spectrometry for the analysis of pesticide residues and their metabolites* (pp. 161–186). Wiley Series on Mass Spectrometry, Wiley & Sons.

Stephenson, G. R., Ferris, I. G., Holland, P. T., & Nordberg, M. (2006). International Union of pure and applied chemistry, chemistry and the environment division. Glossary of terms relating to pesticides. IUPAC recommendations 2006. *Pure and Applied Chemistry, 78,* 2075−2154.

Stöckl, D., D'Hondt, H., & Thienpont, L. M. (2009). Method validation across the disciplines − critical investigation of major validation criteria and associated experimental protocols. *Journal of Chromatography A, 877,* 2180−2190.

Taverniers, I., De Loose, M., & Van Bockstaele, E. (2004a). Trends in quality in the analytical laboratory. I. Traceability and measurement uncertainty of analytical results. *Trends in Analytical Chemistry, 23*(7), 480−490.

Taverniers, I., De Loose, M., & Van Bockstaele, E. (2004b). Trends in quality in the analytical laboratory. II. Analytical method validation and quality assurance. *Trends in Analytical Chemistry, 23*(8), 535−552.

Thompson, M., Ellison, S. L. R., Fajgelj, A., Willetts, P., & Wood, R. (2000). *IUPAC, ISO, AOAC International, Eurachem. Harmonised guidelines for the use of recovery information in analytical measurement.* Available from https://www.eurachem.org/images/stories/Guides/pdf/recovery.pdf.

Thompson, M., Ellison, S. L. R., & Wood, R. (2002). Harmonized guidelines for single laboratory validation of methods of analysis. (IUPAC technical report). *Pure and Applied Chemistry, 74*(5), 835−855.

Thorburn Burns, D., Danzer, K., & Townshend, A. (2002). IUPAC (2002) Use terms "Recovery" "apparent recovery" analytical procedures. *Pure and Applied Chemistry, 4*(11), 2201−2205.

Tomizawa, M., & Casida, J. E. (2005). Neonicotinoid insecticide toxicology: Mechanisms of selective action. *Annual Review of Pharmacology and Toxicology, 45,* 247−268.

Turner, J. (Ed.). (2015). *The pesticide manual* (17th ed.). BCPC Publications. On-line version available from http://www.bcpc.org/shop/PM17.html.

USDA. (2016a). *Residue testing: National residue program.* Available from http://www.fsis.usda.gov/wps/portal/fsis/topics/data-collection-and-reports/chemistry/residue-chemistry.

USDA. (2016b). *Pesticide data program.* Available from https://www.ams.usda.gov/datasets/pdp.

USDA. (2016c). *PDP-data.* Available from https://www.ams.usda.gov/sites/default/files/media/PDP%20DATA%20SOP.pdf.

USDA/SOPs. (2016). *PDP standard operating procedures.* Available from https://www.ams.usda.gov/datasets/pdp/pdp-standard-operating-procedures.

Valverde, A., Fernández-Alba, A. R., Ferrer, C., & Aguilera, A. (2016). Laboratory Triple-A rating: a new approach to evaluate performance-underperformance of laboratories participating in EU proficiency tests for multi-residue analysis of pesticides in fruits and vegetables. *Food Control, 63,* 255−258.

Vanatta, L. E., & Coleman, D. E. (2007). Calibration, uncertainty, and recovery in the chromatographic sciences. *Journal of Chromatography A, 1158,* 47−60.

Vessman, J., Stefan, R. I., Van Staden, J. F., Danzer, K., Lindner, W., Thorburn Burns, D., et al. (2001). Selectivity in analytical chemistry. (IUPAC recommendations 2001). *Pure and Applied Chemistry, 73*(8), 1381−1386.

Wang, J., Chow, W., Chang, J., & Wong, J. W. (2014). Ultrahigh-performance liquid chromatography electrospray ionization Q-orbitrap mass spectrometry for the analysis of 451 pesticide residues in fruits and vegetables: Method development and validation. *Journal of Agricultural and Food Chemistry, 62,* 10375−10391.

WTO. (2013). *Agreement on sanitary and phytosanitary measures (SPS agreement).* Available from https://www.wto.org/English/tratop_e/sps_e/sps_e.htm.

Yarita, T., Otake, T., Aoyagi, Y., Kuroiwa, T., Numata, M., & Takatsu, A. (2015). Proficiency testing for determination of pesticide residues in soybean: comparison of assigned values from participants' results and isotope-dilution mass spectrometric determination. *Talanta, 132,* 269−277.

Zhang, F., Wang, H., Zhang, L., Zhang, J., Fan, R., Yu, C., et al. (2014). Suspected-target pesticide screening using gas chromatography-quadrupole time-of-flight mass spectrometry with high resolution deconvolution and retention index/mass spectrum library. *Talanta, 128,* 156−163.

Zomer, P., & Mol, H. G. (2015). Simultaneous quantitative determination, identification and qualitative screening of pesticides in fruits and vegetables using LC-Q-Orbitrap™-MS. *Food Additives and Contaminants, 32*(10), 1628−1636.

Advanced Sample Preparation Techniques for Pesticide Residues Determination by HRMS Analysis

Renato Zanella, Osmar D. Prestes, Gabrieli Bernardi, Martha B. Adaime

Federal University of Santa Maria, Santa Maria, Brazil

5.1 INTRODUCTION

The presence of organic contaminants in the environment is a result of different sources of pollution from anthropogenic activities (Shamsipur, Yazdanfar, & Ghambarian, 2016). Pesticides are one of the most important tools in the protection of agricultural crops against weeds, pests, and diseases (Commission Regulation, 2009). In accordance with Shamsipur et al. (2016), these compounds are toxic, mobile, and capable of bioaccumulation. They can participate in various physical, chemical, and biological processes. Because of their physicochemical properties, as well as inadequate and intensive use, pesticides can leave residues in food and in the environment (Romero-González, Aguilera-Luiz, Plaza-Bolaños, Garrido Frenich, & Martínez Vidal, 2011). In the last years, many pesticide residues were found in food of vegetal (Qin et al., 2015) and animal origins (Wu et al., 2011), surface (Silva, Daam, & Cerejeira, 2015) and groundwater (Dujaković, Grujić, Radišić, Vasiljević, & Laušević, 2010), soil (Cheng et al., 2016), air (Meire et al., 2016), biological samples (Naksen et al., 2016), etc.

The presence of pesticides may cause toxic effects on consumers and environment (Sturza et al., 2016). Pesticide residue analysis on food is essential for human health protection and to guarantee international trade (Villaverde, Sevilla-Morán, López-Goti, Alonso-Prados, & Sandín-España, 2016). For this reason, regulation agencies have set maximum residue limits (MRLs) for pesticides in food (Commission Regulation, 2005; Farré & Barceló, 2013). Probably, pesticides are one of the most extensively regulated chemical compound classes. Currently, more than 1000 pesticides are used worldwide (Tomlin, 2003). The high number of possible residues indicates the need to develop multiresidue methods to cover compounds of different polarities and to allow routine analysis (Fernández-Alba, 2004). Thus, different sensitive and selective chromatographic-based methods have been reported to monitor pesticides in food samples (Kemmerich et al., 2015; Lozano et al., 2016;

Applications in High Resolution Mass Spectrometry. http://dx.doi.org/10.1016/B978-0-12-809464-8.00005-1

Rizzetti et al., 2016). Sample preparation is the first step in a pesticide residue analysis and has an important impact on the following identification and quantification steps. Therefore, sample preparation is always a bottleneck between specificity and the range of chemical properties of the different analytes to be analyzed (Miao, Kong, Yang, & Yang, 2013).

Currently, triple quadrupole (QqQ) is the first choice for pesticide residue analysis in routine laboratories using gas chromatography (GC) and liquid chromatography (LC) systems (Gómez-Ramos, Ferrer, Malato, Agüera, & Fernández-Alba, 2013). These methods permit target confirmation and quantification with a wide concentration range and high precision in complex matrices at low levels (ng or µg/kg). The reduction of chemical noise due to the high specificity is one the most important advantages of tandem mass spectrometry (MS/MS) techniques (Villaverde et al., 2016). The high velocities to change ionization polarity during chromatographic analysis permit the study of different chemical classes of compounds and the development of multiresidue methods (Wang & Cheung, 2016). QqQ methods also fulfill the international regulations criteria for the identification of target compounds (European Commission, 2015), which should be accomplished by monitoring at least two transitions per target analyte and short dwell times that enable the acquisition of a great number of selected reaction monitoring (SRM) transitions (Greulich & Alder, 2008). All these advantages make the simultaneous analysis of a large number of pesticides by LC and by GC coupled to QqQ systems (Wang & Cheung, 2016) in different matrices (Donato et al., 2012). Nevertheless, QqQ-based methods present drawbacks for its application in a cost-effective way because of the large number of pesticides to be analyzed in a single run.

The method development is extensive and time-consuming and, in this way, QqQ suppliers were able to produce more sensitive and faster instruments with user-friendly software. Thus, it becomes questionable to perform targeted analysis with so many analytes. This requires the use of huge number of standards to ensure good quality analysis (Kaufmann, 2014). Some practical problems are related with a huge standard solutions production such as (1) the availability of some pesticides, (2) use of different solvents to dissolve them, and (3) fast degradation of some of them. Recently, to avoid all these disadvantages and to permit the nontarget analysis, the need for screening methods is growing up (Romero-González et al., 2011). These methods provide a qualitative binary response, and samples can be classified as negative and nonnegative, considering nonnegative samples as those containing one or more analytes above a preestablished concentration level. Then, nonnegative samples must be reanalyzed by an identification/quantification method to determine the final concentration in the positive samples (Romero-González et al., 2011).

Most of screening methods used in routine laboratories detect the presence of pesticides as a pretarget screening (previously known compounds) by low-resolution mass spectrometry instruments, such as single quadrupole (Jedziniak, Szprengier-Juszkiewicz, & Olejnik, 2009), QqQ (Martínez-Vidal, Garrido-Frenich, Aguilera-Luiz, & Romero-González, 2010), and ion trap analyzers (Smith, Gieseker, Rimschuessel, Decker, & Carson, 2009), applying different scan modes.

However, pretarget screening methods are often insufficient because only a limited number of analytes are monitored, and, thus, other nontarget analytes, which could be present in the samples, would not be detected (Romero-González et al., 2011).

High-resolution mass spectrometry (HRMS) is a powerful technique to discriminate between possible target compounds without the need of reference standards, unlike QqQ methods. Nowadays, HRMS is increasingly used as a screening tool in several areas (Lehotay, Sapozhnikova, & Mol, 2015). For pesticide screening, considering the large number of compounds authorized for use, methods must be capable of detection of multiple residues in a single commodity and able to quantitate for a wide MRL range. In the literature, different publications reported multiresidue methods to screen, quantify, and confirm pesticides by HRMS coupled to chromatographic techniques (Hernández, Grimalt, Pozo, & Sancho, 2009; Jamin et al., 2013; Lacina, Urbanova, Poustka, & Hajslova, 2010; Pozo et al., 2007; Wang, Chow, Leung, & Chang, 2012). Nevertheless, this technique presents some disadvantages as false negatives, depending on the pesticide, matrix, and sample preparation methods. Otherwise, this can be avoided by an adequate sample preparation step. HRMS equipment provides high resolution, accurate mass, and high full-scan sensitivity and selectivity, allowing the operation for both target and nontarget compounds (Lehotay et al., 2015).

HRMS permits to work at full-scan, at MS/MS, and in sequential or simultaneous modes (Gómez-Ramos et al., 2013; Romero-González et al., 2011). Although a principal advantage of HRMS is chemical formula generation, even under the best MS conditions, formula generation can only be limited to the number of possibilities within the error tolerance of the measurement (Croley, White, Callahan, & Musser, 2012). All these qualities make HRMS an attractive tool for pesticide residue analysis in food. In the past, HRMS instruments had limited use in food analysis laboratories because of some drawbacks: (1) not enough accurate mass; (2) lower resolution; (3) saturation effects possibility; (4) short dynamic range; and (5) the presence of false positives. Nowadays, all these disadvantages have been solved or minimized with improvements, especially combining full-scan with MS/MS modes (Gómez-Ramos et al., 2013). Time-of-flight (TOF), hybrid quadrupole-TOF (QTOF), and Orbitrap are the most common HRMS analyzers (Romero-González et al., 2011). In this way, HRMS data can be acquired without any previous knowledge about the compounds present in the sample (Peters, Bolck, Rutgers, Stolker, & Nielen, 2009).

5.2 MATRIX EFFECTS AND THE INFLUENCE OF COEXTRACTED COMPONENTS

Chromatographic techniques coupled to MS systems have proven to be a powerful tool for determination of pesticide residues in complex matrices. In accordance with Kaufmann (2014), there is no consensus when sample is considered complex. The presence of target compounds at trace levels (μg/kg or μg/L) in an extract containing different chemical groups such as proteins, carbohydrates, lipids, and other miscellaneous compounds may reveal the dimension of food sample complexity (Kaufmann, 2014). The presence of coextractives influences the analyte signals, which can be either suppressed or enhanced when compared to the response in pure solvent (Trufelli, Palma, Famiglini, & Cappiello, 2011). Ion suppression is the most common matrix effect in LC-based methods, especially when MS operated with electrospray (ESI) positive polarity mode, which is the most used for pesticide analysis (Kittlaus, Schimanke, Kempe, & Speer, 2011). As described by Gómez-Ramos, Rajski, Lozano, and Fernández-Alba (2016) various causes are related with the ion suppression effect in ESI such as (1) competition between matrix components and analytes for available charges and access to the droplet surface and (2) change in the viscosity and surface tension of the LC effluent that, consequently, influences the droplet formation and the evaporation process, affecting the number of charged ions. Furthermore, nonvolatile components can form solid analyte inclusion particles, and matrix components may act as ion-pairing reagents with already ionized analytes (Trufelli et al., 2011). However, the most discussed explanation is related to the competition for access to the droplet surface along with the competition for the surface excess charges (Gómez-Ramos et al., 2016).

As explained by Erney, Gillespie, Gilvydis, and Poole (1993) and Gillespie, Daly, Gilvydis, Schneider, and Walters (1995), for GC methods, the matrix-induced response enhancement can be explained based on the competition between coextractives and the pesticides to access the active sites present in the inlet and chromatographic column. The coextractives protect the target analytes from decomposition in the hot injector. Otherwise, nonvolatile coextractives accumulated into the GC system promote the response diminishment because of the generation of new active sites (Gómez-Ramos et al., 2016). The disadvantages of matrix effects are related with the worst influence on both quantitative and qualitative analytical results (Gosetti, Mazzucco, Zampieri, & Gennaro, 2010). The robustness can be affected because of the necessity of extra cleaning steps in the MS compartments as source, ion transfer capillary, and, in certain cases, the analyzer itself. The matrix effect impact is related to different factors such as matrix type, applied sample preparation method (Kittlaus et al., 2011), mobile phase composition (Gosetti et al., 2010), and also the geometry/design of the ionization source (Stahnke, Kittlaus, Kempe, Hemmerling, & Alder, 2012). Matrix coextractives can affect each commodity differently and can be observed in different parts of the chromatogram. Matrix compounds are "unknowns" in targeted analyses. Currently, only few works have studied these compounds with the focus to understand matrix effect (Yang et al., 2015).

In nontarget analysis, one of the major limitations came from the possibility of unexpected matrix effects or a lack of MS parameter optimization, which could lead to false-negative results. Many matrix compounds often have similar masses to target compounds, and in the case of coelution of matrix compounds with analytes, the instrument might not be capable of fully resolving these two slightly different masses. Resolving power is an important factor in the correct identification of these compounds, especially in complex matrices. TOF instruments usually provide resolutions of up to 15,000 full width at half maximum (FWHM) and can even increase to more than 50,000 FWHM in new TOF instruments. On the other hand, Orbitrap (Exactive) technology currently provides resolutions of over 100,000 FWHM (Martínez-Domínguez, Romero-González, & Garrido-Frenich, 2016).

Different strategies can be applied to minimize matrix coextractive interferences including optimization of chromatographic selectivity and modification of sample preparation and cleanup. An intensive cleanup helps to reduce matrix effects, but for some compounds this step can produce negative effects such as low recoveries. Furthermore, strong matrix effects can remain even after applying cleanup steps (Gómez-Ramos et al., 2016). Dilution of 10 times or more is an easy and successful way to reduce matrix coextractive contents, minimizing matrix effect (Ferrer, Lozano, Agüera, Girón, & Fernández-Alba, 2011). Sample dilution decreases the concentration of coextractive molecules per microdroplet, increasing the ionization efficiency and the analytic signal. When the matrix effect cannot be reduced, some methods can compensate the response variation, such as isotopically labeled standards (ILS), matrix-matched calibration (MMCC), postcolumn infusion (PCI), and ECHO peak technique (Jiao et al., 2016). The ECHO peak technique comprises two injections in each analysis: an unknown sample and a standard solution. The peak of the analyte from the standard elutes in close proximity to the peak of the analyte from the sample, thus forming the ECHO peak. It is expected that both peaks elute so closely that they are affected in the same manner by the coeluted matrix components (Alder, Lüderitza, Lindtnera, & Jürgen Stan, 2004). However, these methods have some disadvantages such as the cost and limited availability of ILS or the fact that MMCC requires many blank extracts. The ECHO peak technique cannot completely compensate matrix effects and PCI cannot adequately compensate in all matrices. Stahnke, Kittlaus, Kempe, and Alder (2012) investigated the relationship between the matrix concentration and the matrix effect for four matrices (avocado, black tea, orange, and arugula) and showed that an appropriate dilution factor could reduce the matrix effects to less than 20%.

5.3 SAMPLE PREPARATION TECHNIQUES FOR PESTICIDE RESIDUE DETERMINATION BY CHROMATOGRAPHIC TECHNIQUES COUPLED TO HIGH-RESOLUTION MASS SPECTROMETRY

Sample preparation is the first step applied in the pesticide residue analysis and has a great impact on the following determination by LC or GC with MS. It is an important step within the entire analytical process. Generally, it is the most time-consuming, labor-intensive, and prone-to-error part of the analysis, which can compromise the results. The main objectives of sample preparation are to promote the extraction and enrichment of the analytes and remove interferences as much as possible. The enrichment step is important because the contaminants are present in food at low concentrations. Usually, extraction with organic solvent is applied. Moreover, cleanup procedures have also been performed after extraction to eliminate coextractives before analysis (Tette, Guidi, Glória, & Fernandes, 2016). According to Hercegová, Domotorova, and Matisova (2007), a sample preparation procedure for pesticide residue analysis should have the following properties: include the highest possible number of pesticides (multiresidue assay); have recoveries as close as possible to 100%; remove potential interfering compounds in the sample to improve selectivity; allow increasing concentration of the analytes; have appropriate precision and ruggedness; and be quick, easy, safe, and cost-wise. Related to the characteristics described by Berendsen, Stolker, and Nielen (2013), the composition of the final extract injected into the instrumental system depends on the sample preparation procedure. This step is closely related to the results in terms of limits of detection (LODs), selectivity, precision, ruggedness, and selectivity. This is a critical process in relation to analysis time, sample throughput, and costs (Berendsen et al., 2013). Sample preparation is always linked to the potential matrix interferences and the chromatographic separation technique. In the last years, the improvements in extraction methods and also in the analytical techniques reduce the sample preparation drawbacks. Nowadays, different methods promote a great reduction in time, sample size, and solvent volumes, increasing the accuracy and precision in the field of pesticide residue analysis (Grimalt & Dehouck, 2016).

The necessity of more information related to the presence of residues and contaminants in raw materials, processed food, and environmental samples promoted the fast development of multiresidue methods focusing on the determination of a large number of pesticides with very different physicochemical properties (Garrido-Frenich, Romero-González, & Aguilera-Luiz, 2014).

Since the first pesticide multiresidue method for organochlorine compounds (Mills & Onley, 1963), other procedures have been developed. Luke, Froberg, and Masumoto (1975) proposed an extraction with acetone for more polar pesticides. Acetone was also used as extraction solvent by the Dutch Food and Consumer Products Safety Authority, which was a reference multiresidue method for more than 25 years (General Inspectorate for Health Protection, 1996). Ethyl acetate was employed as an extraction solvent for pesticide multiresidue method, to improve the polar pesticides analysis employed combined with Na_2SO_4 (Pihlström, Blomkvist, Friman, Pagard, & Osterdahl, 2007). Currently, sample preparation techniques focus on the determination of a large number of pesticides with very different physical and chemical properties (Berendsen et al., 2013; Villaverde et al., 2016). To comply with this demand, analytical laboratories should have suitable and reliable methods that allow the simultaneous determination, increase sample throughput, and allow more than one family of residues or contaminants to be analyzed (Garrido-Frenich et al., 2014). Pesticide residue analysis is a continuing challenge mainly because of low analytes concentration as well as the large amounts of interfering substances, which can be coextracted with them. Thus, extensive sample preparation is often required for the pesticide residue analysis for the effective extraction of the analytes and removal of the interferences. These characteristics have some benefits to HRMS screening analysis that requires comprehensive sample preparation strategies to extract a wide range of chemical classes (Knolhoff & Croley, 2016). However, the vast majority of extraction methods for food analyses tend to be for specific groups, such as pesticides, veterinary drugs, or mycotoxins. In this way, the tendency of sample preparation methods is the development of generic extraction method that can extract and obtain acceptable recoveries with high precision for a wide range of compounds (Garrido-Frenich et al., 2014). These nonselective sample preparation methods can produce final extracts with a significant amount of matrix constituents that might interfere in the analytical results (Berendsen et al., 2013). Thus, the sample preparation is still the bottleneck for the effective and accurate analysis (Zhang et al., 2012).

Bearing in mind the issues described previously, this chapter focuses on reviewing the recent developments in sample preparation techniques for pesticide residue analysis by HRMS. Particularly, their advantages, disadvantages, and future perspectives will be discussed.

5.3.1 DILUTE-AND-SHOOT

Dilute-and-shoot (DS) is considered one of the simplest procedures for pesticide residue analysis (Rentsch, 2016). Most of the DS applications were based on the work published by Mol et al. (2008) that proposed a fast and straightforward generic procedure for the simultaneous extraction of pesticides, mycotoxins, plant toxins, and veterinary drugs in various matrices and analysis by ultrahigh-performance liquid chromatography (UHPLC) coupled to MS/MS. Since then, it has been successfully used by several authors mainly because of its speed, simplicity, and the possibility to extract other residues and contaminant classes (Gómez-Pérez, Plaza-Bolaños, Romero-González, Martínez-Vidal, & Garrido-Frenich, 2012; Martínez-Domínguez, Romero-González, & Garrido-Frenich, 2015, 2016). These procedures minimize sample handling and number of extraction steps and are the current trend used in LC methods. Otherwise, DS, also referred as sample dilution, is not applicable in GC methods because the analyzed extracts typically contain nonvolatile coextractives. As discussed before, these coextractives are gradually deposited in the GC inlet and chromatographic column and can originate new active sites, which may be responsible for the decrease in GC signal response (Cajka et al., 2012).

DS combined with HRMS analysis permits the determination of a wide variety of compounds using only one extraction and injection step into the chromatographic system. Gómez-Pérez et al. (2012) optimized a DS method for the determination of more than 350 pesticides and other contaminants by UHPLC-Orbitrap-MS using 2.5 g honey with a 10 mL water/acetonitrile mixture. The pesticides azoxystrobin, coumaphos, dimethoate, and thiacloprid were detected in four samples. Azoxystrobin and coumaphos were found in two different samples of organic honey at 1.5 and 5.1 µg/kg.

Martínez-Domínguez et al. (2015) optimized a UHPLC-Orbitrap-MS to identify and quantify more than 250 toxic substances, such as pesticides, in ginkgo biloba products. Acetonitrile with 1% (v/v) formic acid was applied for the DS extraction of the target compounds. A cleanup step was performed using a mixture of the sorbents: primary secondary amine (PSA), graphitized carbon black (GCB), octadecylsilica (C_{18}), and Supel QuE Z-Sep+. Generic methods extract more interferences, so cleanup steps can be included to reduce coextractives in the final extract. However, the use of UHPLC has some advantages, such as reduction of the matrix effect by injecting lower sample volume, and HRMS can distinguish between interferences and target compounds. Pesticides, such as hymexazol (10 µg/kg) and tebufenozide (55−459 µg/kg) were found in ginkgo biloba samples. The same approach was applied to determine pesticides and other contaminants in green tea and royal jelly by UHPLC-Orbitrap-MS. Cleanup step was performed using C_{18} and Supel QuE Z-Sep+ for green tea and royal jelly, respectively. The limits of quantification (LOQs) were <10 µg/kg. Eight samples of green tea and one sample of royal jelly were found to be positive for pesticides at concentrations ranging from 10.6 (cinosulfuron) to 47.9 µg/kg (paclobutrazol) (Martínez-Domínguez et al., 2016).

Pesticides and other compounds were analyzed in soy-based nutraceutical products by DS sample preparation and HRMS. A cleanup step with Florisil cartridges was applied to improve the cleanness of the extract. LODs and LOQs were below 5 and 10 µg/kg, respectively. The validated method was applied to the analysis of real samples, finding pesticides such as flutolanil (12.2 µg/kg) and etofenprox (48.2 µg/kg) (Martínez-Domínguez, Romero-González, Arrebola, & Garrido-Frenich, 2016).

Generally, the DS method increases the capabilities of routine laboratories with high sample throughput and reliable characterization of any type of food. Thus, when this procedure is combined with UHPLC-HRMS, a wide variety of compounds from several families and classes of residues and contaminants can be determined using only one fast sample dilution before injection into the chromatographic system (Garrido-Frenich et al., 2014).

5.3.2 QUECHERS METHOD

When a wide-scope analysis is applied, a nonselective sample preparation is preferred to extract the highest number of compounds. For pesticide determination in food, there are well-known general extraction procedures, based on solvent extraction (SE), which are very effective for that purpose (Anastassiades, Lehotay, Stajnbaher, & Schenck, 2003; Lehotay, de Kok, Hiemstra, & Van Bodegraven, 2005; Luke et al., 1975; Mol, van Dam, & Steijger, 2003). The most widely used solvents for pesticides multiresidue extraction are acetonitrile, ethyl acetate, and acetone with advantages and disadvantages for each solvent. Ethyl acetate shows universal characteristics because of its ability to extract several classes of pesticides from different samples. However, for pesticides with basic character (pKa >4), the recovery percentages are generally low because of degradation problems, requiring the addition of sodium hydroxide to increase these percentages (Pihlström et al., 2007). Acetonitrile and acetone are miscible with water and, when in contact with the matrix, promote the extraction in a single phase. When acetone is used, generally, the addition of a nonpolar solvent is necessary to achieve the separation between organic and aqueous phases. This is not the case when acetonitrile is used and salts addition step promotes this phase separation (Maštovská & Lehotay, 2004). Based on acetonitrile extraction, the QuEChERS method (an acronym for quick, easy, cheap, effective, rugged, and safe) was proposed by Anastassiades et al. (2003) and has taken the first position in sample preparation, and it can be used also for environmental samples (Picó, 2015).

QuEChERS method was introduced to overcome the practical limitations of existing multiresidue methods, and during its development, great emphasis was given to obtain a dynamic procedure, which could be applied in routine laboratories. The original QuEChERS method was designed to allow the extraction of pesticide residues in matrices with high percentage of water, such as fruits and vegetables. The extraction procedure is based on acetonitrile extraction and subsequent cleanup with dispersive solid-phase extraction (d-SPE) using the sorbent PSA and anhydrous magnesium sulfate ($MgSO_4$) for water removal. Although the original version provided excellent results for different sample types, some applications have shown that certain compounds had stability problems according to the matrix pH, occasioning poor recovery results (Anastassiades, Scherbaum, Tasdelen, & Stajnbaher, 2007; Lehotay et al., 2005; Payá et al., 2007). To solve this problem, two remarkable modifications were proposed. During the optimization, it was noticed that the use of buffers with pH around 5 promotes satisfactory recoveries ($>70\%$) for pH-dependent compounds (e.g., pymetrozine, imazalil, and thiabendazole) independent of the matrix. Based on that, the first modification proposed to the method was the addition of sodium acetate promoting a buffer effect with pH around 4.8 with the acetic acid added in the acetonitrile. This modification has been adopted by the Association of Analytical Communities (AOAC) as official method (AOAC 2007.01) and was denominated QuEChERS acetate (Lehotay et al., 2005). Later, Anastassiades et al. (2007) proposed the QuEChERS citrate method that uses the mixture of sodium citrate dihydrate and hydrogen citrate sesquihydrate as responsible for the buffering effect of pH 5.0–5.5. This modification has been adopted as an official method by the European Committee for Standardization (CEN 15662). These QuEChERS method versions also include the use of sorbents, such as C_{18} or GCB for fatty and pigmented matrices, respectively, to improve the cleanup procedure.

The changes already made in the QuEChERS method demonstrate the flexibility of the procedure that can be used for multiresidue analysis of pesticides in various matrices. Besides that, the method has been applied satisfactorily in combination with chromatographic techniques coupled to MS. In the past few years, other modifications have been performed such as other extraction solvents (Shaoo, Batttu, & Singh, 2011), new sorbents (Cabrera, Caldas, Prestes, Primel, & Zanella, 2016; Cerqueira, Caldas, & Primel, 2014), and low-temperature precipitation cleanup (Sobhanzadeh, Bakar, Abas, & Nemati, 2012).

QuEChERS method, in addition to be applied to extract a large number of pesticides from different classes, has also been used for other compounds such as mycotoxins (Yogendrarajah, Pouckeb, De Meulenaera, & De Saeger, 2013), veterinary drugs (Guo et al., 2016; Kang et al., 2015), polycyclic aromatic hydrocarbons

(Rociek, Surma, & Cieślik, 2014), and dyes (Jia et al., 2014) in different food and environmental matrices (Fernandes, Domingues, Mateus, & Delerue-Matos, 2013). The QuEChERS method fulfills a high demand for multiresidue analysis applicable to the multiple types of matrices. However, some pesticide characteristics such as a wide range of molecular mass, polarity, acid—base behavior, and the matrix composition may interfere with the extraction and cleanup. Matrix effects is one of the major limitations of nontarget approaches, which can lead to unexpected false-negative results when complex samples are analyzed (Gómez-Ramos et al., 2013).

When QuEChERS method is applied, there are some complex matrices that need a different or an additional cleanup. Conventional solid-phase extraction (SPE), that uses more sorbents, may be a good choice to remove some matrix components such as fatty acids and pigments (Shaoo et al., 2011). QuEChERS may also be combined with other cleanup methods in addition to d-SPE, such as disposable pipette extraction (Koesukwiwat, Lehotay, Miao, & Leepipatpiboon, 2010), dispersive liquid—liquid extraction (DLLME) (Cunha & Fernandes, 2013), or gel permeation chromatography (Luo et al., 2015). Besides that, other strategies may be used when QuEChERS method is used such as analytes protectant to avoid compounds degradation when GC is used, and sample dilution, mainly for LC analysis. A compensation approach, such as the use of matrix-matched standards, is considered a useful method to eliminate the influence of matrix effects on the reliability (accuracy and precision) of the results (Zanella, Prestes, Martins, & Adaime, 2015).

The use of HRMS for pesticide determination in combination with the QuEChERS method has been successfully applied by several studies. Lacina et al. (2010) investigated the potential of UHPLC-TOF-MS to enable rapid and comprehensive analysis of 212 pesticide residues in high-moisture, low-fat fruit and vegetable matrices (apple, strawberry, tomato, and spinach). Sample preparation was done by the original QuEChERS method. Extracts were centrifuged, filtered, and injected without a cleanup step.

Rajski, Gómez-Ramos, and Fernández-Alba (2014) evaluated the main operational parameters for a pesticide multiresidue screening using a UHPLC-Orbitrap-MS system working in full-scan mode. Three different resolution settings (17,500; 35,000; and 70,000) were evaluated at various concentration levels. The evaluation was performed using QuEChERS extracts of tomato, pepper, orange, and green tea. The instrument ensured not only high resolution but also very good sensitivity, linear range, and analysis reproducibility. However, out of three tested resolutions, only 35,000 (for all matrices except tea) and 70,000 (for all matrices) were recommended. The signal response in pepper and orange extracts was similar to tomato while signal suppression occurred mostly in the green tea extract.

Recently, Reinholds, Pugajeva, and Bartkevics (2016) stated that UHPLC-Orbitrap-HRMS is a promising technique in high-throughput method development for the routine analysis of pesticide residues and mycotoxins in spices. The citrate QuEChERS method was used, after partition, and an aliquot of the supernatant was frozen at −80°C for 30 min. Then, the obtained organic sample fraction was centrifuged and 3 mL was evaporated to dryness and reconstituted in 0.5 mL of MeOH:water (50:50, v/v) before analysis.

Munaretto, May, Saibt, and Zanella (2016) proposed a strategy to identify and quantify 182 organic contaminants from different chemical classes, such as pesticides, veterinary drugs, and personal care products in fish fillet using LC-QTOF-MS. For this purpose, two different scan methods (full-scan and all ions MS/MS) were evaluated to assess the best option for screening analysis. Sample preparation was based on original QuEChERS method using Na_2SO_4 instead of $MgSO_4$, improving the recovery of polar compounds. Cleanup was done by d-SPE using PSA and C_{18}.

A screening method was developed by Pérez-Ortega et al. (2016) using UHPLC-TOF-MS for multiclass pesticide analysis in jam. Samples were extracted using the acetate QuEChERS method described by Lehotay et al. (2005). After cleanup with PSA, 1 mL of the extract was evaporated near to dryness under a nitrogen stream and reconstituted with 1 mL of MeOH:water (20:80, v/v) before UHPLC-TOF-MS analysis. The method was based on an experimental database with retention time and accurate mass data for the 353 selected analytes. The LODs were below 10 µg/kg for 90% of the studied compounds.

5.3.3 MATRIX SOLID-PHASE EXTRACTION

Matrix solid-phase extraction (MSPD) was first introduced in 1989 by Barker for the extraction of drug residues from animal tissue (Barker, Long, & Short, 1989). However, since its introduction, MSPD has been widely used for the isolation of a wide range of compounds from a large variety of matrices (Barker, 2007). The main advantages of MSPD are the simplicity and flexibility because it does not require any extra instrumentation-specific equipment. This fact has been contributing to MSPD being chosen over more classical methods for multiresidue pesticides determination in different matrices (Cao, Tang, Chen, & Li, 2015; Ramos, Rial-Otero, Ramos, & Capelo, 2008). A particular application of this technique is the sample preparation for pesticides determination in solid, semisolid, and highly viscous samples. MSPD extraction procedure is based on the application of mechanical forces to the sample, aiming a complete sample disruption and promoting the interactions between sample matrix and a solid bonded sorbent, or the surface chemistry of other solid support materials (Capriotti et al., 2015). Usually, after obtaining a homogeneous mixture, the material is placed in a cartridge and compressed before the elution with a suitable organic solvent. The solid phase and the organic solvent that will be used depend on the polarity of the analyte and the type of sample (Picó, 2015). For pesticide analyses, generally the most used sorbents are C_8, C_{18}, and Florisil. One of the most critical parameters is the ratio of sample-to-solid material (generally 1:1 to 1:4), which depends on sample type and material physicochemical features. The choice of eluent depends on the strength of the interactions between solid support and target analytes (Capriotti et al., 2015). Because there are many parameters that can influence the MSPD technique, such as sample size, type and amount of dispersant material, possible cleanup step, and composition of eluent, experimental design is often used for method optimization (Rallis, Sakkas, Boumba, Vougiouklakis, & Albanis, 2012). Ramos et al. (2008) combined MSPD with ultrasound irradiation using an ultrasonic bath or a probe to decrease the time and to increase significantly the efficiency of the liquid—solid extraction step. Also, the SPE elution step has been successfully substituted by vortex agitation followed by centrifugation (Caldas et al., 2013). Zhan et al. (2016) developed a method using MSPD for the determination of organochlorine pollutants including 16 organochlorine pesticides (OCPs) and seven polychlorinated biphenyls in oil seeds (peanuts and soybeans) by GC with electron capture detector. Under the optimized condition, good recoveries were obtained in the range of 68.9%—103.3% with relative standard deviation (RSD) values below 16.1% in all cases. Low LOD and LOQ values were achieved in the range of 0.1—2.0 µg/kg and 0.4—6.7 µg/kg, respectively. MSPD extraction technique has been used in combination with HRMS method for pesticides determination in food matrices. García-Reyes, Ferrer, Thurman, Fernandez-Alba, and Ferrer (2006) have explored the potential usefulness of LC-TOF-MS for the quantitative analysis of herbicide residues in olive oil samples. The authors compared the values achieved by several analytical parameters evaluated with the values generally obtained with QqQ instruments in SRM mode, illustrating, thus, the potential of LC-TOF-MS for routine quantitative analyses of herbicides in olive oil. Table 5.1 shows DS, QuEChERS, and MSPD applications for pesticide residue analysis in food matrices by chromatographic techniques coupled to HRMS.

5.3.4 SOLID-PHASE EXTRACTION

SPE is one of the most frequently used procedures for cleanup, extraction, and preconcentration of pesticide residues from different samples such as environmental (Donato et al., 2012), food (Gan et al., 2016), and beverage (Wang & Cheung, 2016). Disposable cartridges for SPE were introduced more than 30 years ago, but the development was slow for many years until replacing the classic liquid—liquid extraction (LLE) (Płotka-Wasylka, Szczepańska, Guardia, & Namieśnik, 2016). In addition to cartridges, SPE can be done in disc, pipette tips, and 96-well plates and is recognized as beneficial alternative to LLE, because it overcomes many drawbacks. The advantages of this technique are related to low solvent consumption, low costs, and reduction of processing time. The SPE can be performed in off-line or online mode (automated process). During SPE sample preparation, samples or extracts are percolated through the cartridge and the compounds are sorbed on the solid phase materials, which have been previously conditioned and activated (Zhang et al., 2012). Then the interferences are removed by prewashing by organic solvents while the analytes are retained on the sorbents. More frequently, the interferences are retained in the cartridge and the analytes pass through and are collected for analysis.

Nowadays, a large number of sorbents are available with a wide polarity range, and the most frequently used are (1) chemically modified silica gel, (2) polymeric

Table 5.1 Dilute-and-Shoot (DS), QuEChERS, and Matrix Solid-Phase Extraction (MSPD) Applications for Pesticides Residue Analysis in Food Matrices by Chromatography Techniques Coupled to High-Resolution Mass Spectrometry

Matrix	Compounds	Extraction Procedure	Detection Mode	Resolution (FWHM)	Sensitivity	Mass Accuracy (ppm)	References
Honey	Over 350 pesticides and veterinary drugs	DS: 2.5 g of sample and 2.5 mL of water [50 mL polypropylene (PP) tube] were mixed using a vortex. 7.5 mL of acetonitrile with 1% of formic acid (v/v) was added and shaken for 1 h. Tube was centrifuged and analyzed	UHPLC-Orbitrap-MS ESI(+) and ESI(−)	25,000 (MS) 10,000 (MS/MS)	LOQ <50 µg/kg	<5 ppm	Gómez-Pérez et al. (2012)
Ginkgo biloba nutraceutical products	Over 250 toxic substances, including pesticides	DS: 2.5 g of sample and 2.5 mL of water (50 mL PP tube) were shaken for 30 s; 7.5 mL of acetonitrile with 1% of formic acid (v/v) was added, shaken for 1 h, centrifuged, and analyzed	UHPLC-Orbitrap-MS ESI(+) and ESI(−)	25,000 (MS) 10,000 (MS/MS)	LOQ <10 µg/kg	5 ppm	Martínez-Domínguez et al. (2015)
Green tea and royal jelly	Multiclass pesticides and mycotoxins	DS: 2.5 g of sample and 2.5 mL of water (50 mL PP tube) were shaken for 30 s; 7.5 mL of acetonitrile	UHPLC-Orbitrap-MS ESI(+) and ESI(−)	25,000 (MS) 10,000 (MS/MS)	LOQ <10 µg/kg	<5 ppm	Martínez-Domínguez et al. (2016)

Continued

Table 5.1 Dilute-and-Shoot (DS), QuEChERS, and Matrix Solid-Phase Extraction (MSPD) Applications for Pesticides Residue Analysis in Food Matrices by Chromatography Techniques Coupled to High-Resolution Mass Spectrometry—cont'd

Matrix	Compounds	Extraction Procedure	Detection Mode	Resolution (FWHM)	Sensitivity	Mass Accuracy (ppm)	References
		with 1% of formic acid (v/v) was added and shaken for 2 h. The tube was centrifuged and 1 mL of extract was used for analysis					
Soy-based nutraceutical products	Over 257 toxic substances, including pesticides	DS: 2.5 g of sample and 2.5 mL of water (50 mL PP tube) were shaken for 30 s; 7.5 mL of acetonitrile with 1% of formic acid (v/v) was added and shaken for 1 h, centrifuged, and analyzed	UHPLC-Orbitrap-MS ESI(+) and ESI(−)	25,000 (MS) 10,000 (MS/MS)	LOQ <10 μg/kg	<5 ppm	Martínez-Domínguez et al. (2016)
Wine	50 pesticides	Wine samples were directly injected after filtering with 0.22 μm PTFE filter	LC-QTOF-MS ESI(+)	n.a.	n.a.	<2 ppm	He et al. (2016)
Fruits and vegetables	150 pesticides	QuEChERS acetate—buffered with d-SPE cleanup using a combination	GC-TOF-MS	n.a.	LOQ 10 μg/kg	n.a.	Koesukwiwat et al. (2010)

Sample	Analytes	Method	Instrument	Resolution	LOQ		Reference
Apple, strawberry, tomato, and spinach	212 pesticides	of MgSO$_4$, PSA, C$_{18}$, and GCB QuEChERS original with no cleanup	UHPLC-TOF-MS ESI(+) and ESI(−)	11,000	LOQ <10 µg/kg	n.a.	Lacina et al. (2010)
Orange, tomato, leek	53 pesticides	QuEChERS citrate	LC-QTOF-MS ESI(+)	15,000	n.a.	n.a.	Ferrer et al. (2011)
Fish fillet	182 organic contaminants	QuEChERS original with d-SPE cleanup using a combination of Na$_2$SO$_4$, PSA, and C$_{18}$	LC-QTOF-MS ESI(+)	10,000	LOQ <25 µg/kg	<5 ppm	Munaretto et al. (2016)
Mushrooms	Nicotine	QuEChERS modified using acetonitrile (pH = 9) with no cleanup step	LC-TOF/MS ESI(+)	7500	LOQ 10 µg/kg	n.a.	Lozano et al. (2012)
Fruits and vegetables	170 pesticides	QuEChERS citrate	UHPLC-Orbitrap-MS	35,000 and 70,000	n.a.	5 ppm	Rajski et al. (2014)
Leek, wheat, and tea	389 multiclass food contaminants	QuEChERS original and d-SPE cleanup using C$_{18}$ and MgSO$_4$	HPLC-Orbitrap-MS	70,000 (MS) 17,500 (MS/MS)	LOQ <10 µg/kg	5 ppm	Dzuman, Zachariasova, Veprikova, Godula, and Hajslova (2015)
Paprika	134 pesticides and 11 mycotoxins	QuEChERS citrate	UHPLC-Orbitrap-MS	70,000 mycotoxins 35,000 pesticides	n.a.	n.a.	Reinholds et al. (2016)

Continued

Table 5.1 Dilute-and-Shoot (DS), QuEChERS, and Matrix Solid-Phase Extraction (MSPD) Applications for Pesticides Residue Analysis in Food Matrices by Chromatography Techniques Coupled to High-Resolution Mass Spectrometry—cont'd

Matrix	Compounds	Extraction Procedure	Detection Mode	Resolution (FWHM)	Sensitivity	Mass Accuracy (ppm)	References
Olive oil	Simazine, atrazine, diuron, and terbuthylazine	MSPD: The extract from the previous LLE step was mixed with NH_2-silica until a fine powder was obtained. The mixture was transferred to a minicolumn containing florisil and eluted with acetonitrile. The extract was evaporated, redissolved, and analyzed	LC-TOF/MS ESI(+)	n.a.	LOD <5 µg/kg	<2 ppm	García-Reyes et al. (2006)

DS, dilute-and-shoot; d-SPE, dispersive solid-phase extraction; ESI, electrospray ionization; FWHM, full width at half maximum; GCB, graphitized carbon black; GC–TOF-MS, gas chromatography–time-of-flight-mass spectrometry; LC–QTOF, liquid chromatography–quadrupole time-of-flight; LLE, liquid–liquid extraction; LOQ, limit of quantification; MS, mass spectrometry; MS/MS, tandem mass spectrometry; MSPD, matrix solid-phase extraction; n.a., not applied; PSA, primary secondary amine; PTFE, polytetrafluoroethylene; QuEChERS, quick, easy, cheap, effective, rugged, and safe; UHPLC, ultrahigh-performance liquid chromatography.

sorbents, and (3) graphitized or porous carbon. Gilbert-López et al. (2010) developed a method for the determination of 33 multiclass pesticides in fruit-based soft drinks. The proposed method consists of SPE using hydrophilic—lipophilic balanced polymer-based reverse-phase cartridges, followed by identification and quantification of the target pesticides by LC-TOF-MS. Fruit-based soft drinks spiked at 10 and 50 μg/L yielded average recoveries in the range of 66%—124% with RSD <14%. The obtained LOQs were from 0.02 to 2 μg/L.

Rodríguez-Cabo, Rodríguez, Ramil, Silva, and Cela (2016) evaluated for the first time the quantitative and screening possibilities of GC-HRMS for multiclass determination of semivolatile compounds in wine. SPE was used in combination with DLLME, with previous acetylation to phenolic compounds, for determination of 39 semivolatile compounds, including pesticides, related to wine quality by GC coupled to a QTOF MS system. LOQ values at μg/L were attained for pesticides and the combination of narrow-mass extraction windows, typically 5 mDa, with retention times data permits the high selective detection of target compounds.

5.3.5 SOLID-PHASE MICROEXTRACTION AND STIR BAR SORPTIVE EXTRACTION

Solid-phase microextraction (SPME) is a solvent-free sample preparation technique developed by Arthur and Pawliszyn (1990). SPME eliminates the need for using solvents and combines sampling, isolation, concentration, and enrichment in one step. Besides, it offers the benefit of short sample preparation step, solvent-free extraction, small sample volumes, and analyte concentration from solid, liquid, or gaseous samples. Its applications have been optimized and automated (Arthur, Killam, Buchholz, Pawliszyn, & Berg, 1992; Chai and Tan, 2009). The main advantage of SPME is the good analytical performance combined with simplicity and low cost, resulting in relatively clean and concentrated extracts, which is ideal for MS applications (Abdulra'uf, Sirhan, & Tan, 2012). SPME uses a very simple apparatus similar to a modified syringe, and the device comprises a fiber holder and a fiber assembly that contains a 1- to 2-cm-long retractable SPME fiber (Vas & Vékey, 2004). The SPME fiber consists in a thin fused silica fiber coated with a thin polymer film such as polydimethylsiloxane (PDMS), conventionally used as a coating material in GC.

The extraction process involves two basic steps: partition of the analytes between the sample matrix and the fiber coating followed by desorption of these analytes from the fiber before injection into analytical instrumentation (Abdulra'uf et al., 2012; Vas & Vékey, 2004). SPME can be used for extraction by direct immersion of the fiber into the sample (DI-SPME) or by suspending the fiber into the sample headspace (HS-SPME). According to Abdulra'uf et al. (2012), HS-SPME has advantage over DI-SPME because of shorter equilibrium time, the higher diffusion coefficients in gaseous phase, and higher concentration of the volatile analytes in the HS before the extraction, and also the variation of sample matrix properties has no effect on the fiber. Mmualefe, Torto, Huntsman-Mapila, and Mbongwe (2009) compared the peak areas of both SPME modes employing a 65 μm PDMS/

divinylbenzene (DVB) fiber, for pesticides extraction in water samples. As reported by the authors, HS-SPME was more sensitive than the DI mode because HS sampling eliminates competition for adsorption sites on the fiber coating by nonvolatile compounds present in the liquid sample. In both cases, after sampling, the fiber is retracted into the metal needle for mechanical protection and the next step is the transference of the analytes from the fiber into the chromatographic system. When the analysis is performed by GC, thermal desorption (TD) is used inserting the fiber directly into the injector. However, when the analysis is by LC, commercial interfaces for online desorption are available. A liquid desorption (LD) can also be applied as an alternative approach, using a few milliliters of an organic solvent and injecting the resulted extract into an LC or GC system (Abdulra'uf et al., 2012; Vas & Vékey, 2004).

More recently, a new system of SPME, denominated in-tube SPME, was developed using a capillary instead of an SPME fiber. In this system, the analytes are desorbed either by mobile phase flow or by aspirating a desorption solvent of choice from a second vial (Masiá, Moliner-Martinez, Munoz-Ortuno, Pico, & Campíns-Falcó, 2013). The most common capillaries have coatings similar to common commercially available SPME fibers, such as polyethylene glycol or polystyrene (PS)/DVB (Balasubramanian & Panigrahi, 2011). In-tube SPME is suitable for automation, which not only shortens analysis time but also often provides accuracy and precision relative to manual techniques (Vas & Vékey, 2004).

The SPME fiber has been widely used for the extraction of pesticide residues in different matrices. Usually, an optimization step is required when this technique is used, to evaluate parameters that may influence the analytes extraction such as fiber coatings, extraction temperature and time, ionic strength, and stirring. The PDMS/DVB fiber is the most used for fruits and vegetables analysis and has been employed in the extraction of pyrethroid pesticides from tomato and strawberry (Beltran, Peruga, Pitarch, & Lopez, 2003) and of 70 multiclass pesticides from cucumber, pepper, and tomato (Cortes-Aguado, Sanchez-Morito, Grenido, Vidal, & Arrebola, 2007).

Souza-Silva, Lopez-Avila, and Pawliszyn (2013) developed a fast and robust method for the determination of 10 triazole fungicides in grape and strawberry pulps using DI-SPME coupled to GC-TOF-MS. Potential factors affecting the extraction efficiency were investigated and optimized, including extraction temperature, sample pH, ionic strength, agitation speed, and extraction and desorption times. Under optimized conditions, the method was linear for over four orders of magnitude in concentration, with determination coefficients (r^2) greater than 0.99 for all test compounds in both matrices. The LOQ values ranged from 0.25 to 5 µg/kg, which are well below the maximum residue levels allowed for those compounds in both matrices. The proposed DI-SPME method demonstrates potential for the automated screening of triazole pesticides in fruit samples with no sample pretreatment needed.

Stir bar sorptive extraction (SBSE) was introduced in 1999 as a solventless sample preparation method for the extraction and enrichment of organic compounds from aqueous matrices (Baltussen, Sandra, David, & Cramers, 1999). SBSE emerged from the SPME technique, because it has been found that the inclusion of a polymeric phase in a conventional stir bar promotes a greater extraction capacity, giving a better enrichment factor when compared to SPME (Tan & Abdulra'uf, 2012). The enrichment factor for SBSE, which is determined by the analyte recovery in the extraction phase, is higher than that of SPME because of the 50−250 times larger volume of extraction phase on the stir bar (David & Sandra, 2007). Usually, in the SBSE a 10- to 40-mm-long magnetic stir bar coated with 50−300 µL of PDMS is used as the extracting phase (Kataoka, 2010).

The extraction mechanisms are very similar to SPME, where the analytes are adsorbed on a PDMS-coated stir bar by stirring the sample solution for a given time. The stir bar can then be submitted to a TD when a GC is used or to a LD with an organic solvent to be subsequently injected into an LC or GC systems (Prieto et al., 2010; Tan & Abdulra'uf, 2012). When a GC is used, the stir bar cannot be directly injected into the split/splitless injection port as in the SPME, and a thermal-desorption unit (TDU) is required. According to Tan and Abdulra'uf (2012), the need for a TDU or the extraction in a solvent before chromatographic analysis is one of the limitations of the technique. However, after the extraction by SBSE, the analytes can be introduced quantitatively into the analytical instrumentation by TD or LD.

The available studies demonstrate that the SBSE technique provides good recovery results and extremely low LOD values, mainly for analytes with hydrophobic characteristics. Stir bar can also be frozen after extraction, given more flexibility to routine laboratories (Margoum, Guillemain, Yang, & Coquery, 2013). Ochiai et al. (2011) developed an SBSE method for determination of ultratrace amounts of OCPs in water using 20-mm-long stir bars coated with a 0.5 mm layer (47 µL) of PDMS and an extraction time of 3 h. The stir bar was thermally desorbed and the compounds were analyzed by GC × GC-TOF-MS, achieving very low LOD values in the range of 10−44 pg/L.

5.3.6 MICROWAVE-ASSISTED EXTRACTION

Microwave-assisted extraction (MAE) was introduced by Ganzler, Salgó, and Valkó (1986) and has been successfully applied for extraction of organic compounds from various solid and liquid matrices (Fang, Lau, Law, Fong, & Li, 2012; Haroune et al., 2015; Wu et al., 2014). MAE extraction is based on the changes in the solute cell structure caused by electromagnetic waves. The acceleration and high extraction yield may be the result of a synergistic combination of heat and mass gradients (Chemat, Abert-Vian, & Zill-e-Huma, 2009). Generally, MAE devices comprise a closed extraction vessel under controlled pressure and temperature or a focused microwave oven at atmospheric pressure. These two technologies are commonly named pressurized MAE or focused MAE, respectively (Wang, Ding, & Ren, 2016). MAE has many advantages over other classical extraction techniques such as reduction of extraction time and solvent consumption as well as the possibility of running multiple samples (Merdassa, Liu, & Megersa, 2013). However, the instrumentation required is relatively expensive. Besides, when MAE is used, the analysts have to deal with thermal degradation for some labile pesticides (Smalling and Kuivila, 2008).

The potential of HRMS to confirm and quantify pesticides at trace levels, allied to MAE efficiency, has not yet well explored. Coscollà, León, Pastor, and Yusà (2014) and López, Yusà, Millet, and Coscollà (2016) used the MAE method developed by Coscollà, Yusà, Beser, and Pastor (2009), as a generic extraction method, aiming to achieve the highest number of compounds. In both works, MAE was allied to UHPLC-Orbitrap-MS for the determination of several pesticides and metabolites in ambient air using quantitative target analysis, postrun target screening analysis, and nontarget screening strategies.

5.3.7 PRESSURIZED LIQUID EXTRACTION

Pressurized liquid extraction (PLE) is a technique performed to extract solid or semisolid samples employing liquid solvents at elevated temperatures (70−200°C) and pressures (1500−3000 psi). The elevated temperature enhances the solubility of the analytes, breaking matrix−analyte interactions and achieving a higher diffusion rate, while elevated pressure keeps the solvent below its boiling point (Vazquez-Roig & Picó, 2015). Similar to MAE, PLE procedures are performed rapidly with reduced solvent use when compared with traditional extraction techniques. However, an expensive lab equipment is required (Marazuela & Bogialli, 2009). PLE can be performed in static, dynamic, or a combination of both these modes. Usually, samples are dried (e.g., lyophilization, oven desiccation, air dried, or reduction of water using sorbents) before the extraction procedure, which increases the solvent penetration into the sample matrix providing high extraction efficiency (Vazquez-Roig & Picó, 2015). PLE being an official US Environmental Protection Agency method for persistent organic pollutants (POPs) in many environmental solid samples is already a routine technique for trace analysis of organic contaminants (Aguilar, Williams, Brooks, & Usenko, 2014).

The main parameters that influence PLE extraction efficiency are temperature, extraction time, flush volume, number of extraction cycles, sorbent type, and solvents (Vazquez-Roig & Picó, 2015). These parameters, as in other extractions techniques, are generally optimized to gain acceptable recoveries for most analytes in a multiresidue method. The extraction procedure by PLE is usually conducted using a mixture between the sample with a dispersing/drying agent to reduce particle clumping and solvent channeling, resulting in higher extraction efficiency. The most common dispensing agents applied for pesticides analysis are Hydromatrix, an inert diatomaceous earth that absorbs 2 mL of water per gram, sodium sulfate, and sea sand (Vazquez-Roig & Picó, 2015).

Many works reported the use of several sorbents in the extraction cell, which provides an integration of extraction and in-cell cleanup in a single step, known as selective PLE (SPLE) (Canosa et al., 2007; Rodrigues, Pardal, Salgueiro-González, Muniategui-Lorenzo, & Alpendurada, 2016). Most of the published works that employ PLE extraction for pesticides determinations have low resolution as detection system. However, PLE has been applied in combination to HRMS in some specific applications, mainly for the determination of POPs (Do, Lundstedt, & Haglund, 2013). Soler et al. (2007) successfully used PLE extraction and LC-QTOF-MS in positive ion mode for the determination of carbofuran, 3-hydroxycarbofuran, 3-ketocarbofuran, 3-hydroxy-7-phenol carbofuran, 3-keto-7-phenolcarbofuran, 7-phenolcarbofuran, and dibutylamine in food matrices. PLE recoveries ranged from 55% to 94% with LOQs from 10 (for carbosulfan, carbofuran, 3-hydroxycarbofuran, and dibutylamine) to 70 µg/kg (3-keto-7-phenolcarbofuran). The method was used to monitor the presence and fate of the target compounds in orange, potato, and rice crops treated with a commercial product containing carbosulfan. Table 5.2 shows SPE, microextraction solid phase extraction, and PLE applications for pesticide residues determination in food matrices by chromatographic techniques coupled to HRMS.

5.4 PERSPECTIVES AND CONCLUSIONS

Over the past few years much progress has been made in the pesticide residue analysis field. In this time, hyphenated techniques promoted great efficient separation and highly sensitive detection. The methods based on chromatographic separation with MS detection resulted in a powerful tool during identification and quantification of pesticide residues. However, even with powerful HRMS, the risk of interference increases with the complexity of the matrix studied, so an adequate sample preparation before instrumental analysis is still required in many food analysis applications. For the development of new methods one interesting issue is the simultaneous determination of several classes of residues and contaminants in a single run. In this sense, UHPLC has allowed comprehensive analyses of complex matrices in short running times. Moreover, the development of generic extraction methods, using sample preparation procedures such as QuEChERS and "DS," combined with chromatographic techniques coupled to HRMS, permits the determination of a wide number of residues and contaminants in complex matrices.

Nowadays, ambient mass spectrometry becomes a future trend with swiftness and simplicity as advantages. In these techniques samples are analyzed at room temperature and atmospheric pressure. They present an alternative technique with high salt and matrix tolerances, shortcutting many steps and therefore allowing high-throughput analysis (Shiea et al., 2015). Different ambient mass spectrometry techniques including desorption electrospray ionization, direct analysis in real time, electrospray laser desorption ionization, and low-temperature plasma probe coupled to HRMS have been used for pesticides residue determination.

As discussed by Li et al. (2013) different efforts have been made to solve some drawbacks related to validation steps and practical applications of these techniques. These efforts aim to improve the throughput and dissemination of ambient mass spectrometry. Other perspective is related to the association of microextraction

Table 5.2 Different Extraction Procedures Applied for Pesticide Residues Determination in Food Matrices by Chromatographic Techniques Coupled to High-Resolution Mass Spectrometry

Matrix	Compounds	Extraction Procedure	Detection Mode	Resolution (FWHM)	Sensitivity	Mass Accuracy	References
Fruit-based soft drinks	33 multiclass pesticides	SPE: samples were passed through the polymer-based hydrophilic–lipophilic SPE cartridges and the analytes were eluted with methanol. This elute was then evaporated and redissolved with methanol:water, filtered through a PTFE filter, and analyzed	LC-TOF-MS	9700	LOQ <2 µg/L	<3 ppm	Gilbert-López et al. (2010)
Wine	Multiclass semivolatile compounds	SPE: samples were passed through an OASIS MAX sorbent aiming to remove very polar and acidic compounds and methanol was used for elution. A subsequent step was conducted using DLLME	GC-TOF-MS	4900–7500	LOQ <1.4 µg/L	<5 ppm	Rodríguez-Cabo et al. (2016)

Fruits	Triazole fungicides	DI-SPME: 9 g of whole fruit pulp was weighed in a 10 mL vial. An SPME PDMS/DVB fiber was used with 5 min preextraction incubation with agitation at 500 rpm and 50 C, followed by a 15 min extraction in DI mode. After, the fiber was rinsed in water, and then desorbed in the GC system	GC-TOF-MS	n.a.	LOQ <5 µg/kg	n.a.	Souza-Silva et al. (2013)
Food matrices	Carbofuran (CF), 3-hydroxy-CF, 3-keto-CF, 3-hydroxy-7-phenol-CF, 3-keto-7-phenol-CF, 7-phenol-CF, dibutylamine	PLE: 5 g of sample was packed into a 22 mL stainless steel vessel with anhydrous sodium sulfate. Extraction conditions: dichloromethane at 40°C and 2000 psi; heating time of 2 min with two cycles, 5 min each static extraction (40 mL). The extract was concentrated to 10 mL in a rotary evaporator and evaporated to dryness. Sample was redissolved in 1 mL of methanol and analyzed	LC-QTOF-MS ESI(+)	n.a.	LOQ <70 µg/kg	<20 ppm	Soler et al. (2007)

DI-SPME, *direct immersion-solid-phase microextraction*; DLLME, *dispersive liquid–liquid extraction*; ESI, *electrospray ionization*; FWHM, *full width at half maximum*; GC–TOF-MS, *gas chromatography–time-of-flight-mass spectrometry*; LC–TOF-MS, *liquid chromatography–time-of-flight-mass spectrometry*; LOQ, *limit of quantification*; n.a., *not applied*; PDMS/DVB, *polydimethylsiloxane/divinylbenzene*; PLE, *pressurized liquid extraction*; PTFE, *polytetrafluoroethylene*; SPE, *solid-phase extraction*.

techniques based on solventless and solvent-minimized sample preparation techniques combined with GC and LC coupled to HRMS. This is a powerful option for the development of rapid, selective, and ultrasensitive analytical methods. Currently, the most common sample preparation techniques for pesticide residue analysis still focused on chromatographic methods coupled to low-resolution mass spectrometry. However, HRMS applications are increasingly being used as a promising technique in high-throughput method development for the pesticide residues control.

REFERENCES

Abdulra'uf, L. B., Sirhan, A. H., & Tan, G. H. (2012). Recent developments and applications of liquid phase microextraction in fruits and vegetables analysis. *Journal of Separation Science, 35*, 3540–3553.

Aguilar, L., Williams, E. S., Brooks, B. W., & Usenko, S. (2014). Development and application of a novel method for high-throughput determination of PCDD/Fs and PCBs in sediments. *Environmental Toxicology Chemistry, 33*, 1529–1536.

Alder, L., Lüderitza, S., Lindtnera, K., & Jürgen Stan, H. (2004). The ECHO technique — the more effective way of data evaluation in liquid chromatography–tandem mass spectrometry analysis. *Journal of Chromatography A, 1058*, 67–79.

Anastassiades, M., Lehotay, S. J., Stajnbaher, D., & Schenck, F. J. (2003). Fast and easy multiresidue method employing acetonitrile extraction/partitioning and "dispersive solid-phase extraction" for the determination of pesticide residues in produce. *Journal of AOAC International, 86*, 412–431.

Anastassiades, M., Scherbaum, E., Tasdelen, B., & Stajnbaher, D. (2007). *Crop protection, public health, environmental safety.* Weinheim, Germany: Wiley-VCH.

Arthur, C. L., Killam, L. M., Buchholz, K. D., Pawliszyn, J., & Berg, J. R. (1992). Automation and optimization of solid-phase microextraction. *Analytical Chemistry, 64*, 1960–1966.

Arthur, C. L., & Pawliszyn, J. (1990). Solid phase microextraction with thermal desorption using fused silica optical fibers. *Analytical Chemistry, 62*, 2145–2148.

Balasubramanian, S., & Panigrahi, S. (2011). Solid-phase microextraction (SPME) techniques for quality characterization of food products: A review. *Food Bioprocess Technology, 4*, 1–26.

Baltussen, E., Sandra, P., David, F., & Cramers, C. (1999). Stir bar sorptive extraction (SBSE), a novel extraction technique for aqueous samples: Theory and principles. *Journal of Microcolumn Separations, 11*, 737–747.

Barker, S. A. (2007). Matrix solid phase dispersion (MSPD). *Journal of Biochemistry and Biophysics Methods, 70*, 151–162.

Barker, S. A., Long, A. R., & Short, C. R. (1989). Isolation of drug residues from tissues by solid phase dispersion. *Journal of Chromatography A, 475*, 353–361.

Beltran, J., Peruga, A., Pitarch, E., & Lopez, F. J. (2003). Application of solid-phase microextraction for the determination of pyrethroid residues in vegetable samples by GC-MS. *Analytical and Bioanalytical Chemistry, 376*, 502–511.

Berendsen, B. J. A., Stolker, L. A. A. M., & Nielen, M. W. F. (2013). Selectivity in the sample preparation for the analysis of drug residues in products of animal origin using LC-MS. *TrAC Trends in Analytical Chemistry, 43*, 229–239.

Cabrera, L. D. C., Caldas, S. S., Prestes, O. D., Primel, E. G., & Zanella, R. (2016). Evaluation of alternative sorbents for dispersive solid-phase extraction clean-up in the QuEChERS method for the determination of pesticide residues in rice by liquid chromatography with tandem mass spectrometry. *Journal of Separation Science, 39,* 1945–1954.

Cajka, T., Sandy, C., Bachanova, V., Drabova, L., Kalachova, K., Pulkrabova, J., et al. (2012). Streamlining sample preparation and gas chromatography–tandem mass spectrometry analysis of multiple pesticide residues in tea. *Analytica Chimica Acta, 743,* 51–60.

Caldas, S. S., Bolzan, C. M., de Menezes, E. J., Escarrone, A. L. V., Martinez, C. G. M., Bianchini, A., et al. (2013). A vortex-assisted MSPD method for the extraction of pesticide residues from fish liver and crab hepatopancreas with determination by GC-MS. *Talanta, 112,* 63–68.

Canosa, P., Perez-Palacios, D., Garrido-Lopez, A., Tena, M. T., Rodríguez, I., Rubí, E., et al. (2007). Pressurized liquid extraction with in-cell clean-up followed by gas chromatography–tandem mass spectrometry for the selective determination of parabens and triclosan in indoor dust. *Journal of Chromatography A, 1161,* 105–112.

Cao, Y., Tang, H., Chen, D., & Li, L. (2015). A novel method based on MSPD for simultaneous determination of 16 pesticide residues in tea by LC–MS/MS. *Journal of Chromatography B, 998–999,* 72–79.

Capriotti, A. L., Cavaliere, C., Foglia, P., Samperi, R., Stampachiacchiere, S., Ventura, S., et al. (2015). Recent advances and developments in matrix solid-phase dispersion. *TrAC Trends in Analytical Chemistry, 71,* 186–193.

Cerqueira, M. B. R., Caldas, S. S., & Primel, E. G. (2014). New sorbent in the dispersive solid phase extraction step of quick, easy, cheap, effective, rugged, and safe for the extraction of organic contaminants in drinking water treatment sludge. *Journal of Chromatography A, 1336,* 10–22.

Chai, M. K., & Tan, G. H. (2009). Validation of a headspace solid-phase microextraction procedure with gas chromatography-electron capture detection of pesticide residues in fruits and vegetables. *Food Chemistry, 117,* 561–567.

Chemat, F., Abert-Vian, M., & Zill-e-Huma, Y.-J. (2009). Microwave assisted separations: Green chemistry in action. In J. T. Pearlman (Ed.), *Green chemistry research trends.* New York: Nova Science Publishers.

Cheng, Z., Dong, F., Xu, J., Liu, X., Wu, X., Chen, Z., et al. (2016). Atmospheric pressure gas chromatography quadrupole-time-of-flight mass spectrometry for simultaneous determination of fifteen organochlorine pesticides in soil and water. *Journal of Chromatography A, 1435,* 115–124.

Commission Regulation (EC). (2005). No 396/2005 of the European parliament and of the council of 23 February 2005 on maximum residue levels of pesticides in or on food and feed of plant and animal origin and amending Council Directive 91/414/EEC. *Official Journal of the European Union, 70,* 1–16.

Commission Regulation (EC). (2009). No 1107/2009 of the European parliament and of the council of 21 October 2009 concerning the placing of plant protection products on the market and repealing Council Directives 79/117/EEC and 91/414/EEC. *Official Journal of the European Union, 309,* 1–50.

Cortes-Aguado, S., Sanchez-Morito, N., Grenido Frenich, A., Vidal, J. L. M., & Arrebola, L. F. J. (2007). Screening method for the determination at parts per trillion levels of pesticide residues in vegetables combining solid-phase microextraction and gas chromatography-tandem mass spectrometry. *Analytical Letters, 40,* 2886–2914.

Coscollà, C., León, N., Pastor, A., & Yusà, V. (2014). Combined target and post-run target strategy for a comprehensive analysis of pesticides in ambient air using liquid chromatography-Orbitrap high resolution mass spectrometry. *Journal of Chromatography A, 1368*, 132–142.

Coscollà, C., Yusà, V., Beser, M. I., & Pastor, A. (2009). Multi-residue analysis of 30 currently used pesticides in fine airborne particulate matter (PM 2.5) by microwave-assisted extraction and liquid chromatography–tandem mass spectrometry. *Journal of Chromatography A, 1216*, 8817–8827.

Croley, T. R., White, K. D., Callahan, J. H., & Musser, S. M. (2012). The chromatographic role in high resolution mass spectrometry for non-targeted analysis. *Journal of the American Society for Mass Spectrometry, 23*, 1569–1578.

Cunha, S. C., & Fernandes, J. O. (2013). Assessment of bisphenol A and bisphenol B in canned vegetables and fruits by gas chromatography–mass spectrometry after QuEChERS and dispersive liquid–liquid microextraction. *Food Control, 33*, 549–555.

David, F., & Sandra, P. (2007). Stir bar sorptive extraction for trace analysis. *Journal of Chromatography A, 1152*, 54–69.

Do, L., Lundstedt, S., & Haglund, P. (2013). Optimization of selective pressurized liquid extraction for extraction and in-cell clean-up of PCDD/Fs in soils and sediments. *Chemosphere, 90*, 2414–2419.

Donato, F. F., Kemmerich, M., Facco, J. F., Friggi, C. A., Prestes, O. D., Adaime, M. B., et al. (2012). Simultaneous determination of pesticide and antibiotic residues at trace levels in water samples by SPE and LC-MS/MS. *Brazilian Journal of Analytical Chemistry, 2*, 331–340.

Dujaković, N., Grujić, S., Radišić, M., Vasiljević, T., & Laušević, M. (2010). Determination of pesticides in surface and ground waters by liquid chromatography-electrospray-tandem mass spectrometry. *Analytica Chimica Acta, 678*, 63–72.

Dzuman, Z., Zachariasova, M., Veprikova, Z., Godula, M., & Hajslova, J. (2015). Multi-analyte high performance liquid chromatography coupled to high resolution tandem mass spectrometry method for control of pesticide residues, mycotoxins, and pyrrolizidine alkaloids. *Analytica Chimica Acta, 863*, 29–40.

Erney, D. R., Gillespie, A. M., Gilvydis, D. M., & Poole, C. F. (1993). Explanation of the matrix-induced chromatographic response enhancement of organophosphorus pesticides during open tubular column gas chromatography with splitless or hot on-column injection and flame photometric detection. *Journal of Chromatography A, 638*, 57–63.

European Commission (EC) 2015. Guidance document on analytical quality control and method validation procedures for pesticides residues analysis in food and feed SANTE/11945/2015

Fang, G., Lau, H. F., Law, W. S., Fong, S., & Li, Y. (2012). Systematic optimisation of coupled microwave-assisted extraction-solid phase extraction for the determination of pesticides in infant milk formula via LC–MS/MS. *Food Chemistry, 134*, 2473–2480.

Farré, M., & Barceló, D. (2013). Analysis of emerging contaminants in food. *TrAC Trends in Analytical Chemistry, 43*, 240–253.

Fernandes, V. C., Domingues, V. F., Mateus, N., & Delerue-Matos, C. (2013). Multiresidue pesticides analysis in soils using modified QuEChERS with disposable pipette extraction and dispersive solid-phase extraction. *Journal of Separation Science, 36*, 376–382.

Fernández-Alba, A. R. (Ed.). (2004). *Comprehensive analytical chemistry* (Vol. XLIII). Amsterdam: Elsevier.

Ferrer, C., Lozano, A., Agüera, A., Girón, A. J., & Fernández-Alba, A. R. (2011). Overcoming matrix effects using the dilution approach in multiresidue methods for fruits and vegetables. *Journal of Chromatography A, 1218*, 7634–7639.

Gan, J., Lv, L., Peng, J., Li, J., Xiong, Z., Chen, D., et al. (2016). Multi-residue method for the determination of organofluorine pesticides in fish tissue by liquid chromatography triple quadrupole tandem mass spectrometry. *Food Chemistry, 207*, 195–204.

Ganzler, K., Salgó, A., & Valkó, K. (1986). Microwave extraction: A novel sample preparation method for chromatography. *Journal of Chromatography A, 371*, 299–306.

García-Reyes, J. F., Ferrer, C., Thurman, E. M., Fernandez-Alba, A. R., & Ferrer, I. (2006). Analysis of herbicides in olive oil by liquid chromatography time-of-flight mass spectrometry. *Journal of Agriculture and Food Chemistry, 54*, 6493–6500.

Garrido-Frenich, A., Romero-González, R., & Aguilera-Luiz, M. M. (2014). Comprehensive analysis of toxics (pesticides, veterinary drugs and mycotoxins) in food by UHPLC-MS. *TrAC Trends in Analytical Chemistry, 63*, 158–169.

General Inspectorate for Health Protection. (1996). *Analytical methods for pesticide residues in foodstuffs* (6th ed.) Part 1, The Hague, The Netherlands.

Gilbert-López, B., García-Reyes, J. F., Mezcua, M., Ramos-Martos, N., Fernández-Alba, A. R., & Molina-Díaz, A. (2010). Multi-residue determination of pesticides in fruit-based soft drinks by fast liquid chromatography time-of-flight mass spectrometry. *Talanta, 81*, 1310–1321.

Gillespie, A. M., Daly, S. L., Gilvydis, D. M., Schneider, F., & Walters, S. M. (1995). Multicolumn solid-phase extraction cleanup of organophosphorus and organochlorine pesticide residues in vegetable oils and butterfat. *Journal of AOAC International, 78*, 431–437.

Gómez-Pérez, M. L., Plaza-Bolaños, P., Romero-González, R., Martínez-Vidal, J. L., & Garrido-Frenich, A. (2012). Comprehensive qualitative and quantitative determination of pesticides and veterinary drugs in honey using liquid chromatography-Orbitrap high resolution mass spectrometry. *Journal of Chromatography A, 1248*, 130–138.

Gómez-Ramos, M. M., Ferrer, C., Malato, O., Agüera, A., & Fernández-Alba, A. R. (2013). Liquid chromatography-high-resolution mass spectrometry for pesticide residue analysis in fruit and vegetables: Screening and quantitative. *Journal of Chromatography A, 1287*, 24–37.

Gómez-Ramos, M. M., Rajski, Ł., Lozano, A., & Fernández-Alba, A. R. (2016). *Analytical Methods.* http://dx.doi.org/10.1039/C6AY00436A (online).

Gosetti, F., Mazzucco, E., Zampieri, D., & Gennaro, M. C. (2010). Signal suppression/enhancement in high-performance liquid chromatography tandem mass spectrometry. *Journal of Chromatography A, 1217*, 3929–3939.

Greulich, K., & Alder, L. (2008). Fast multiresidue screening of 300 pesticides in water for human consumption by LC-MS/MS. *Analytical and Bioanalytical Chemistry, 391*, 183–197.

Grimalt, S., & Dehouck, P. (2016). Review of analytical methods for the determination of pesticide residues in grapes. *Journal of Chromatography A, 1433*, 1–23.

Guo, C., Wang, M., Hui, X., Binbin, H., Wang, F., Pan, G., et al. (2016). Development of a modified QuEChERS method for the determination of veterinary antibiotics in swine manure by liquid chromatography tandem mass spectrometry. *Journal of Chromatography B, 1027*, 110–118.

Haroune, L., Cassoulet, R., Lafontaine, M., Belisle, M., Garant, D., Pelletier, F., et al. (2015). Liquid chromatography-tandem mass spectrometry determination for multiclass pesticides from insect samples by microwave-assisted solvent extraction followed by a salt-out effect and micro-dispersion purification. *Analytica Chimica Acta, 891*, 160–170.

He, Z., Xu, Y., Wang, L., Peng, Y., Luo, M., Cheng, H., et al. (2016). Wide-scope screening and quantification of 50 pesticides in wine by liquid chromatography/quadrupole time-of-flight mass spectrometry combined with liquid chromatography/quadrupole linear ion trap mass spectrometry. *Food Chemistry, 196*, 1248–1255.

Hercegová, A., Domotorova, M., & Matisova, E. (2007). Sample preparation methods in the analysis of pesticide residues in baby food with subsequent chromatographic determination. *Journal of Chromatography A, 1153*, 54–73.

Hernández, F., Grimalt, S., Pozo, O., & Sancho, J. (2009). Use of ultra-high-pressure liquid chromatography-quadrupole time-of-flight MS to discover the presence of pesticide metabolites in food samples. *Journal of Separation Science, 32*, 2245–2261.

Jamin, E., Bonvallot, N., Tremblay-Franco, M., Cravedi, J., Chevrier, C., Cordier, S., et al. (2013). Untargeted profiling of pesticide metabolites by LC-HRMS: An exposomics tool for human exposure evaluation. *Analytical and Bioanalytical Chemistry, 406*, 1149–1161.

Jedziniak, P., Szprengier-Juszkiewicz, T., & Olejnik, M. (2009). Determination of benzimid-azoles and levamisole residues in milk by liquid chromatography—mass spectrometry: Screening method development and validation. *Journal of Chromatography A, 1216*, 8165–8172.

Jia, W., Chu, X., Ling, Y., Huang, J., Lin, Y., & Chang, J. (2014). Simultaneous determination of dyes in wines by HPLC coupled to quadrupole orbitrap mass spectrometry. *Journal of Separation Science, 37*, 782–791.

Jiao, W., Xiao, Y., Qian, X., Tong, M., Hu, Y., Hou, R., et al. (2016). Optimized combination of dilution and refined QuEChERS to overcome matrix effects of six types of tea for determination eight neonicotinoid insecticides by ultra performance liquid chromatography-electrospray tandem mass spectrometry. *Food Chemistry, 210*, 26–34.

Kang, J., Park, H. C., Gedi, V., Park, S. J., Kim, M. A., & Kim, M. K. (2015). Veterinary drug residues in domestic and imported foods of animal origin in the Republic of Korea. *Journal Food Additives and Contaminants: Part B, 8*, 106–112.

Kataoka, H. (2010). Recent developments and applications of microextraction techniques in drug analysis. *Analitycal Bioanalytical Chemistry, 396*, 339–364.

Kaufmann, A. (2014). Combining UHPLC and high-resolution MS: A viable approach for the analysis of complex samples? *TrAC Trends in Analytical Chemistry, 63*, 113–128.

Kemmerich, M., Rizzetti, T. M., Martins, M. L., Prestes, O. D., Adaime, M. B., & Zanella, R. (2015). Optimization by central composite design of a modified QuEChERS method for extraction of pesticide multiresidue in sweet pepper and analysis by ultra-high-performance liquid chromatography—tandem mass spectrometry. *Food Analytical Methods, 8*, 728–739.

Kittlaus, S., Schimanke, J., Kempe, G., & Speer, K. (2011). Assessment of sample cleanup and matrix effects in the pesticide residue analysis of foods using postcolumn infusion in liquid chromatography—tandem mass spectrometry. *Journal of Chromatography A, 1218*, 8399–8410.

Knolhoff, A. M., & Croley, T. R. (2016). Non-targeted screening approaches for contaminants and adulterants in food using liquid chromatography hyphenated to high resolution mass spectrometry. *Journal of Chromatography A, 1428*, 86–96.

Koesukwiwat, U., Lehotay, S. J., Miao, S., & Leepipatpiboon, N. (2010). High throughput analysis of 150 pesticides in fruits and vegetables using QuEChERS and low-pressure gas chromatography—time-of-flight mass spectrometry. *Journal of Chromatography A, 1217*, 6692—6703.

Lacina, O., Urbanova, J., Poustka, J., & Hajslova, J. (2010). Identification/quantification of multiple pesticide residues in food plants by ultra-high-performance liquid chromatography-time-of-flight mass spectrometry. *Journal of Chromatography A, 1217*, 648—659.

Lehotay, S. J., de Kok, A., Hiemstra, M., & Van Bodegraven, P. (2005). Validation of a fast and easy method for the determination of residues from 229 pesticides in fruits and vegetables using gas and liquid chromatography and mass spectrometric detection. *Journal of AOAC International, 88*, 595—614.

Lehotay, S. J., Sapozhnikova, Y., & Mol, H. G. J. (2015). Current issues involving screening and identification of chemical contaminants in foods by mass spectrometry. *TrAC Trends in Analytical Chemistry, 69*, 62—75.

Li, L. P., Feng, B. S., Yang, J. W., Chang, C. L., Bai, Y., & Liu, H. W. (2013). Applications of ambient mass spectrometry in high-throughput screening. *Analyst, 138*, 3097—3103.

López, A., Yusà, V., Millet, M., & Coscollà, C. (2016). Retrospective screening of pesticide metabolites in ambient air using liquid chromatography coupled to high-resolution mass spectrometry. *Talanta, 150*, 27—36.

Lozano, A., Kiedrowska, B., Scholten, J., de Kroon, M., de Kok, A., & Fernández-Alba, A. R. (2016). Miniaturisation and optimisation of the Dutch mini-Luke extraction method for implementation in the routine multi-residue analysis of pesticides in fruits and vegetables. *Food Chemistry, 192*, 668—681.

Lozano, A., Martínez-Uroz, M. A., Gómez-Ramos, M. J., Gómez-Ramos, M. M., Mezcua, M., & Fernández-Alba, A. R. (2012). Determination of nicotine in mushrooms by various GC/MS- and LC/MS-based methods. *Analytical and Bioanalytical Chemistry, 402*, 935—943.

Luke, M. A., Froberg, J. E., & Masumoto, H. T. (1975). Extraction and cleanup of organochlorine, organophosphate, organonitrogen, and hydrocarbon pesticides in produce for determination by gas-liquid chromatography. *Journal of the Association of Official Analytical Chemists, 58*, 1020—1026.

Luo, Y., Zheng, H., Jiang, X., Li, X., Zhang, H., Zhu, Z., et al. (2015). Determination of pesticide residues in tobacco using modified QuEChERS procedure coupled to on-line gel permeation chromatography-gas chromatography/tandem mass spectrometry. *Chinese Journal of Analytical Chemistry, 43*, 1538—1544.

Marazuela, M. D., & Bogialli, S. (2009). A review of novel strategies of sample preparation for the determination of antibacterial residues in foodstuffs using liquid chromatography-based analytical methods. *Analytica Chimica Acta, 645*, 5—17.

Margoum, C., Guillemain, C., Yang, X., & Coquery, M. (2013). Stir bar sorptive extraction coupled to liquid chromatography-tandem mass spectrometry for the determination of pesticides in water samples: Method validation and measurement uncertainty. *Talanta, 116*, 1—7.

Martínez-Domínguez, G., Romero-González, R., Arrebola, F. J., & Garrido-Frenich, A. (2016). Multi-class determination of pesticides and mycotoxins in isoflavones supplements obtained from soy by liquid chromatography coupled to Orbitrap high resolution mass spectrometry. *Food Control, 59*, 218—224.

Martínez-Domínguez, G., Romero-González, R., & Garrido-Frenich, A. (2015). Determination of toxic substances, pesticides and mycotoxins, in ginkgo biloba nutraceutical products by liquid chromatography Orbitrap-mass spectrometry. *Microchemical Journal, 118*, 124–130.

Martínez-Domínguez, G., Romero-González, R., & Garrido-Frenich, A. (2016). Multi-class methodology to determine pesticides and mycotoxins in green tea and royal jelly supplements by liquid chromatography coupled to Orbitrap high resolution mass spectrometry. *Food Chemistry, 197*, 907–915.

Martínez-Vidal, J. L., Garrido-Frenich, A., Aguilera-Luiz, M. M., & Romero-González, R. (2010). Development of fast screening methods for the analysis of veterinary drug residues in milk by liquid chromatography-triple quadrupole mass spectrometry. *Analytical and Bioanalytical Chemistry, 397*, 2777–2790.

Masiá, A., Moliner-Martinez, Y., Munoz-Ortuno, M., Pico, Y., & Campíns-Falcó, P. (2013). Multiresidue analysis of organic pollutants by in-tube solid phase microextraction coupled to ultra-high performance liquid chromatography – Electrospray-tandem mass spectrometry. *Journal of Chromatography A, 1306*, 1–11.

Maštovská, K., & Lehotay, S. J. (2004). Evaluation of common organic solvents for gas chromatographic analysis and stability of multiclass pesticide residues. *Journal of Chromatography A, 1040*, 259–272.

Meire, R. O., Khairy, M., Targino, A. C., Galvão, P. M. A., Torres, J. P. M., Malm, O., et al. (2016). Use of passive samplers to detect organochlorine pesticides in air and water at wetland mountain region sites (S-SE Brazil). *Chemosphere, 144*, 2175–2182.

Merdassa, Y., Liu, J., & Megersa, N. (2013). Development of a one-step microwave-assisted extraction method for simultaneous determination of organophosphorus pesticides and fungicides in soils by gas chromatography – Mass spectrometry. *Talanta, 114*, 227–234.

Miao, Q., Kong, W., Yang, S., & Yang, M. (2013). Comparison of sample preparation methods combined with gas chromatography with electron-capture detection for the analysis of multipesticide residues in lotus seeds. *Journal of Separation Science, 36*, 2010–2019.

Mills, P. A., & Onley, J. H. (1963). Rapid method for chlorinated pesticide residues in nonfatty foods. *Journal Association of Official Analytical Chemists, 46*, 186–191.

Mmualefe, L. C., Torto, N., Huntsman-Mapila, P., & Mbongwe, B. (2009). Headspace solid phase microextraction in the determination of pesticides in water samples from the Okavango Delta with gas chromatography-electron capture detection and time-of-flight mass spectrometry. *Microchemical Journal, 91*, 239–244.

Mol, H. G., van Dam, R. C., & Steijger, O. M. (2003). Determination of polar organophosphorus pesticides in vegetables and fruits using liquid chromatography with tandem mass spectrometry: Selection of extraction solvent. *Journal of Chromatography A, 1015*, 119–127.

Mol, H. G., Plaza-Bolaños, P., Zomer, P., De Rijk, T. C., Stolker, A. A., & Mulder, P. P. (2008). Toward a generic extraction method for simultaneous determination of pesticides, mycotoxins, plant toxins, and veterinary drugs in feed and food matrixes. *Analytical Chemistry, 80*, 9450–9459.

Munaretto, J. S., May, M. M., Saibt, N., & Zanella, R. (2016). Liquid chromatography with high resolution mass spectrometry for identification of organic contaminants in fish fillet: Screening and quantification assessment using two scan modes for data acquisition. *Journal of Chromatography A, 1456*, 205–216.

Naksen, W., Prapamontol, T., Mangklabruks, A., Chantara, S., Thavornyutikarn, P., Robson, M., et al. (2016). A single method for detecting 11 organophosphate pesticides in human plasma and breastmilk using GC-FPD. *Journal of Chromatography B, 1025*, 92–104.

Ochiai, N., Teruyo, I., Sasamoto, K., Yoshikatsu, T., Hashimoto, S., Fushimi, A., et al. (2011). Stir bar sorptive extraction and comprehensive two-dimensional gas chromatography coupled to high-resolution time-of-flight mass spectrometry for ultra-trace analysis of organochlorine pesticides in river water. *Journal of Chromatography A, 1218,* 6851−6860.

Payá, P., Anastassiades, M., Mack, D., Sigalova, I., Tasdelen, B., Oliva, J., et al. (2007). Analysis of pesticide residues using the Quick Easy Cheap Effective Rugged and Safe (QuEChERS) pesticide multiresidue method in combination with gas and liquid chromatography and tandem mass spectrometric detection. *Analytical and Bioanalytical Chemistry, 389,* 1697−1714.

Pérez-Ortega, P., Lara-Ortega, F. J., García-Reyes, J. F., Beneito-Cambra, M., Gilbert-López, B., Martos, N. R., et al. (2016). Determination of over 350 multiclass pesticides in Jams by ultra-high performance liquid chromatography time-of-flight mass spectrometry (UHPLC-TOFMS). *Food Analytical Methods.* http://dx.doi.org/10.1007/s12161-015-0369-2 (online).

Peters, R. J. B., Bolck, Y. J. C., Rutgers, P., Stolker, A. A. M., & Nielen, M. W. F. (2009). Multi-residue screening of veterinary drugs in egg, fish and meat using high-resolution liquid chromatography accurate mass time-of-flight mass spectrometry. *Journal of Chromatography A, 1216,* 8206−8216.

Picó, Y. (2015). Advanced sample preparation techniques for MS analysis. In D. Tsipi, H. Botitsi, & A. Economou (Eds.), *Mass spectrometry for the analysis of pesticides residues and their metabolites.* Hoboken: John Wiley & Sons, Inc.

Pihlström, T., Blomkvist, G., Friman, P., Pagard, U., & Osterdahl, B. G. (2007). Analysis of pesticide residues in fruit and vegetables with ethyl acetate extraction using gas and liquid chromatography with tandem mass spectrometric detection. *Analytical and Bioanalytical Chemistry, 389,* 1773−1789.

Płotka-Wasylka, J., Szczepańska, N., Guardia, M., & Namieśnik, J. (2016). Modern trends in solid phase extraction: New sorbent media. *TrAC Trends in Analytical Chemistry, 77,* 23−43.

Pozo, M., Barreda, O. J., Sancho, J. V., Hernández, F., Liberia, J., Cortés, M. A., et al. (2007). Multiresidue pesticide analysis of fruits by ultra-performance liquid chromatography tandem mass spectrometry. *Analytical and Bioanalytical Chemistry, 389,* 1765−1771.

Prieto, A., Basauri, O., Rodil, R., Usobiaga, A., Fernández, L. A., & Etxebarria, N. (2010). Stir-bar sorptive extraction: A view on method optimization, novel applications, limitations and potential solutions. *Journal of Chromatography A, 1217,* 2642−2666.

Qin, G., Li, Y., Chen, Y., Sun, Q., Zuo, B., He, F., et al. (2015). Pesticide residues determination in China vegetables in 2010−2013 applying gas chromatography with mass spectrometry. *Food Research International, 72,* 161−167.

Rajski, Ł., Gómez-Ramos, M., & Fernández-Alba, A. R. (2014). Large pesticide multiresidue screening method by liquid chromatography-Orbitrap mass spectrometry in full scan mode applied to fruit and vegetables. *Journal of Chromatography A, 1360,* 119−127.

Rallis, G. N., Sakkas, V. A., Boumba, V. A., Vougiouklakis, T., & Albanis, T. A. (2012). Determination of organochlorine pesticides and polychlorinated biphenyls in post-mortem human lung by matrix solid-phase dispersion with the aid of response surface methodology and desirability function. *Journal of Chromatography A, 1227,* 1−9.

Ramos, J. J., Rial-Otero, R., Ramos, L., & Capelo, J. L. (2008). Ultrasonic-assisted matrix solid-phase dispersion as an improved methodology for the determination of pesticides in fruits. *Journal of Chromatography A, 1212,* 145−149.

Reinholds, I., Pugajeva, I., & Bartkevics, V. (2016). A reliable screening of mycotoxins and pesticide residues in paprika using ultra-high performance liquid chromatography coupled to high resolution Orbitrap mass spectrometry. *Food Control, 60,* 683—689.

Rentsch, K. M. (2016). Knowing the unknown-state of the art of LCMS in toxicology. *TrAC Trends in Analytical Chemistry.* http://dx.doi.org/10.1016/j.trac.2016.01.028 (online).

Rizzetti, T. M., Kemmerich, M., Martins, M. L., Prestes, O. D., Adaime, M. B., & Zanella, R. (2016). Optimization of a QuEChERS based method by means of central composite design for pesticide multiresidue determination in orange juice by UHPLC-MS/MS. *Food Chemistry, 196,* 25—33.

Rociek, A. S., Surma, M., & Cieślik, W. (2014). Comparison of different modifications on QuEChERS sample preparation method for PAHs determination in black, green, red and white tea. *Environmental Science and Pollution Research, 21,* 1326—1338.

Rodrigues, E. T., Pardal, M. A., Salgueiro-González, N., Muniategui-Lorenzo, S., & Alpendurada, M. F. (2016). A single-step pesticide extraction and clean-up multi-residue analytical method by selective pressurized liquid extraction followed by on-line solid phase extraction and ultra-high-performance liquid chromatography-tandem mass spectrometry for complex matrices. *Journal of Chromatography A, 1452,* 10—17.

Rodríguez-Cabo, T., Rodríguez, I., Ramil, M., Silva, A., & Cela, R. (2016). Multiclass semi-volatile compounds determination in wine by gas chromatography accurate time-of-flight mass spectrometry. *Journal of Chromatography A, 1442,* 107—117.

Romero-González, R., Aguilera-Luiz, M. M., Plaza-Bolaños, P., Garrido Frenich, A., & Martínez Vidal, J. L. (2011). Food contaminant analysis at high resolution mass spectrometry: Application for the determination of veterinary drugs in milk. *Journal of Chromatography A, 1218,* 9353—9365.

Shamsipur, M., Yazdanfar, N., & Ghambarian, M. (2016). Combination of solid-phase extraction with dispersive liquid—liquid microextraction followed by GC-MS for determination of pesticide residues from water, milk, honey and fruit juice. *Food Chemistry, 204,* 289—297.

Shaoo, K. S., Batttu, R. S., & Singh, B. (2011). Development and validation of QuEChERS method for estimation of propamocarb residues in tomato (*Lycopersicon esculentum* mill) and soil. *American Journal of Analytical Chemistry, 2,* 26—31.

Shiea, C., Huang, Y., Liu, D. L., Chou, C. C., Chou, J. H., Chen, P. Y., et al. (2015). Rapid screening of residual pesticides on fruits and vegetables using thermal desorption electrospray ionization mass spectrometry. *Rapid Communications in Mass Spectrometry, 29,* 163—170.

Silva, E., Daam, M. A., & Cerejeira, M. J. (2015). Aquatic risk assessment of priority and other river basin specific pesticides in surface waters of Mediterranean river basins. *Chemosphere, 135,* 394—402.

Smalling, K. L., & Kuivila, K. M. (2008). Multi-residue method for the analysis of 85 current-use and legacy pesticides in bed and suspended sediments. *Journal of Chromatography A, 1210,* 8—18.

Smith, S., Gieseker, C., Rimschuessel, R., Decker, C. S., & Carson, M. C. (2009). Simultaneous screening and confirmation of multiple classes of drug residues in fish by liquid chromatography-ion trap mass spectrometry. *Journal of Chromatography A, 1216,* 8224—8232.

Sobhanzadeh, E., Bakar, N. K. A., Abas, M. R. B., & Nemati, K. (2012). A simple and efficient multi-residue method based on QuEChERS for pesticides determination in palm oil by liquid chromatography time-of-flight mass spectrometry. *Environmental Monitoring and Assessment, 184*, 5821−5828.

Soler, C., Hamilton, B., Furey, A., James, K. J., Manes, J., & Picó, Y. (2007). Liquid chromatography quadrupole time-of-flight mass spectrometry analysis of carbosulfan, carbofuran, 3-hydroxycarbofuran, and other metabolites in food. *Analytical Chemistry, 79*, 1492−1501.

Souza-Silva, E. A., Lopez-Avila, V., & Pawliszyn, J. (2013). Fast and robust direct immersion solid phase microextraction coupled with gas chromatography−time-of-flight mass spectrometry method employing a matrix compatible fiber for determination of triazole fungicides in fruits. *Journal of Chromatography A., 1313*, 139−146.

Stahnke, H., Kittlaus, S., Kempe, G., Hemmerling, C., & Alder, L. (2012). The influence of electrospray ion source design on matrix effects. *Journal of Mass Spectrometry, 47*, 875−884.

Stahnke, H., Kittlaus, S., Kempe, G. N., & Alder, L. (2012). Reduction of matrix effects in liquid chromatography−electrospray ionization−mass spectrometry by dilution of the sample extracts: How much dilution is needed? *Analytical Chemistry, 84*, 1474−1482.

Sturza, J., Silver, M. K., Xu, L., Li, M., Mai, X., Xia, Y., et al. (2016). Prenatal exposure to multiple pesticides is associated with auditory brainstem response at 9 months in a cohort study of Chinese infants. *Environment International, 92−93*, 478−485.

Tan, G. H., & Abdulra'uf, L. B. (2012). Developments and applications of microextraction techniques for the analysis of pesticide residues in fruits and vegetables. In R. P. Soundararajan (Ed.), *Pesticides − Recent trends in pesticide residue assay*. Rijeka, Croatia: InTech.

Tette, P. A. S., Guidi, L. R., Glória, M. B. A., & Fernandes, C. (2016). Pesticides in honey: A review on chromatographic analytical methods. *Talanta, 149*, 124−141.

Tomlin, C. (2003). *The pesticide manual − A World Compendium* (13th ed.). Alton, Hampsire: British Crop Protection Council (BCPC).

Trufelli, H., Palma, P., Famiglini, G., & Cappiello, A. (2011). An overview of matrix effects in liquid chromatography-mass spectrometry. *Mass Spectrometry Reviews, 30*, 491−509.

Vas, G., & Vékey, K. (2004). Solid-phase microextraction: A powerful sample preparation tool prior to mass spectrometric analysis. *Journal of Mass Spectrometry, 39*, 233−254.

Vazquez-Roig, P., & Picó, Y. (2015). Pressurized liquid extraction of organic contaminants in environmental and food samples. *TrAC Trends in Analytical Chemistry, 71*, 55−64.

Villaverde, J. J., Sevilla-Morán, B., López-Goti, C., Alonso-Prados, J. L., & Sandín-España, P. (2016). Trends in analysis of pesticide residues to fulfill the European Regulation (EC) No. 1107/200. *TrAC Trends in Analytical Chemistry, 80*, 568−580.

Wang, H., Ding, J., & Ren, N. (2016). Recent advances in microwave-assisted extraction of trace organic pollutants from food and environmental samples. *TrAC Trends in Analytical Chemistry, 75*, 197−208.

Wang, J., & Cheung, W. (2016). UHPLC/ESI-MS/MS determination of 187 pesticides in wine. *Journal of AOAC International, 99*, 539−557.

Wang, J., Chow, W., Leung, D., & Chang, J. (2012). Application of ultrahigh-performance liquid chromatography and electrospray ionization quadrupole Orbitrap high-resolution mass spectrometry for determination of 166 pesticides in fruits and vegetables. *Journal of Agricultural and Food Chemistry, 60*, 12088−12104.

Wu, G., Bao, X., Zhao, S., Wu, J., Han, A., & Ye, Q. (2011). Analysis of multi-pesticide residues in the foods of animal origin by GC—MS coupled with accelerated solvent extraction and gel permeation chromatography cleanup. *Food Chemistry, 126*, 646—654.

Wu, L., Song, Y., Xu, X., Li, N., Shao, M., Zhang, H., et al. (2014). Medium-assisted nonpolar solvent dynamic microwave extraction for determination of organophosphorus pesticides in cereals using gas chromatography-mass spectrometry. *Food Chemistry, 162*, 253—260.

Yang, P., Chang, J. S., Wong, J. W., Zhang, K., Krynitsky, A. J., Bromirski, M., et al. (2015). Effect of sample dilution on matrix effects in pesticide analysis of several matrices by liquid chromatography-high-resolution mass spectrometry. *Journal of Agricultural and Food Chemistry, 63*, 5169—5177.

Yogendrarajah, P., Pouckeb, C., De Meulenaera, B., & De Saeger, S. (2013). Development and validation of a QuEChERS based liquid chromatography tandem mass spectrometry method for the determination of multiple mycotoxins in spices. *Journal of Chromatography A, 1297*, 1—11.

Zanella, R., Prestes, O. D., Martins, M. L., & Adaime, M. B. (2015). Quantitative analysis and method validation. In T. Tuzimski, & J. Sherma (Eds.), *High performance liquid chromatography in pesticide residue analysis*. Boca Raton: CRC Press.

Zhan, J., Li, J., Liu, D., Liu, C., Yang, C., Zhou, Z., et al. (2016). A simple method for the determination of organochlorine pollutants and the enantiomers in oil seeds based on matrix solid-phase dispersion. *Food Chemistry, 194*, 319—324.

Zhang, L., Liu, S., Cui, X., Pan, C., Zhang, A., & Chen, F. (2012). A review of sample preparation methods for the pesticide residue analysis in foods. *Central European Journal of Chemistry, 10*, 900—925.

Applications of Liquid Chromatography Coupled With High-Resolution Mass Spectrometry for Pesticide Residue Analysis in Fruit and Vegetable Matrices

P. Sivaperumal

National Institute of Occupational Health, Ahmedabad, Gujarat, India

6.1 INTRODUCTION

Pesticides belong to a distinct group of chemicals, which have been used to control pests and weeds in the agricultural crops. They are polar, semipolar, and ionic molecules and are characterized with varying degrees of thermal stability and volatility. Pesticide residues in fruit and vegetables are traditionally monitored mainly by gas chromatography (GC) technique with multiresidue methods (Cajka, Hajslova, Lacina, Mastovska, & Lehotay, 2008; Lehotay, de Kok, Hiemstra, & Bodegraven, 2005; Pihlström, Blomkvist, Friman, Pagard, & Osterdahl, 2007). To ensure food safety, maximum residue limits (MRLs) have been laid down by various countries for a wide range of pesticide residues in food samples. To control and monitor the residues, accurate analytical methodology is necessary for the determination of residues at trace levels, and the methodology should be able to determine a maximum number of pesticide residues in a single run with short run time.

The primary advantage of high-resolution mass spectrometry (HRMS) is its suitability for the determination of pesticide residues with adequate sensitivity, mass accuracy, and resolution (or mass resolving power). The preamble of more advanced instruments led to debate as to how much mass resolving power is required for HRMS. van der Heeft et al. (2008) considered that 50,000—60,000 [full width at half maximum (FWHM)] is adequate to ensure reliable mass assignment of the analytes. In the case of complex matrices, inadequate mass resolution of the instrument may lead to inaccurate mass measurements of the analytes in the samples. Ferrer and Thurman (2007) reported that a resolving power of

Applications in High Resolution Mass Spectrometry. http://dx.doi.org/10.1016/B978-0-12-809464-8.00006-3

6000–10,000 (FWHM) is sufficient for many vegetable samples. For example, instruments with resolving power (FWHM) of 10,000 at m/z 200 could separate or resolve masses that differ by 0.020 mass units. In complex matrices, adequate mass resolution is required to provide sufficient selectivity to enable detection of residues at trace levels. Otherwise, it will provide false-positive results in the sample. Accurate mass measurements of targeted small molecules are obtained from the molecular formula of the respective molecules and nontargeted molecules are confirmed from possible elemental compositions (Brenton & Godfrey, 2010). The newly upgraded time-of-flight (TOF) and Orbitrap-based mass analyzers offer a mass accuracy of <1 ppm. The earlier TOF instruments exhibited poor performance of mass accuracy (>5 ppm). These analyzers currently use a time-to-digital converter technology. Thus they could produce more accurate mass measurements (Lacina, Urbanova, Poustka, & Hajslova, 2010). Furthermore, excellent mass accuracy is recorded while using internal mass calibration, i.e., simultaneously the sample and the calibrant solution are introduced into the ion source (Hird, Lau, Schuhmacher, & Krska, 2014). Across the world, development of multiresidue analytical methods in food samples is mainly based on mass analyzers coupled with GC and high-performance liquid chromatography (HPLC). The liquid chromatography (LC) technique has been chosen for the determination of polar, less volatile, or thermally stable pesticides in food samples. These molecules are not suitable for GC analysis; hence they have to be derivatized before GC analysis. Moreover, the LC system has become the most popular technique for the multiresidue/multiclass determination of agrochemical residues in food (Ferrer & Thurman, 2007). In routine analysis, the analysis time plays a key role for determination of residues in the sample. For fast separations in the LC system, several approaches have been applied, viz., operating at high temperatures, monolithic columns, fused-core columns, narrow columns, and sub-2 μm particle size packed columns. Among the different approaches, sub-2 μm columns provide a threefold increase in chromatographic resolution and 2- to 10-fold increase in high-speed analysis with similar efficiency (Ferrer, García-Reyes, Mezcua, Thurman, & ernánndez-Alba, 2005). Using small columns with fine particles could reduce solvent consumption, increase the flow rate, and reduce the run time considerably. As a consequence, smaller sized stationary phases became popular since 2004, as they can withstand high pressures and are compatible with ultrahigh-pressure liquid chromatography (UHPLC) systems. The UHPLC system produces narrow peaks (usually 1–3 s, or even lower), and hence it requires fast-scan and sensitive mass analyzers for the determination of hundreds of analytes at trace levels with short run time. In food sample analysis, UHPLC coupled with various mass analyzers such as triple quadrupole (QqQ), Q-trap, QqTOF, and Orbitrap are currently used for the determination of pesticide residues in food samples (Taylor, Keenan, Reid, & Fernández, 2008). Studies have focused on the issue related to the sensitivity and selectivity of pesticide residues by LC tandem mass spectrometry (Ortelli, Edder, & Corvi, 2005). The high-resolution mass analyzers (TOF and Orbitrap) are independent of specified ion abundances, which leads to improved mass accuracy, higher selectivity, and subsequently improves the limit of detection (LOD). This greatly reduces the number of false results of the analytes (Kellmann, Muenster, Zomer, & Mol, 2009). The HRMS techniques, such as TOF and Orbitrap, possess several advantages in the function of full-scan operation mode for multiresidue analysis (Mol, Zomer, & de Koning, 2012). The Orbitrap mass analyzer provides high mass resolution and accuracy for the multiresidue screening of the analytes (Kellmann et al., 2009; Moulard et al., 2011). De Dominicis, Commissati, and Suman (2012) described theoretically that an unlimited number of nontarget organic molecules can be monitored without reinjecting the samples. The TOF and Orbitrap analyzers possess similar mass accuracy; however, the Orbitrap has a higher resolving power (up to 100,000), and also the new generation of TOF analyzers could provide a suitable resolving power for the measurement of analytes in complex matrices.

Del Mar Gómez-Ramos, Rajski, Heinzen, & Fernández-Alba (2015) described the use of LC/Q-Orbitrap for the analysis of 139 pesticide residues in tomato, pepper, orange, and green tea samples followed by the QuEChERS (quick, easy, cheap, effective, rugged, and safe) extraction method. Q-Orbitrap is used along with a full-scan mode and a resolution power of 70,000. The method showed very good reproducibility, peak shape, and linearity with similar retention time. For minimizing false results, the authors suggested the use of appropriate control systems, viz., automatic gain control, maximum injection time and underfill ratio, and also suggested that for identifying the pesticide residues at trace level, the intensity of the threshold value to be set at lower counts. The results obtained from LC/Q-Orbitrap mass spectrometry (MS) were compared with those from LC−QqQ MS/MS. Finally, they concluded that these techniques were consistent with similar quantitative detection capabilities.

Zomer and Mol (2015) developed simultaneous quantitative determination, identification, and qualitative screening of pesticide restudies (184) in fruit and vegetable samples by LC-Q-Extractive Orbitrap−based MS. QuEChERS extraction and detection was performed in an improved nontargeted data acquisition with and without fragmentation. The method was validated as per European Commission's Directorate General for Health and Consumer Protection (DG SANCO) guidelines. The screening detection limit (SDL) was evaluated for 184 pesticide residues in the samples. Of these, the SDL obtained was 10 ng/g for 134 pesticides, 50 ng/g for 39 pesticides, and 200 ng/g for 2 pesticides, and the SDLs of 9 pesticides were not established. The obtained results showed that the pesticide residues met the criteria for trueness and precision set in the DG SANCO guidance document.

As a novel approach, HRMS has been introduced for the analysis of pesticide residues in food samples. This technique is potentially useful for the determination of target and nontarget analytes in food samples (Ferrer & Thurman, 2007; Ferrer, Thurman, & Fernández-Alba, 2005; Gilbert-López, García-Reyes, Ortega-Barrales, Molina-Díaz, & Fernández-Alba, 2007; Taylor et al., 2008). Hence, this chapter describes HRMS capabilities for the identification, confirmation, and quantification of pesticide residues in fruit and vegetable samples. Furthermore, it elaborates the QuEChERS method coupled with UHPLC-QTOF/MS (UHPLC-HRMS).

6.2 APPLICATIONS OF PESTICIDE RESIDUE ANALYSIS IN FRUIT AND VEGETABLE SAMPLES BY LC-HRMS

Traditionally, most of the pesticide residues have been quantified by GC combined with nitrogen—phosphorus, flame photometric, and electron capture detectors. In case of LC, photo diode array detector was used in the majority of the analytical techniques. The identification and confirmation of pesticide residues in fruit and vegetable samples are difficult at lower concentrations, due to the complex matrix.

Chromatography coupled with MS was the most popular technique for the analysis of pesticide residues in food samples. The quadrupole mass analyzers were used successfully, due to their wide linear range, specificity, and sensitivity. GC-single quadrupole mass spectrometry has been widely used for quantitative analysis of pesticide residues in food samples (Cajka et al., 2008; van Leeuwen & de Boer, 2008). Since 2000, LC coupled with QqQ mass analyzer has been widely used for the determination of multiclass pesticides. In addition, the LC technique is more suitable for thermolabile, polar, and nonvolatile organic compounds.

The LC-MS technique finds wide usage for the analysis of pesticide residues in food samples (Lambropoulou & Albanis, 2007; Niessen, Manini, & Andreoli, 2006; Picó, Font, Ruiz, & Fernández, 2006). QqQ mass analyzers provide good sensitivity and specificity but limited structural information and do not identify the nontarget compounds (Picó et al., 2006; Soler, Hamilton, et al., 2007). The quadrupole-time-of-flight (QTOF)-MS has a wide range of applications and is mainly used for screening purposes (Hernández, Ibáñez, Sancho, & Pozo, 2004; van Leeuwen & de Boer, 2008; Núñez, Moyano, & Galceran, 2004).

The development of QTOF-MS took place during 2005—2010. Technological improvement in ionization sources and detectors has resulted in better sensitivity compared with earlier instruments. A number of studies have been reported for the quantitative determination of pesticide residues using QTOF-MS (Picó, Blasco, & Font, 2003; Taylor et al., 2008; Williamson & Bartlett, 2007). The advantage of using the full-scan operation mode in high-resolution mass spectrometers, such as TOF and Orbitrap, is it provides high specificity without limiting the number of molecules in a single run. HRMS (QTOF-MS) is more flexible and rapid than LC-QqQ-MS, while developing the methods for complex matrices. Thus HRMS can be an attractive tool for identifying both target and nontarget analytes in complex matrices. Development and validation of the analytical method is of supreme importance during pesticide residue analysis in fruit and vegetable samples by HRMS. The analytical methodology must be within the acceptable level of mass accuracy, mass resolution, selectivity, LOD, limit of quantification (LOQ), linearity, accuracy, precision, recovery, ruggedness, and robustness for every step of sample preparation and quantitation of the analytes. Therefore Table 6.1 briefly describes the various essential developments and validation parameters, viz., matrices, number of target analysts, sample preparation, and

Table 6.1 Analytical Methods for Determination of Pesticide Residues in Fruit and Vegetable Samples by LC-HRMS

Matrix	No. of Pesticides	Extraction Methods	LC Parameters	Detection Mode	TOF-MS Parameters	Method Validation Parameters	References
Apple, Pepper, Broccoli, Cucumber, Tomato, Grapefruit, Orange, Lemon, Pear, Carrot, Eggplant, Lettuce, Grapes, Pineapple, Strawberry, and Melon	3	15 g sample + 15 mL ACN + 1.5 g NaCl + 4 g MgSO₄, centrifuged at 3700 rpm for 1 min; 5 mL of supernatant transferred to 250 mg PSA + 750 mg MgSO₄, centrifuged at 3700 rpm for 1 min; 2 mL sample evaporated to dryness and reconstituted with 2 mL of 10% MeOH. Extract filtered through 0.45 μm PTFE and injected to the instrument.	Column: C₈ (150 mm × 4.6 mm; 5 μm) Mobile phase: A: 0.1% HCOOH in H₂O; B:ACN Flow rate: 600 μL/min	HPLC-TOF-MS (ESI+)	Capillary voltage: 4000 V Nebulizer pressure: 40 psig Drying gas: 9 L/min Gas temperature: 300°C Skimmer voltage: 60 V Octapole DC 1: 37.5 V Octapole RF: 250 V Resolution: 9500 ± 500	Mass accuracy: <8.2 ppm	Garcia-Reyes, Ferrer, Thurman, Molina-Díaz, and Fernández-Alba (2005)
Apple and Orange	350	15 g sample + 45 mL EtOAc + 1 mL NaOH (6.5 M); sample blended in a Polytron (high-speed blender) at 21,000 rpm for 30 s; extract filtered through 20 g of anh. Na₂SO₄; the solid washed with 50 mL EtOAc; extract evaporated to dryness and reconstituted with 15 mL of MeOH. Extract filtered through 0.2 μm PTFE and injected to the instrument.	Column: C₈ (150 mm × 4.6 mm; 5 μm) Mobile phase: A: ACN; B: 0.1% HCOOH in H₂O Flow rate: 600 μL/min	LC-TOF-MS (ESI+)	Capillary voltage: 4000 V Nebulizer pressure: 40 psig Drying gas: 9 L/min Gas temperature: 300°C Fragmentor: 190 V Skimmer voltage: 60 V Octapole DC 1: 37.5 V Octapole RF: 250 V Resolution: 9500 ± 500	Mass accuracy: <2 ppm	Thurman, Ferrer, Malato, and Fernández-Alba (2006)
Pepper, Broccoli, Tomato,	15	15 g sample + 90 mL EtOAc + 1 mL NaOH (6.5 M); sample blended in a Polytron	Column: C₈ (150 mm × 4.6 mm; 5 μm)	HPLC-TOF-MS (ESI+)	Capillary voltage: 4000 V Nebulizer pressure:	LOD: 0.5 –50 μg/kg Linearity:	Ferrer, Garcia-Reyes, et al. (2005)

Continued

Table 6.1 Analytical Methods for Determination of Pesticide Residues in Fruit and Vegetable Samples by LC-HRMS—cont'd

Matrix	No. of Pesticides	Extraction Methods	Instrumentation			Method Validation Parameters	References
			LC Parameters	Detection Mode	TOF-MS Parameters		
Orange, Lemon, Apple, and Melon		(high-speed blender) at 21,000 rpm for 30 s; extract filtered through 20 g of anh. Na₂SO₄; the solid washed with 50 mL EtOAc; extract evaporated to dryness and reconstituted with 15 mL of MeOH. Extract filtered through 0.45 μm PTFE and injected to the instrument.	Mobile phase: A: ACN; B: 0.1% HCOOH in H₂O Flow rate: 600 μL/min		40 psig Drying gas: 9 L/min Gas temperature: 300°C Skimmer voltage: 60 V Octapole DC 1: 37.5 V Octapole RF: 250 V Resolution: 9500 ± 500	>0.992 RSD: <11% Mass accuracy: <2 ppm	
Orange, Potato, and Rice	8	5 g sample with anh. Na₂SO₄ (PLE); accelerated solvent extraction conditions; DCM at 40°C and 2000 psi heating time 2 min and 2 cycles 5 min each static extraction (40 mL); extract purged up to ~ 10 mL and evaporated to dryness; reconstituted with 1 mL MeOH. Extract injected to the instrument.	Column: (150 mm × 2.1 mm; 5 μm) Mobile phase: A: 1.0 mM NH₄OAc in H₂O; B: 1.0 mM NH₄OAc in MeOH; C: 1.0 mM NH₄OAc in ACN Flow rate: 200 μL/min	LC-QqTOF-MS (ESI+)	Ion spray voltage: 5000 V Nebulizer gas flow: 10 L/min Curtain gas (N₂): 12 L/min Collision gas (N₂): 3 L/min Focusing potential: 150 V Entrance potential: 4 V Cell exit potential: 12 V Declustering potential: 30 V Declustering potential-2: 15 V	LOD: 3–18 μg/kg LOQ: 10–70 μg/kg Linearity: ≥0.99 Recovery: 55% –94% RSD: 3%–13% Mass accuracy: <4 mDa	Soler, Hamilton, et al. (2007)
Orange peel and flesh, Banana skin and flesh, Strawberry, and Pear	2	10 g sample + 70 mL acetone, blended with high-speed blender at 8000 rpm for 2 min; extract filtered through 25–30 μm filter paper and washed with 10 mL of acetone; volume adjusted to 100 mL with acetone. Extract filtered through 0.45 μm nylon syringe filter and injected to the instrument.	Column: C₁₈ (50 mm × 2.1 mm; 5 μm) Mobile phase: A: 0.01% HCOOH in MeOH; B: 0.01% HCOOH in H₂O Flow rate: 300 μL/min	HPLC-QTOF-MS (ESI+)	Nebulizer gas flow: 15 L/h Desolvation gas flow: 700–800 L/h Cone voltage: 30 V Capillary voltage: 3.5 kV Nitrogen Desolvation temp.: 350°C Source temp.: 120°C Resolution: 5000	LOD: 7.5 & 1 μg/L Linearity: >0.99 Recovery: 51% –130% RSD: <20% Mass accuracy: <5 ppm	Grimalt, Pozo, Sancho, and Hernández (2007)

Matrix	Pesticides/No.	Sample preparation	Technique	Column/Mobile phase	Instrument conditions	Validation/Results	Reference
Pears	3	5 g sample + 30 mL EtOAc + 15 g anh. Na$_2$SO$_4$ warring blender for 2 min and allow to set; solid residue again homogenized with 30 mL of EtOAc then through anh. Na$_2$SO$_4$ and collect with first fraction. Rinsed twice with 15 mL EtOAc and filter. Extract evaporated to dryness and reconstituted with 0.5 mL MeOH and injected to the instrument.	UPLC-QqTOF-MS and UPLC-QqTOF-MS/MS (ESI+)	Column: C$_{18}$ (5 cm × 2.1 mm; 1.7 µm) Mobile phase: A: 10 mM AF in H$_2$O; B: MeOH Flow rate: 400 µL/min	Desolvation gas flow: 500 L/h Desolvation temp.: 350°C Cone gas: 50 L/h Source temp.: 120°C Capillary voltage: 3000 V Cone voltage: 20 V TOF flight tube voltage: 5630 V Reflection voltage: 1780 V Pusher voltage: 815 V Puller voltage: 640 V MCP detector voltage: 815 V Resolution: 5000	UPLC-QqTOF-MS LOD: 0.4 µg/kg LOQ: 2 µg/kg Linearity: >0.99 Recovery: 73%–78% RSD: 7%–17% Mass accuracy: <24.2 ppm UPLC-QqTOF-MS/MS LOD: 0.4 µg/kg LOQ: 2 µg/kg Linearity: >0.99 Recovery: 74%–79% RSD: 6%–19%	Picó, Farré, Soler, and Barceló (2007) and Picó, la Farré, Soler, and Barceló (2007)
Orange	Fenthion and its metabolites	5 g sample + 30 mL EtOAc + 15 g anh. Na$_2$SO$_4$, mix thoroughly by using warring blender for 2 min and allow to set and supernatant passed through filter paper; solid residue again homogenized with 30 mL EtOAc; filter through anh. Na$_2$SO$_4$; collected with first extraction; twice 15 mL of EtOAc used to rinse and passed through filter and collected; extract evaporated to dryness and reconstituted with 0.5 mL MeOH and injected to the instrument.	UPLC-QqTOF-MS (ESI+)	Column: C$_{18}$ (5 cm × 2.1 mm; 1.7 µm) Mobile phase: A: MeOH; B: 10 mM AF in H$_2$O Flow rate: 400 µL/min	Desolvation gas flow: 500 L/h Desolvation temp.: 350°C Source temp.: 120°C Cone gas: 50 L/h Capillary voltage: 3000 V Cone voltage: 20 V TOF flight tube voltage: 5630 V Reflection voltage: 1780 V Pusher voltage: 840 V Puller voltage: 645 V MCP detector voltage: 2200 V Resolution: 5000	LOD: 0.4 –5.6 µg/kg LOQ: 3–15 µg/kg Linearity: >0.99 Recovery: >70% RSD: 2.8% –10% Mass accuracy: <5 ppm	Picó, Farré, et al. (2007) and Picó, la Farré, et al. (2007)
Oranges, Strawberries, Cherries, Peaches, Apricots, and Pears	12	Ethyl Acetate Extraction (EtOAc): 15 g sample + 100 mL EtOAc + 50 g anh. Na$_2$SO$_4$ mix thoroughly by using warring blender for 2 min and allow to set supernatant passed through filter paper; solid residue again homogenized with 30 mL EtOAc passed through filter paper through anh. Na$_2$SO$_4$; solid residue again	LC-QqTOF-MS/MS (ESI+)	Column: C$_{18}$ (150 mm × 4.6 mm; 5 µm) proceed by security guard cartridge C$_{18}$ (4 mm × 2 mm) Mobile phase: A: 70% of 10 mM AF in	Ion spray source temp.: 450°C Ion spray voltage: 5000 V Nebulizer gas flow: 10 L/min Curtain gas (N$_2$): 12 L/min	EtOAc Extraction Linearity: >0.99 Recovery: 58% –99% RSD: 7%–19% PLE LOQ: 10	Soler, James, and Picó (2007)

Continued

Table 6.1 Analytical Methods for Determination of Pesticide Residues in Fruit and Vegetable Samples by LC-HRMS—cont'd

Matrix	No. of Pesticides	Extraction Methods	Instrumentation		TOF-MS Parameters	Method Validation Parameters	References
			LC Parameters	Detection Mode			
		homogenized with 100 mL EtOAc; filter through anh. Na₂SO₄ and collected with first extraction fraction; twice 25 mL of EtOAc used to rinse and passed through filter and collected; extract evaporated less than 5 mL; reconstituted with 10 mL EtOAc in volumetric flask. Pressurized Liquid Extraction (PLE): 2.5 g sample blended with 20 g of acidic alumina for 5 min using pestle to obtain homogeneous mixture; mixture introduced into a stainless steel extraction cell; N₂ at pressure of 10,000 mbars supplied to PLE system and EtOAc at 75°C for 7 min; the extract transferred to 15-mL tube and rinsed twice with 0.5 mL MeOH; the extract evaporated to dryness and reconstituted with 0.5 mL of MeOH and injected to the instrument.	MeOH; B: 30% of 10 mM AF in H₂O Flow rate: 600 µL/min		Declustering potential: 30 V Focusing potential: 150 V Declustering potential-2, 15 V Collision gas (N₂): 3 L/min	~400 µg/kg Linearity: >0.99 Recovery: 72% –90% RSD: 10% –19%	
Green pepper, Tomato, Cucumber, and Orange	101	15 g sample +15 mL ACN; Shake well for 1 min; Add 4 g MgSO₄ + 1.5 g NaCl and shake for 1 min; centrifuged at 3700 rpm for 1 min. 5-mL aliquot transferred to tube containing 250 mg PSA + 750 mg MgSO₄; shake well for 20 s; centrifuged at 3700 rpm for 1 min; Extract of the sample evaporated to near dryness and reconstituted with mobile phase. Extract filtered through 0.45 µm PTFE and injected to the instrument.	Column: C₈ (150 mm × 4.6 mm; 5 µm) Mobile phase: A: ACN; B: 0.1% HCOOH in H₂O Flow rate: 600 µL/min	HPLC-TOF-MS (ESI+)	Capillary voltage: 4000 V Nebulizer pressure: 40 psig Drying gas: 9 L/min Gas temp.: 300°C Fragmentor voltage: 190 V Skimmer voltage: 60 V Octapole DC 1: 37.5 V Octapole RF: 250 V Resolution: 9700 ± 500	LOD: 0.04 ~150 µg/kg Linearity: >0.99 RSD: 0.9%–9% Mass accuracy: <0.05 Da	Ferrer and Thurman (2007)

Pears	1 and 4 degradation products	5 g sample + 30 mL EtOAc + 15 g anh. Na$_2$SO$_4$ mix thoroughly by using warring blender for 2 min and allow to set; supernatant (EtOAc) passed through filter paper that contains anh. Na$_2$SO$_4$; collect and solid residue again homogenized with 30 mL EtOAc and filter through anh. Na$_2$SO$_4$; collect first extraction fraction; twice 15 mL of EtOAc used to rinse, extract evaporated to dryness and reconstituted with 0.5 mL ACN and injected to the instrument.	Column: C$_{18}$ (5 cm × 2.1 mm; 1.7 μm) Mobile phase: A: MeOH; B: 10 mM AF in H$_2$O Flow rate: 400 μL/min	UPLC-Qq-TOF-MS (ESI+) and UPLC-QqTOF-MS/MS	Desolvation gas flow: 500 L/h Desolvation temp.: 350°C Cone gas: 50 L/h Source temp: 120°C Capillary voltage: 3000 V Cone voltage: 20 V TOF flight tube voltage: 5630 V Reflection voltage: 1780 V Pusher voltage: 840 V Puller voltage: 645 V MCP detector voltage: 2200 V Resolution: 5000	UHPLC-QqTOF-MS LOD: 2–7 μg/kg LOQ: 5–20 μg/kg Linearity: >0.99 Recovery: 83%–101% RSD: 9%–19% Mass accuracy: <5 ppm UHPLC-QqTOF-MS/MS LOD: 5–10 μg/kg LOQ: <30 μg/kg Linearity: >0.99 Recovery: 80%–93% RSD: ≤19%	Picó, Farré, Tokman, and Barceló (2008)
Organic Strawberry	100	10 g sample + 50 g Na$_2$SO$_4$ + 2 g NaHCO$_3$ + 50 mL EtOAc; mixture homogenized for 1 min using tissue disperser; liquid layer decanted through Whatman filter paper No. 1; the residue rehomogenized using 50 mL EtOAc for 1 min and filtered as above; rinsed with EtOAc; concentrated the sample at low temp. and make up with 10 mL EtOAc.1 mL aliquot transferred and 10 mL of MeOH added; evaporated ~1 mL and transferred to 2-mL volumetric flask and make up with 10 mM NH$_4$Ac. Extract filtered through 0.45 μm PTFE and injected to the instrument.	Column: C$_{18}$ (50 mm × 2.1 mm; 1.7 μm) Mobile phase: A: 5 mM NH$_4$OAc in H$_2$O:MeOH (95:5 v/v); B: 5 mM NH$_4$OAc in MeOH Flow rate: 480 μL/min	UPLC-qTOF-MS (ESI+/−)	Cone voltage: 30 V Source temp.: 120°C Capillary voltage: 3000/ 2800 V ± ionization Desolvation Gas (N$_2$) Temp: 300/ 150°C ± ionization Desolvation gas flow rate: 60/1000 L/h ± ionization Cone gas flow rate: 50 L/h Resolution: 11,500	Linearity: ≥0.95 Recovery: 70%–120% RSD: ≤30% Mass accuracy: ≤5 ppm	Taylor et al. (2008)

Continued

Table 6.1 Analytical Methods for Determination of Pesticide Residues in Fruit and Vegetable Samples by LC-HRMS—cont'd

Matrix	No. of Pesticides	Extraction Methods	Instrumentation		TOF-MS Parameters	Method Validation Parameters	References
			LC Parameters	Detection Mode			
Pepper, Rice, Garlic, and Cauliflower	300	15 g sample + 15 mL ACN shaken well; add 6 g MgSO₄ + 1.5 g NaCl and shaken for 1 min; centrifuged at 3700 rpm for 1 min; 5 mL supernatant transferred to tube containing 250 mg PSA + 750 mg of MgSO₄; shaken well for 20 s and centrifuged at 3700 rpm for 1 min; 2 mL of extract of sample evaporated to near dryness and reconstituted with 2 mL 10% ACN. Extract filtered through 0.45 μm PTFE and injected to the instrument.	Column: C₁₈ (50 × 4.6 mm; 1.8 μm) Mobile phase: A: 0.1% HCOOH in H₂O: ACN (95:5 v/v); B: 0.1% HCOOH in ACN:H₂O (95:5 v/v) with flow rate: 600 μL/min	HPLC-TOF-MS (ESI+)	HPLC-TOF-MS-(1) Capillary voltage: 4000 V Nebulizer pressure: 40 psig Drying gas flow rate: 9 L/min Gas temp.: 325°C Skimmer voltage: 60 V Octapole DC 1: 37.5 V Octapole RF: 250 VFragmentor Voltage (CID): 190 V Resolution: 9700 ± 500 HPLC-TOF-MS-(2) Superheated N₂ sheath gas temp.: 400°C Flow rate: 12 L/h Capillary voltage: 4000 V Nebulizer pressure: 40 psi Drying gas flow rate: 10 L/min Gas temperature: 350°C Skimmer voltage: 65 V Octapole RF: 750 V Fragmentor voltage (CID): 90 V Resolution: 19,500 ± 500	LOQ: 0.6 –1.1 mg/kg	Mezcua et al. (2011)

Fruits and Vegetables	297	15 g sample + 15 mL ACN shaken well; add 6 g MgSO₄ + 1.5 g NaCl and shaken for 1 min; centrifuged at 3700 rpm for 1 min; 5 mL supernatant transferred to tube containing 250 mg PSA + 750 mg of MgSO₄; shaken well for 20 s; centrifuged at 3700 rpm for 1 min; 2 mL of extract of sample evaporated to near dryness and reconstituted with 2 mL 10% ACN. Extract filtered through 0.45 µm PTFE and injected to the instrument.	Column: C₁₈ (50 × 4.6 mm; 1.8 µm) Mobile phase: A: 0.1% HCOOH in H₂O:ACN (95:5 v/v); B: 0.1% HCOOH in ACN:H₂O (95:5 v/v) with flow rate: 600 µL/min	HPLC-TOF-MS (ESI+)	Capillary voltage: 4000 V Nebulizer pressure: 40 psig Drying gas flow rate: 9 L/min Gas temperature: 325°C Skimmer voltage: 60 V Octapole DC 1: 37.5 V Octapole RF: 250 V Fragmentor voltage (CID): 190, 210, and 230 V Resolution: 9700 ± 500	LOD: 5–50 µg/kg Mass accuracy: <6 ppm	Mezcua, Malato, García-Reyes, Molina-Díaz, and Fernández-Alba (2009)
Pears and Apples	4	In 500-mL beaker, sample covered with EtOAc and sonicated for 20 min; EtOAc filter through filter paper; sample washed with 30 mL of EtOAc which is filtered and collected in first extraction fraction; EtOAc evaporated to less than 10 mL then transferred tube; evaporated to almost dryness and adjusted to 10 mL with MeOH and injected to the instrument.	Column: C₁₈ (15 cm × 2.1 mm; 1.7 µm) Mobile phase: A: 10 mM AF in H₂O; B: 10 mM AF in MeOH Flow rate: 200 µL/min	UPLC-QqTOF-MS (ESI+)	Capillary voltage: 3000 V Cone voltage: 20 V Desolvation gas flow rate: 500 L/h Desolvation gas (N₂) temp.: 350°C Cone gas flow rate: 50 L/h Source temp.: 120°C Resolution: 5000	LOD: 0.02 –0.34 µg LOQ: 0.05 –0.60 µg Linearity: >0.99 Recovery: >72% RSD: <19% Mass accuracy: <6 ppm	Picó, la Farré, Segarra, and Barceló (2010)
Apple, Strawberry, Tomato, and Spinach	212	10 g sample + 10 mL ACN shake well for 1 min; add 4 g MgSO₄ + 1 g NaCl and shaken for 1 min; centrifuged at 11,000 rpm for 5 min; supernatant filtered through 0.2 mm PTFE filter and injected to the instrument.	Column: HSS T3 (100 mm × 2.1 mm; 1.8 µm) Mobile phase: A: MeOH; B: 5 mM AF Flow rate: 300 –600 µL/min	UHPLC-TOF-MS (ESI+)	ESI (+) Capillary voltage: 3500 V Cone voltage: 30 V Source temp.: 120 °C Desolvation gas (N₂) temp.: 350°C Desolvation gas flow rate: 700 L/h Cone gas flow rate: 10 L/h	LOQ: ≤10 µg/kg Linearity: >0.99 RSD: 1.28% –16.28%	Lacina et al. (2010)

Continued

Table 6.1 Analytical Methods for Determination of Pesticide Residues in Fruit and Vegetable Samples by LC-HRMS—cont'd

Matrix	No. of Pesticides	Extraction Methods	Instrumentation		TOF-MS Parameters	Method Validation Parameters	References
			LC Parameters	Detection Mode			
Orange peel and flesh, Banana peel and flesh, Strawberry, Tomato, Grapefruit, Cucumber, and Pepper	11	10 g sample + 25 mL MeOH: H_2O (80:20 v/v) extraction for 2 min using high-speed blender at 8000 rpm, extract filtered through filter paper and washed with 10 mL of extraction solvent, adjusted with 50 mL of same extraction solvent. Extract filtered through 0.22 μm Nylon syringe filter and injected to the instrument.	Column: BEH C_{18} (50 mm × 2.1 mm; 1.7 μm) Mobile phase: A: 0.5 mM NH_4OAc in methanol; B: 0.5 mM NH_4OAc in H_2O Flow rate: 300 μL/min	UHPLC-TOF and QTOF-MS/MS (ESI+)	ESI (−) Capillary voltage: −2000 V Cone voltage: −30 V Source temp: 120°C Desolvation gas (N_2) temp: 350°C Desolvation gas flow rate: 700 L/h Cone gas flow rate: 10 L/h Resolution: 11,000 Desolvation gas flow rate: 600 L/h Cone gas flow rate: 50 L/h Desolvation gas (N_2) temp: 350°C Source temp.: 120°C Capillary voltage: 3.5 kV Collision gas (Ar) flow rate: 0.50 mL/min Resolution: 10,000	Recovery: 70% –110 % RSD: <20 % TOF LOD: 0.28 –8.42 pg LOC: 2.0 –140 μg/kg Linearity: >0.995 QTOF LOD: 0.31 –12.5 pg LOC: 4.2 –229.2 μg/kg Linearity: > 0.995 Mass accuracy: <2.6 ppm	Grimalt, Sancho, Pozo, and Hernández (2010)
Orange, Tomato, and Leek	53	10 g sample + 10 mL ACN shake well for 1 min; add 4 g $MgSO_4$ + 1 g NaCl and shaken for 1 min; add 50 μL ISTD solution, mix on a vortex for 30 s, and centrifuged for 5 min at 5000 rpm; Transfer 1 mL aliquot of to 1.5 mL microcentrifuge vial containing 25 mg PSA + 150 mg anh. $MgSO_4$; vortexed for 30 s; centrifuged for 1 min at 6000 rpm to separate solids from solution.	Column: C_8 (150 mm × 4.6 mm; 5 μm) Mobile phase: A: 0.1% HCOOH in ACN; B: 0.1% HCOOH in high purity H_2O Flow rate: 600 μL/min	HPLC-Q-TOF-MS (ESI+)	Superheated N_2 sheath gas temp: 400°C Flow rate: 12 L/min Capillary voltage: 4000 V Nebulizer pressure: 40 psig Drying gas flow rate: 10 L/min Gas temperature: 300°C	Not reported	Ferrer, Lozano, Agüera, Girón, and Fernández-Alba (2011)

Matrix	No.	Sample preparation	Instrument	MS conditions	Performance	Reference
Fruits and Vegetables	850	15 g sample + 15 mL 1% acetic acid in ACN; shaken well; add 6 g anh. MgSO$_4$ + 2.5 g C$_2$H$_9$NaO$_5$ shaken for 4 min; centrifuged at 4000 rpm for 4 min; 5 mL supernatant transferred to tube containing 250 mg PSA + 750 mg anh. MgSO$_4$ vortexed for 20 s; centrifuged at 4000 rpm for 4 min; 800 μL of extract evaporated under argon steam and reconstituted with 800 μL initial eluent composition used in separation (ACN/H$_2$O with 0.1% v/v HCOOH 10/90% v/v). Extract filtered through 0.45 μm PTFE and injected to the instrument. Extract filtered through 0.45 μm PTFE and injected to the instrument.	Column: C$_{18}$ (50 mm × 4.6 mm; 1.8 μm) Mobile phase: A: 0.1% HCOOH in H$_2$O; B: ACN Flow rate: 500 μL/min	Skimmer voltage: 65 V Octapole RF: 750 V Fragmentor voltage (CID): 90 V Resolution: 19,500 ± 500 Capillary voltage: 4000 V Nebulizer pressure: 40 psi Drying gas flow rate: 9 L/min Gas temperature: 325°C Skimmer voltage: 65 V Octapole RF: 250 V Fragmentor voltage (CID): 190 V	Recovery: 56% –94% RSD: <4.8% Mass accuracy: <3.5 ppm	Polgár et al. (2012)
HPLC-TOF-MS (ESI+)						
Lemon and Grapes	2	10 g sample + 25 mL MeOH: H$_2$O (80:20 v/v) extraction for 2 min using high-speed blender at 8000 rpm. Extract filtered through filter paper and washed with 10 mL of extraction solvent, adjusted with 50 mL same solvent. Extract filtered through 0.22-μm Nylon syringe filter and injected to the instrument.	Column: BEH C$_{18}$ (100 mm × 2.1 mm; 1.7 μm) Mobile phase: A: 0.5 mM NH$_4$OAc in MeOH; B: 0.5 mM NH$_4$OAc in H$_2$O Flow rate: 300 μL/min	Desolvation gas flow rate: 600 L/h Cone gas flow rate: 60 L/h Desolvation gas (N$_2$) temp.: 350°C Source temp.: 120°C Capillary voltage: 3.5 kV Resolution: 10,000	Not reported	Hernández, Grimalt, Pozo, and Sancho (2009)
UPLC-QTOF-MS (ESI+)						
Brinjal, Cabbage, Cauliflower, Guava, Okra, Onion, Potato, Apple, Banana, Grape, Mango,	60	10 g sample + 25 mL ACN: MeOH (90:10 v/v) vortexed for 3 min; add 5 g NaCl vortexed for 3 min; centrifuged at 5000 rpm for 5 min; supernatant transferred to round bottom flask and evaporated to dryness. Clean-up process: 0.5 g	Column: C$_{18}$ (50 mm × 2.1 mm; 1.7 μm) Mobile phase: A: 0.1% HCOOH in MeOH; B: 0.1% HCOOH in H$_2$O Flow rate: 500 μL/min	Capillary voltage: 80 V Cone voltage: 30 V Source temp.: 115°C Desolvation gas flow rate: 600 L/h Desolvation temp.:	MDL: 0.3 –3.8 μg/kg LOQ: 0.8 –11.8 μg/kg Linearity: >0.99 Recovery: 74% –111% RSD: <15%	Sivaperumal et al. (2015)
UHPLC-TOF-MS (ESI+)						

Continued

Table 6.1 Analytical Methods for Determination of Pesticide Residues in Fruit and Vegetable Samples by LC-HRMS—cont'd

Matrix	No. of Pesticides	Extraction Methods	Instrumentation		TOF-MS Parameters	Method Validation Parameters	References
			LC Parameters	Detection Mode			
Orange, and Pomegranate		GCB + 0.5 g PSA in a 3.0 mL cartridge. A layer of anh. Na_2SO_4 was added to the column to remove traces of H_2O from the eluent. The column was washed with 5 mL of ACN:MeOH (95:5 v/v); The extracted sample was reconstituted in 2 mL ACN: MeOH (95:5 v/v) and loaded onto the column and passed through the column at 1 mL/min flow rate and analyte eluted with 10 mL of ACN:MeOH (95:5 v/v) at a rate of 2 mL/min. This eluent was collected in test tube and evaporated to near dryness. The residues were reconstituted in 1 mL of MeOH and injected to the instrument.			250 °C Cone gas flow rate: 50 L/h Resolution: 11,000	Mass accuracy: <2 ppm	
Apple, Lemon, Orange, and Tomato	54	10 g sample + 10 mL ACN shaken vigorously for 1 min; Add 4 g anh. $MgSO_4$ + 1 g NaCl +1 g $C_6H_9Na_3O_9$ + 0.5 g $C_6H_9NaO_8$ shaken vigorously for 1 min; centrifuged at 600 rpm for 5 min; 1 mL supernatant transferred to tube containing 125 mg PSA + 750 mg $MgSO_4$ + 15 mg of GCB shaken vigorously for 2 min; centrifuged for 5 min at 600 rpm. Extract filtered through membrane filter (PVDF, 0.22 µm) and injected to the instrument.	Column: C_{18} (5 mm × 2.10 mm; 1.7 µm) equipped with security guard ULTRA Cartridges UHPLC C_{18} for 2.1 mm Mobile phase: A: 0.1% HCOOH in H_2O; B: 0.1% HCOOH in MeOH Flow rate: 300 µL/min	UHPLC/ESI-LTQ-Orbitrap-MS	Capillary temp.: 275°C Source voltage: 3 kV Resolution: 30,000	LOD: ≤2 µg/kg Linearity: ≥0.99 Recovery: 58% –120% RSD: <22% Mass accuracy: ≤4 ppm	Farré, Picó, and Barceló (2014)
Apple, Broccoli, Celery, Leek, Melon, Nectarine, Onion, Pear, Pepper, Tomato	199	10 g sample + 10 mL ACN shaken for 5 min using a mechanical horizontal shaker; add mixture of salts; 4 g $MgSO_4$ + 1 g NaCl + 0.5 g $C_6H_8Na_2O_8$ + 1 g $C_6H_5Na_3O_9$ then shaken by hand immediately; centrifuged at 3500 rpm for 3 min; 1.5 mL supernatant transferred to 2 mL dSPE tube containing 150 mg $MgSO_4$ + 50 mg PSA; shaken	Column: C_{18} (100 mm × 2.1 mm; 1.7 µm) Mobile phase: A: 10 mM NH_4OAc in H_2O; B: 10 mM NH_4OAc in MeOH Flow rate: 450 µL/min	UPLC-QTOF-MS (ESI+)	Sample cone voltage: 25 V Nebulizer gas (N_2) flow rate: 50 L/h Nebulizer Gas temp: 120°C Desolvation gas flow: 1000 L/h Desolvation temp.: 550°C Resolution: 19,000	SDLs: 10 –50 µg/kg Mass accuracy: <5 ppm	López et al. (2014)

Lemon	2	vigorously for 1 min; centrifuged at 3500 rpm for 1 min; 1 mL supernatant evaporated to near dryness and reconstituted with 1 mL of ACN:H$_2$O (1:3 v/v) then injected to the instrument. 15 g sample + 15 mL ACN shaken well for 1 min; add 4 g MgSO$_4$ + 1.5 g NaCl and vigorously shaken for 1 min; centrifuged at 3700 rpm for 1 min; 5 mL supernatant transferred to tube containing 250 mg PSA + 750 mg MgSO$_4$; energetically shaken for 20 s and centrifuged at 3700 rpm for 1 min; extract of sample evaporated to near dryness and reconstituted with initial mobile phase. Extract filtered through 0.2 µm PTFE and injected to the instrument.	Column: C$_8$ (150 mm × 4.6 mm; 5 µm) Mobile phase: A: 0.1% HCOOH in H$_2$O; B: ACN Flow rate: 600 µL/min	Capillary voltage: 4000 V Nebulizer pressure: 40 psig Vaporizer temperature: 350°C Drying gas flow rate: 9 L/min Gas temperature: 300°C Fragmentor voltage (CID): 190 V Skimmer voltage: 60 V Octapole DC 1: 37.5 V Octapole RF: 250 V Resolution: 9500 ± 500	Mass accuracy: ≤9.5 ppm	Thurman, Ferrer, and Fernández-Alba (2005)
Tomato, Lettuce, Pepper, and Cucumber	3	15 g sample + 30 mL EtOAc blended with high-speed blender for 30 s at 21,000 rpm; this step repeated two times; obtained extract filtered through 20 g of anh. Na$_2$SO$_4$; the solid washed with 50 mL EtOAC; extract evaporated to dryness and reconstituted with 15 mL of MeOH by sonication. Extract filtered through 0.2 µm PTFE and injected to the instrument.	Column: C$_8$ (150 mm × 4.6 mm; 5 µm) Mobile phase: A: ACN; B: 0.1% HCOOH in H$_2$O Flow rate: 600 µL/min	Capillary voltage: 4000 V Nebulizer pressure: 40 psig Drying gas flow rate: 9 L/min Gas temperature: 300°C Fragmentor voltage: 190 V Skimmer voltage: 60 V Octapole DC 1: 37.5 V Octapole RF: 250 V Resolution: 9500 ± 500	LOD: 5–10 µg/kg Linearity: >0.991 RSD: 2%–5% Mass accuracy: <3 ppm	Thurman et al. (2005)

The instrument/method for row 2 is HPLC-TOF-MS (ESI+) and for row 3 is HPLC-TOF-MS.

Continued

Table 6.1 Analytical Methods for Determination of Pesticide Residues in Fruit and Vegetable Samples by LC-HRMS—cont'd

Matrix	No. of Pesticides	Extraction Methods	Instrumentation — LC Parameters	Detection Mode	TOF-MS Parameters	Method Validation Parameters	References
Apple, Pear, Tomato, Potato, Pepper, and Cucumber	100	15 g sample + 15 mL ACN shaken well for 1 min; add 4 g MgSO$_4$ + 1.5 g NaCl and vigorously shaken for 1 min; centrifuged at 3700 rpm for 1 min; 5 mL supernatant transferred to tube containing 250 mg PSA + 750 mg MgSO$_4$; energetically shaken for 20 s and centrifuged at 3700 rpm for 1 min; extract evaporated to near dryness and reconstituted with initial mobile phase. Extract filtered through 0.45 µm PTFE and injected to the instrument.	Column: C$_8$ (150 mm × 4.6 mm; 5 µm) Mobile phase: A: 0.1% HCOOH in H$_2$O; B: ACN Flow rate: 600 µL/min	HPLC- TOF-MS (ESI+)	Capillary voltage: 4000 V Nebulizer pressure: 40 psig Drying gas flow rate: 9 L/min Gas temperature: 300°C Fragmentor voltage: 190 and 230 V Skimmer voltage: 60 V Octapole DC 1: 37.5 V Octapole RF: 250 V Resolution: 9500 ± 500	LOD: 10 –100 µg/kg Mass accuracy: ±5 ppm	Ferrer, Fernandez-Alba, Zweigenbaum, and Thurman (2006)
Pear and Pepper	8	15 g sample + 15 mL ACN shaken well for 1 min; add 4 g MgSO$_4$ + 1.5 g NaCl and vigorously shaken for 1 min; centrifuged at 3700 rpm for 1 min; 5 mL supernatant transferred to tube containing 250 mg PSA + 750 mg MgSO$_4$; energetically shaken for 20 s and centrifuged at 3700 rpm for 1 min; 2 mL extract evaporated to near dryness and reconstituted with initial mobile phase (10% MeOH). Extract filtered through 0.45 µm PTFE and injected to the instrument.	Column: C$_8$ (150 mm × 4.6 mm; 5 µm) Mobile phase: A: 0.1% HCOOH in H$_2$O; B: ACN Flow rate: 600 µL/min	HPLC- TOF-MS (ESI+)	Capillary voltage: 4000 V Nebulizer pressure: 40 psig Drying gas flow rate: 9 L/min Gas temperature: 300°C Skimmer voltage: 60 V Octapole DC 1: 37.5 V Octapole RF: 250 V Fragmentor voltage (CID): 190 V Resolution: 9500 ± 500	Mass accuracy: <4.65 ppm	Garcia-Reyes, Molina-Diaz, and Fernandez-Alba (2007)
Fruit and Vegetable	100	15 g sample + 15 mL ACN shaken well for 1 min; add 4 g MgSO$_4$ + 1.5 g NaCl and vigorously shaken for 1 min; centrifuged at 3700 rpm for 1 min; 5 mL supernatant transferred to tube containing 250 mg PSA + 750 mg MgSO$_4$;	Column: C$_8$ (150 mm × 4.6 mm; 5 µm) Mobile phase: A: 0.1% HCOOH in H$_2$O; B: ACN Flow rate: 600 µL/min	HPLC-TOF-MS(ESI+) and HPLC-TOF-MS/MS (ESI+)	Capillary voltage: 4000 V Nebulizer pressure: 40 psig Drying gas flow rate: 9 L/min Gas temperature: 325°C	LOD: 0.06 –35 µg/kg LOQ: 0.2 –100 µg/kg Linearity: >0.997 RSD: <0.45%	Garcia-Reyes, Hernando, Ferrer, Molina-Diaz, and Fernández-Alba (2007)

Matrix	No. of pesticides	Sample preparation	Chromatographic conditions	MS	MS conditions	Method performance	Reference
(continued from previous page)		energetically shaken for 20 s and centrifuged at 3700 rpm for 1 min; 2 mL extract evaporated to near dryness and reconstituted with 2 mL of 10% MeOH. Extract filtered through 0.45 μm PTFE and injected to the instrument.			Skimmer voltage: 60 V Octapole DC 1: 37.5 V Octapole RF: 250 V Fragmentor voltage (CID): 190 V Resolution: 9500 ± 500	Mass accuracy: ≤2.8 ppm	
Peppers	3	15 g sample + 15 mL ACN shaken well for 1 min; add 4 g MgSO$_4$ + 1.5 g NaCl and vigorously shaken for 1 min; centrifuged at 3700 rpm for 1 min; 5 mL aliquot transferred to tube containing 250 mg PSA + 750 mg MgSO$_4$; centrifuged at 3700 rpm for 1 min; 2 mL extract evaporated to near dryness and reconstituted with 2 mL of 10% MeOH. Extract filtered through 0.45 μm PTFE and injected to the instrument.	Column: C$_8$ (150 mm × 4.6 mm; 5 μm) Mobile phase: A: 0.1% HCOOH in H$_2$O; B: ACN Flow rate: 600 μL/min	HPLC-TOF-MS (ESI+)	Capillary voltage: 4000 V Nebulizer pressure: 40 psig Vaporizer temperature: 325°C Drying gas flow rate: 9 L/min Fragmentor voltage (CID): 160 V Skimmer voltage: 60 V Octapole DC 1: 37.5 V Octapole RF: 250 V Resolution: 9700 ± 500	LOD: 0.06–0.6 μg/kg LOQ: 0.1–5 μg/kg Linearity: >0.997 Recovery: 76%–109% RSD: <10% Mass accuracy: ≤2 ppm	Mezcua et al. (2008)
Apple, Banana, Pear, Apple Juice, Peas, Sweet Potato, Creamed Corn, Squash, and Carrots	138	10 g sample + 10 mL ACN + acetic acid (99:1 v/v) + 1 g anh. NaOAc + 4 g anh. MgSO$_4$ shaken for 45 s by hand followed by centrifuge at 3000 rpm for 2 min; 6–8 mL supernatant transferred to tube containing 400 mg PSA + 1200 mg MgSO$_4$ shaken for 45 s; centrifuged at 3000 rpm for 2 min; 5 mL extract evaporated to 0.2–0.3 mL and made up to 1 mL with MeOH; vortexed for 30 s then made up to 2 mL with 0.1 M NH$_4$OAc and vortexed for 30 s. Extract filtered through PVDF, 0.22-μm membrane and injected to the instrument.	Column: C$_{18}$ (100 mm × 2.1 mm; 1.7 μm) Mobile phase: A: ACN; B: 10 mM ammonium Flow rate: 400 μL/min	UPLC-QqTOF-MS (ESI+)	Capillary voltage: 3.20 kV Source temp.: 120°C Desolvation temp.: 300°C Nebulizer gas N$_2$ flow: 50 L/h Desolvation gas (N$_2$) flow: 800 L/h Collision gas (Ar) pressure: 5.3×10^{-3} mbar Collision energy: 5 eV Cone voltage: 20 V Resolution: 15,000	LOD: <1 μg/kg LOQ: <10 μg/kg Linearity: ≥0.97 Recovery: 17.6%–139.9% RSD: 6.7%–47.3% Mass accuracy: 2–10 ppm	Wang and Leung (2009)

Continued

Table 6.1 Analytical Methods for Determination of Pesticide Residues in Fruit and Vegetable Samples by LC-HRMS—cont'd

Matrix	No. of Pesticides	Extraction Methods	Instrumentation			Method Validation Parameters	References
			LC Parameters	Detection Mode	TOF-MS Parameters		
Tomato, Zucchini, and Watermelon	1	15 g sample + 15 mL 1% acetic acid in ACN + 6 g anh. MgSO$_4$ + 1.5 g anh. NaOAc; shaken vigorously for 1 min by hand and centrifuged at 3700 rpm for 2 min; 5 mL supernatant transferred to tube containing 250 mg PSA + 750 mg anh. MgSO$_4$; shaken for 20 s and centrifuged at 3700 rpm for 2 min; 1 mL extract evaporated to dryness and reconstituted with 1 mL of H$_2$O:MeOH (9:1) and injected to the instrument.	Column: C$_8$ (150 mm × 4.6 mm; 5 μm) Mobile phase: A: ACN; B: 0.1% HCOOH in H$_2$O Flow rate: 600 μL/min	HPLC-TOF-MS (ESI+)	Capillary voltage: 4000 V Nebulizer pressure: 40 psig Drying gas flow rate: 9 L/min Drying gas temperature: 300°C Fragmentor voltage (CID): 190 V Skimmer voltage: 60 V Octapole DC 1: 37.5 V Octapole RF: 250 V Resolution: 9500 ± 500	LOD: 0.5 μg/kg LOQ: 10 μg/kg Linearity: >0.999 Recovery: 65%–87% RSD: <14% Mass accuracy: ≤0.5 ppm	Valverde, Aguilera, Ferrer, Camacho, and Cammarano (2010)
Apple, Banana, Cantaloupes, Orange and Orange Juice, Carrot, Corn, Onion, Pear, Potato, Spinach, and Tomato	148	15 g sample + 15 mL ACN + acetic acid (99:1 v/v) + 1.5 g anh. NaOAc + 6 g anh. MgSO$_4$ and shaken for 45 s; centrifuged at 3000 rpm for 3 min. 9 mL supernatant transferred to tube containing 600 mg PSA + 1800 mg MgSO$_4$; shaken for 45 s and centrifuged at 3000 rpm for 3 min; 1 mL supernatant evaporated to 0.1 −0.2 mL and made up to 0.5 mL with MeOH; vortexed for 30 s then made up to 1 mL with 0.1 M NH$_4$OAc and vortexed for 30 s; 1 mL extract transferred to vial and add 500 μL mixture of 0.1 M NH$_4$OAc + MeOH (50:50, v/v) and vortexed for 30 s then filtered and injected to the instrument.	Column: C$_{18}$ (100 mm × 2.1 mm; 1.7 μm) Mobile phase: A: ACN; B: 10 mM NH$_4$OAc in H$_2$O Flow rate: 400 μL/min	UHPLC-QqTOF-MS (ESI+)	Capillary voltage: 3.20 kV Source temp.: 120°C Desolvation gas (N$_2$) temp.: 300°C Nebulizer gas (N$_2$) flow rate: 50 L/h Desolvation gas (N$_2$) flow rate: 800 L/h Collision gas (Ar) pressure: 5.3 × 10^{-3} mbar Resolution: 15,000	Recovery: 81%–110% RSD: <20%	Wang, Chow et al. (2010)
Berries	148	15 g sample + 15 mL ACN + acetic acid (99:1 v/v) + 1.5 g anh. NaOAc + 6 g anh. MgSO$_4$ and shaken for 45 s; centrifuged at 3000 rpm for	Column: C$_{18}$ (100 mm × 2.1 mm; 1.7 μm) Mobile phase: A: ACN; B: 10 mM	UHPLC-QqTOF-MS (ESI+)	Capillary voltage: 3.20 KV Source temp.: 120°C Desolvation gas (N$_2$)	Recovery: 81%–110% RSD: ≤20%	Wang, Leung, and Chow (2010)

Matrix	No.	Sample preparation	Column / Mobile phase	Instrument	MS parameters	Accuracy	Reference
Tomato, Pepper, Zucchini, Orange, and Leek	97	3 min; 9 mL aliquot transferred to tube containing 600 mg PSA + 1800 mg MgSO$_4$; shaken for 45 s and centrifuged at 3000 rpm for 3 min; 1 mL supernatant evaporated to 0.1 −0.2 mL and made up to 0.5 mL with MeOH; vortexed for 30 s then made up to 1 mL with 0.1 M NH$_4$OAc and vortexed for 30 s; 1 mL extract transferred to vial and added 500 µL mixture of 0.1 M NH$_4$OAc + MeOH (50: 50 v/v) and vortexed for 30 s and injected to the instrument.	Column: C$_{18}$ (50 mm × 4.6 mm; 1.8 µm) Mobile phase: A: 0.1% HCOOH + 5% Milli Q H$_2$O in ACN; B: 0.01% HCOOH in H$_2$O Flow rate: 600 µL/min	LC-QTOF-MS (ESI+)	temp.: 300°C Nebulizer Gas (N$_2$) flow rate: 50 L/h Desolvation gas (N$_2$) flow rate: 800 L/h Collision gas (Ar) pressure: 5.3 × 10^{-3} mbar Resolution: 15,000 Superheated N$_2$ sheath gas temp.: 400°C Flow rate: 12 L/min Capillary voltage: 4000 V Nebulizer pressure: 40 psig Drying gas flow rate: 10 L/min Gas temperature: 350°C Skimmer voltage: 65 V Octapole RF: 750 V Fragmentor voltage (CID): 90 V Resolution: 19,500 ± 500	Mass accuracy: <2.37 ppm	Malato, Lozano, Mezcua, Agüera, and Fernandez-Alba (2011)
Tomato skin	Unknown pesticides	10 g sample + 10 mL ACN shaken for 8 min using automatic axial extractor; add 4 g MgSO$_4$ + 1 g NaCl + 1 g C$_6$H$_5$Na$_3$O$_9$ + 0.5 g C$_6$H$_9$NaO$_8$ and shaken by automatic axial extractor for 8 min; centrifuged at 3700 rpm for 5 min; 5 mL supernatant transferred to tube containing 750 mg MgSO$_4$ + 125 mg PSA + 125 mg C$_{18}$; The extract vortexed for 30 s; centrifuged at 3700 rpm for 5 min; extract acidified with 10 µL HCOOH 5% per mL of extract. This treatment 1 mL of sample extract represents 1 g of sample and injected to the instrument. Wash the skin of tomato three times with 2–5 mL MeOH; collect the solvent in 150 mL Pyrex beaker; transfer the MeOH to 5 mL syringe and filter through PTFE filter and 0.3 mL supernatant diluted with 0.6 mL deionized H$_2$O then injected to the instrument.	NH$_4$Ac in H$_2$O Flow rate: 400 µL/min Column: C$_8$ (150 mm × 4.6 mm; 5 µm) Mobile phase: A: ACN; B: 0.1% HCOOH in H$_2$O Flow rate: 600 µL/min	LC-TOF-MS (ESI+)	Capillary voltage: 4000 V Nebulizer pressure: 40 psig Drying gas flow rate: 9 L/min Drying gas temperature: 300°C Fragmentor voltage (CID): 190 V Skimmer voltage: 60 V Octapole DC 1: 37.5 V Octapole RF: 250 V Resolution: 9500 ± 500	Mass accuracy: ≥2.4 ppm	Thurman et al. (2006)

Continued

Table 6.1 Analytical Methods for Determination of Pesticide Residues in Fruit and Vegetable Samples by LC-HRMS—cont'd

Matrix	No. of Pesticides	Extraction Methods	Instrumentation		Detection Mode	TOF-MS Parameters	Method Validation Parameters	References
			LC Parameters					
Wheat, Lettuces, Avocado, and Orange	30	0.5 g sample + 0.5 g C$_{18}$ blended using pestle; this mix introduced to a glass column; add 10 mL DCM; The sample allowed to elute by applying slight vacuum; The eluent collected in a conical tube; add 3 mL MeOH to avoid evaporation to dryness then evaporated up to 0.5 mL. Extract filtered through 0.45 μm PTFE and injected to the instrument.	Column: C$_{18}$ (5 cm × 2.1 mm; 1.7 μm) Mobile phase: A: 10 mM AF in H$_2$O; B: 10 mM AF in MeOH Flow rate: 400 μL/min		UPLC-QqTOF-MS (ESI+)	Collision energy: 4 eV Desolvation gas (N$_2$) flow rate: 500 L/h Desolvation gas (N$_2$) temp: 350°C Cone gas: 50 L/h Source temp.: 120°C Capillary voltage: 3000 V Cone voltage: 20 V Resolution: 5000	Linearity: >0.995 Recovery: 74% –99% RSD: <19% Mass accuracy: ≤13.5	Picó, Blasco, Farré, and Barceló (2009)
Peach and Pear	71	10 g sample + 10 mL ACN + 4 g MgSO$_4$ + 1 g NaCl + 1 g C$_6$H$_9$Na$_3$O$_9$ + 0.5 g C$_6$H$_9$NaO$_8$; adjust the pH 5–5.5 using 5 M NaOH and shaken sample vigorously for 1 min; centrifuged at 3000 rpm for 5 min; 6 mL supernatant transferred to tube containing 150 mg PSA + 900 mg MgSO$_4$ and shaken for 30 s; centrifuged at 3000 rpm for 5 min. Adjust the pH to 5.0 and injected to the instrument.	Mobile phase: A: 0.1% HCOOH in H$_2$O; B: 100% ACN Flow rate: 300 μL/min		LC-QTOF-MS (ESI+)	Drying gas flow rate: 10 L/min Drying gas temperature: 325°C Nebulizer pressure: 50 psig Vcap: 4000 V Fragment voltage (CID): 175 V	Not reported	Meng, Zweigenbaum, Fürst, and Blanke (2010)
Fruits and Vegetables	184	10 g sample + 10 mL 1% acetic acid in ACN + 1 g anh. NaOAc + 4 g anh. MgSO$_4$; rotate by end-over-end for 30 min; supernatant was diluted 1:1 with H$_2$O then filtered and injected to the instrument.	Column: T3 (100 mm × 3 mm; 3 μm) Mobile phase: A: H$_2$O with 2 mM AF and 20 μL HCOOH/L; B: MeOH:H$_2$O (95:5 v/v) with 2 mM AF and 20 μL HCOOH/L Flow rate: 300 –450 μL/min		LC-Orbitrap MS (ESI + and ESI −)	Capillary voltage (+): 3.5 kV Capillary voltage (+): –2.5 kV Sheath gas flow rate: 47.5 AU Auxiliary gas flow rate: 11.25 AU Cone gas: 2.25 AU Capillary temperature: 256°C Heater temperature: 413°C Resolution: 70,000	Recovery: ≤139 % RSD: 1% –147% SDL: 10 –200 ng/g	Zomer and Mol (2015)

Sample	No.	Sample preparation	Column/Mobile phase	Instrument	MS conditions	Performance	Reference
Tomato, Pepper, and Orange	170	10 g + 10 mL ACN and shaken for 4 min. Afterward 4 g MgSO$_4$, 1 g NaCl, 1 g C$_6$H$_9$Na$_3$O$_9$, and 0.5 g C$_6$H$_9$NaO$_8$ added and shaken 4 min; The extract was centrifuged (3700 rpm) for 5 min; 5 mL supernatant transferred to a 15 mL PTFE centrifuge tube containing 750 mg of MgSO$_4$; 125 mg of PSA and 125 mg of C$_{18}$; shaken for 30 s and centrifuged at 3700 rpm for 5 min; The extracts transferred into amber vial and acidified with 10 μL formic acid 5% per mL of extract. Injected to the instrument.	Column: C$_{18}$ (150 mm × 2.1 mm; 2.6 μm) Mobile phase: A: H$_2$O: MeOH (98:2%) with 5 mM AF and 0.1% HCOOH; B: MeOH: H$_2$O (98:2%) with 5 mM AF and 0.1% HCOOH	UHPLC Orbitrap MS (HESI II$^+$ and HESI II$^-$)	Sheath gas flow rate: 40 Auxiliary gas flow rate: 5 Sweep gas flow rate: 1 Spray voltage: 3.00 kV Capillary temperature: 280°C S-lens RF level: 55 Heater temperature: 350°C Resolution: 17,500, 35,000, 70,000	Linearity: ≥0.995 Recovery: 83% –91% Mass accuracy: ≤2.5 ppm	Rajski, Gomez-Ramos Mdel, and Fernandez-Alba (2014)

ACN, acetonitrile; AF, ammonium formate; anh., anhydrous; Ar, argon; BEH, ethylene bridged hybrid; C$_2$H$_9$NaO$_5$, sodium acetate trihydrate; C$_6$H$_9$NaO$_8$, disodium hydrogen citrate sesquihydrate; C$_6$H$_9$Na$_3$O$_9$, trisodium citrate dehydrate; CID, collision-induced dissociation; DCM, dichloromethane; dSPE, dispersive solid-phase extraction; ESI, electrospray ionization; EtOAc, ethyl acetate; GCB, graphite carbon black; H$_2$O, water; HCOOH, formic acid; HPLC, high-performance liquid chromatography; HRMS, high-resolution mass spectrometry; HSS, high-strength silica; LCHRMS, liquid chromatography high-resolution mass spectrometry; LC-TOF MS, liquid chromatography time-of-flight mass spectrometry; LOC, limit of confirmation; LOD, limit of detection; LOQ, limit of quantitation; MCD, multichannel detector; MeOH, methanol; MgSO$_4$, magnesium sulfate; MS, mass spectrometry; m/z, mass-to-charge ratio; N$_2$, nitrogen; Na$_2$SO$_4$, sodium sulfate; NaCl, sodium chloride; NaOAc, sodium acetate; NaOH, sodium hydroxide; NH$_4$OAc, ammonium acetate; PLE, pressurized liquid extraction; PSA, primary secondary amine; Psig, pounds per square inch gage; PTFE, polytetrafluoroethylene; PVDF, polyvinylidene difluoride; QqTOF, quadrupole-time-of-flight; QuEChERS, quick, easy, cheap, effective, rugged, and safe; RSD, relative standard deviation; s, second; SDL, screen detection limit; temp., temperature; UHPLC, ultrahigh-performance liquid chromatography.

optimized instrument condition and method accuracy parameters for the estimation of pesticide residues in fruit and vegetable samples by HRMS. The optimization of sample extraction and instrumentation conditions is a tedious procedure. The optimized extracting solvents, LC column, mobile phase, and mass analyzer parameters, which are required for analysis of the analytes, are emphasized in the table. The table provides the simplified procedure for the estimation of pesticide residues in fruit and vegetable samples by HRMS. An overview of the recent applications of pesticide residues in fruit and vegetable samples by LC-TOF-MS is showed in Table 6.1.

6.3 OPTIMIZED SAMPLE PREPARATION AND CHROMATOGRAPHIC CONDITIONS FOR MASS ANALYZERS

Food samples are complex matrices; therefore sample preparation is very essential to achieve accurate results of the analytes and maintain the performance of the instrument. In the past years, conventional extraction procedures were used for the determination of pesticides in food samples. The extraction techniques, viz., solid-phase extraction, solid-phase microextraction, pressurized liquid extraction, and solid–liquid extraction were mainly used (van Leeuwen & de Boer, 2008). In the solid–liquid extraction technique many organic solvents, viz., methanol, acetonitrile, and ethyl acetate were used for the extraction of pesticides in different matrices (Kotretsou & Koutsodimou, 2006). In addition, a cleanup step is necessary to remove the coextractant from the sample. For the cleanup process, several sorbents such as C_8, C_{18}, polymeric, primary secondary amine or Bondesil-NH_2, etc., are used. The selection of suitable sorbent depends upon the polarity of the analytes (van Leeuwen & de Boer, 2008). Nowadays many generic extraction and cleanup procedures have been developed, which are useful in determining a wide range of pesticide residues in food samples. The QuEChERS is one such method for the extraction of pesticide residues in fruit and vegetable samples (Picó et al., 2006). The QuEChERS method was adopted for the determination of pesticide residues in plant matrices. The principle involves an initial extraction with organic solvents, viz., acetone, ethyl acetate, and acetonitrile and liquid–liquid partitioning after the addition of $MgSO_4$ and NaCl, followed by dispersive solid-phase extraction for the sample cleanup step (Rodriguez-Aller, Gurny, Veuthey, & Guillarme, 2012). The generic extraction methods had some disadvantages, i.e., coextraction of matrix components along with analytes interferes during the analysis. The matrix effect (ME) depends upon the sample size, extraction and cleanup procedure, and instrument technique. The ME has a significant role during the analysis and is to be minimized. The UHPLC-MS technique is less susceptible to ME compared with conventional HPLC-MS. However, the ME varies from sample to sample and analyte to analyte (O'Mahony et al., 2013). There are various options available to minimize MEs, viz., matrix-matched calibration or the standard-addition method, atmospheric pressure chemical ionization, cleanup procedure, use of isotopically labeled internal standards, and dilution of the samples (Beltrán et al., 2013; O'Mahony et al., 2013).

UHPLC-MS is a fast analysis technique and a large number of pesticide residues can be determined from different matrices. Generally, C_{18}-based stationary phases are used and aqueous and organic phases are used as mobile phase for the chromatographic separation (Geis-Asteggiante et al., 2012). Methanol or acetonitrile is used as the organic phase and formic acid is used as an additive, which may enhance ionization and chromatographic separation of the pesticides. However, some specific conditions are applied for the determination of pesticides in fruit and vegetable samples as described in Table 6.1. For example, Polgár et al. (2012) used a mobile phase consisting of acetonitrile and 0.1% formic acid in water with gradient conditions for determination of pesticide residues in fruit and vegetable samples.

UHPLC coupled with QqQ and HRMS analyzers, viz., TOF or Orbitrap, are increasingly being used for the quantitation and confirmation of a number of analytes (Geis-Asteggiante et al., 2012). UHPLC-QqTOF has been used for the determination of 138 pesticides in five infant foods based on fruits and vegetables. The detection level of target analytes was achieved at 1 and 10 µg/kg as the confirmation and quantification levels (Williamson & Bartlett, 2007). Wang, Chow, and Leung (2010) analyzed 148 pesticides in five fruits and seven vegetables by the UHPLC-QqTOF technique. Lacina et al. (2010) used UHPLC-TOF for the analysis of 212 pesticide residues in strawberry, spinach, tomato, and apple by the QuEChERS method. The UHPLC-HRMS (UHPLC-Q-TOF/MS) was optimized for the determination of pesticide residues in fruits and vegetables using the QuEChERS method. Fig. 6.1 shows extracted ion chromatogram of pesticide residues in tomato matrix.

6.4 ANALYTICAL METHOD VALIDATION
6.4.1 MATRIX EFFECT

UHPLC coupled with HRMS with electrospray ionization (UHPLC−ESI-TOF/MS) is the most powerful analytical technique for qualitative and quantitative determination of pesticide residue in fruit and vegetable samples. A high degree of recovery, precision, and mass accuracies are a perfect choice for the multiresidue method for food analysis. However, despite all these advantages, ME is one of the most important drawbacks while using the ESI technique. MEs are a major concern in food samples, which can create serious problems and severely affect the determination of residue analysis at trace levels and also affect the reproducibility of the analytes (Hernández et al., 2006; Pihlström et al., 2007). The evaluation of MEs is done by comparing peak areas of the matrix matched standards with corresponding peak areas of standards in the solvent. The quantification of MEs is generally based on signal enhancement or suppression in the corresponding peak area by using the

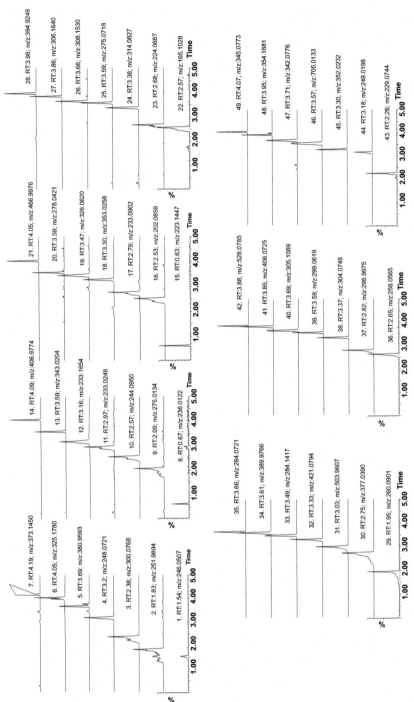

FIGURE 6.1 Ultrahigh-Performance Liquid Chromatography–High-Resolution Mass Spectrometry Extracted Ion Chromatogram of Pesticides Spiked at 10 µg/kg in Tomato Matrix.

1. Monocrotophos; 2. Dimethoate; 3. Phosphamidon; 4. Methiocarb; 5. Chlorfenvinphos; 6. Allethrin; 7. Propagite; 8. Omethoate; 9. Thiocloprid; 10. Carbofuran; 11. Diruon; 12. Siduron; 13. Phenthoate; 14. Ethion; 15. Mexacarbate; 16. Simazin; 17. Fluometuron; 18. Malathion; 19. Etaconazole; 20. Imidacloprid; 21. Abate; 22. Fenuron; 23. Carbaryl; 24. Hexaconazole; 25. Neburon; 26. Tebuconazole; 27. Buprofezin; 28. Profenofos; 29. 3-Hydroxycarbofurn; 30. Thiodicarb; 31. Chlorantraniliprole; 32. Ethoxysulfuron; 33. Metolachlor; 34. Phosalone; 35. Penconazole; 36. Carboxin; 37. Phorate sulfoxide; 38. Propetamphos; 39. Quinalphos; 40. Diazinon; 41. Difenoconazole; 42. Indoxacarb; 43. Metoxuran; 44. Linuron; 45. Iprodion; 46. Flubendiamide; 47. Tilt; 48. Tetramethrin; 49. Oxadiazon.

multilevel with and without matrix-matched calibration. The values of MEs give the signal strength, i.e., a negative value indicates matrix-induced signal suppression, whereas a positive value indicates an enhancement in signal intensity.

6.4.2 EVALUATION OF THE MATRIX INTERFERENCES BY UHPLC-HR/MS

In fruit and vegetable samples, signal enhancement and/or signal suppression have been observed when ESI is used in UHPLC-Q-TOF/MS. Comparison of analyte recovery with and without matrix-matched calibration serves a better procedure for the evaluation of ME. For the evaluation of ME four representative pesticides, viz., dimethoate, chlorantraniliprole, phosphamidon, and 3-hydroxycarbofuran, were selected in green chili samples. Dimethoate was eluted at 2.07 min, with mass of m/z 251.9894. The percentage concentration deviations observed without and with matrix-matched calibrations were 47.6%−89.8% and −0.1% to 9.3%, respectively (Fig. 6.2). Chlorantraniliprole was eluted at 3.21 min, with accurate mass of m/z 503.9606. The percentage concentration deviations observed without and with matrix-matched calibrations were −3.5% to 52.1% and −13.02% to 8.3%, respectively. Phosphamidon was observed at 2.65 min, with accurate mass of m/z 300.0768. The percentage concentration deviations observed without and with matrix-matched calibration were −1.7% to 30.4% and −8.5% to 7.2%, respectively. 3-Hydroxycarbofuran was observed at 2.12 min, with accurate mass of m/z 260.0899. The percentage concentration deviations observed without and with matrix-matched calibration were −55% to 160% and −9.3% to 5.1%, respectively. The observations indicate that, using the matrix-matched calibration curve and dual nebulizer technique significantly reduces the ME. The representative matrix-matched calibration curve is shown in Fig. 6.3. For the investigation of MEs, three matrices, viz., orange, tomato, and leek, with 53 pesticides were selected. In the three matrices studied, signal suppression was the effect observed. In tomato matrix, 68% of the pesticides did not show the relevant effect and one pesticide showed less ME. Thirty percent of the pesticides showed medium signal suppression, and 2% of the compounds had a strong ME. In the leek samples, 9% did not show signal suppression, 54% showed medium ME, and 37% showed strong signal suppression. In case of orange matrix, 43% of the pesticides showed no ME, 29% showed medium signal suppression, and 28% showed strong signal suppression. For the reduction of signal suppression, several dilutions were tested to study the evolution of signal suppression. Finally, 15 times dilution of the matrix eliminated most of the MEs and the authors also suggested "if signal suppression could not be reduced, stable isotope-labelled internal standards are to be used for quantification of problematic pesticides." Sivaperumal, Anand, and Riddhi (2015) evaluated the ME in fruit and vegetable samples by UHPLC-TOF/MS. The study revealed that, using the ESI and TOF/MS technique, relatively strong ME was observed. To reduce the ME, dual-nebulizer and matrix-matched calibration techniques were performed. Finally, it was concluded that by using these techniques no significant ME was observed.

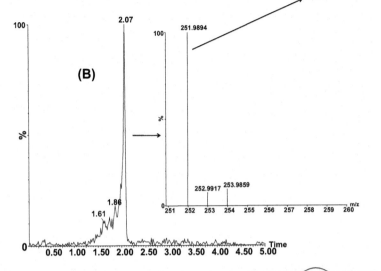

	#	Name	Type	Std. Conc	RT	Area	IS Area	Response	Primary Flags	ppb	%Dev	Adj.RT	Quan Trace
1	1	6397	QC	10.000	2.07	5.636		5.636	MM	10.6	6.1	2.05	251.989
2	2	6398	QC	25.000	2.07	12.080		12.080	MM	27.3	9.3	2.05	251.989
3	3	6399	QC	50.000	2.07	22.190		22.190	MM	53.6	7.1	2.05	251.989
4	4	6400	QC	100.000	2.07	40.059		40.059	bb	99.9	-0.1	2.05	251.989

(B)

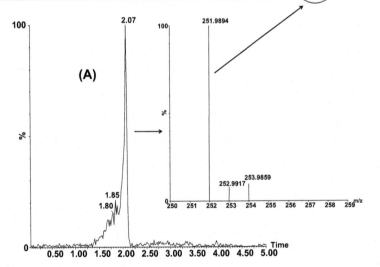

	#	Name	Type	Std. Conc	RT	Area	IS Area	Response	Primary Flags	ppb	%Dev	Adj.RT	Quan Trace
1	1	6397	QC	10.000	2.07	15.378		15.378	MM	14.8	47.6	2.05	251.989
2	2	6398	QC	25.000	2.07	22.826		22.826	MM	36.0	43.8	2.05	251.989
3	3	6399	QC	50.000	2.07	35.887		35.887	MM	73.1	46.3	2.05	251.989
4	4	6400	QC	100.000	2.07	76.879		76.879	MM	189.8	89.8	2.05	251.989

(A)

FIGURE 6.2

Evidence of matrix interference reduction from chili matrix by ultrahigh-performance liquid chromatography—high-resolution mass spectrometry; (A) percentage concentration deviation of dimethoate (RT 2.07 and *m/z* 251.9894) in chili matrix using without matrix-matched calibration; (B) percentage concentration deviation of dimethoate (RT 2.07 and *m/z* 251.9894) in chili matrix using with matrix-matched calibration.

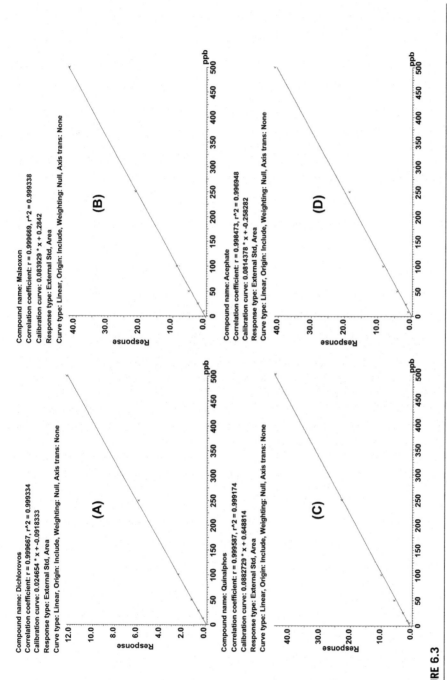

FIGURE 6.3

Matrix-matched calibration of representative pesticide residues in chili matrix. (A) Dichlorovos; (B) malaoxon; (C) quinalphos; (D) acephate.

6.4.3 LIMIT OF DETECTION, LIMIT OF QUANTITATION, ACCURACY, AND PRECISION

LODs and LOQs were also investigated to evaluate the performance of the QuEChERS method coupled with Q-TOF MS. LODs and LOQs were calculated based on signal-to-noise ratio of 3 ($S/N \geq 3$) and 10 ($S/N \geq 10$), respectively. LODs and LOQs ranged from 0.1 to 3 μg/kg and from 0.4 to 10 μg/kg, respectively. The LOQs comply with the regulatory limit of 10 μg/kg prescribed for pesticide residues in fruits and vegetables. Only two compounds (acephate, dichlorvos) showed higher LOQs due to the lower efficiency of electrospray ionization. A recovery study was performed to confirm the extraction efficiency of pesticide residues in fruit and vegetable samples. The mean recovery (%) and coefficient of variation (%) were 68% −122% and 4%−20%, respectively, except carboxin (24.5%) and phenthoate (35.5%). The optimized method meets all the acceptability criteria laid down by SANCO 11945/2015. The quantitation process of representative tomato spiked sample is shown in Fig. 6.4.

6.5 ACCURATE MEASUREMENT OF PESTICIDE RESIDUES IN FRUIT AND VEGETABLE SAMPLES

6.5.1 DETERMINATION OF CHLORINE ISOTOPE IN FOOD SAMPLES

Clomazone, difenoconazole, and chlorfenvinphos were selected for evaluation of chlorine isotope ratio in mango matrix. Clomazone has been chosen for evaluation of single chlorine present in the molecule. Clomazone m/z $[C_{12}H_{14}ClNO_2]^+$ accurate mass $[M]^+$ is at 240.0791 and ^{37}Cl atom isotope m/z is at 242.0765. The mass difference is 1.9974 mass units, which is the mass defect of chlorine isotope in clomazone. The mass intensity of $[M+2]^+$ peak is about 30%, compared with that $[M]^+$ peak, which is consistent with the reference spectrum of clomazone (Fig. 6.5). Difenoconazole has been chosen for evaluation of two chlorine atoms in the molecule. The difenoconazole m/z $[C_{19}H_{17}Cl_2 N_3O_3]^+$ accurate mass $[M]^+$ is at 406.0725 and ^{37}Cl atom isotope m/z is at 408.0700. The mass difference is 1.9975 mass units, which is the mass defect of chlorine isotope in difenoconazole (Fig. 6.5). The mass intensity of $[M+2]^+$ peak is about 66%, compared with $[M]^+$ peak, which is consistent with the reference spectrum of difenoconazole. Chlorfenvinphos has been selected for evaluation of three chlorine atoms present in the molecule. The chlorfenvinphos m/z $[C_{12}H_{14}Cl_3O_4P]^+$ accurate mass $[M]^+$ is at 380.9593 and ^{37}Cl atom isotope m/z is at 382.9565. The mass difference is 1.9972 mass units, which is the mass defect of chlorine isotope in chlorfenvinphos (Fig. 6.5). The mass intensity of $[M+2]^+$ peak is about 99%, compared with $[M]^+$ peak, which is consistent with the reference spectrum of chlorfenvinphos. These results indicate that the $^{35}Cl/^{37}Cl$ isotopic mass ratio increases with increase in number of chlorine atoms present in a single molecule and specified matrix.

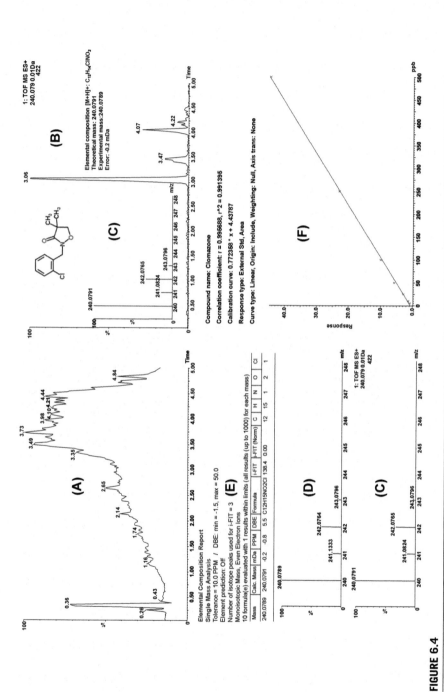

FIGURE 6.4

Quantification process of pesticide residues in fruit and vegetable samples by ultrahigh-performance liquid chromatography–high-resolution mass spectrometry; (A) total ion chromatogram of chlomazone in tomato matrix spiked at 10 μg/kg; (B) extracted ion chromatogram of chlomazone (RT: 3.06 min); (C) mass spectrum of chlomazone (m/z 240.0791); (D) reference spectrum of chlomazone; (E) mass spectrum of chlomazone with chemical composition; (F) matrix-matched calibration of chlomazone.

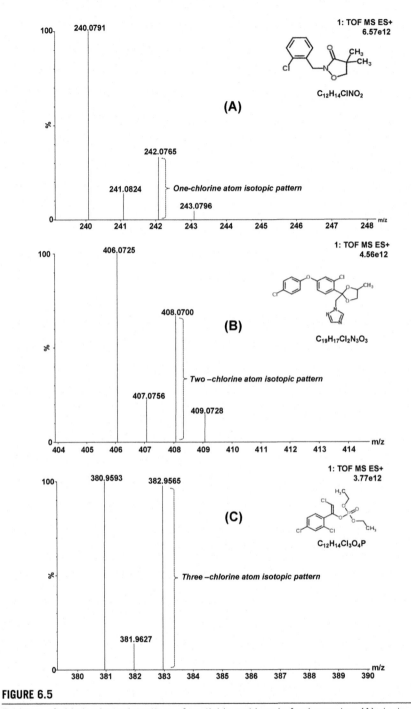

FIGURE 6.5

Evidence of chlorine isotopic pattern of pesticide residues in food samples; (A) single chlorine isotopic pattern in single molecule—clomazone (*m/z* Cl 240.0791 and ^{37}Cl 242.0765) spiked at 25 µg/kg in mango matrix; (B) two chlorine isotopic pattern in single molecule—difenoconazole (*m/z* 406.0725 and ^{37}Cl 408.0700) spiked at 25 µg/kg in mango matrix; (C) three chlorine isotopic patterns in single molecule—clorfenvinphos (*m/z* 380.9593 and ^{37}Cl 382.9565) spiked at 25 µg/kg in mango matrix.

6.5.2 DETERMINATION OF CARBON, CHLORINE, AND BROMINE ISOTOPE IN FOOD SAMPLES

Profenophos $[C_{11}H_{15}BrClO_3PSNa]^+$ has been chosen for evaluation of carbon, chlorine, and bromine isotopes in mango matrix. HRMS accurate mass of profenophos m/z was at 394.9249 $[M]^+$, ^{13}C atom isotopic mass was 395.9281 $[M+1]^+$, and the isotopic mass intensity was approximately 11%. For chlorine and bromine atoms, the isotopic mass was 396.9227 $[M+2]^+$ and mass defect of chlorine and bromine isotopes was 1.9978 mass units. ^{37}Cl and ^{81}Br mass intensity was approximately 125%, which was compared with $[M]^+$ intensity of profenophos in mango matrix (Fig. 6.6). The intensity ratio could provide additional confirmation of pesticide residues in fruit and vegetable samples.

6.6 EVALUATION OF PESTICIDE RESIDUES IN FRUIT AND VEGETABLE SAMPLES

The QuEChERS method was validated for evolutions of pesticide residues in fruit and vegetable samples. The residues were detected in fruit and vegetable samples

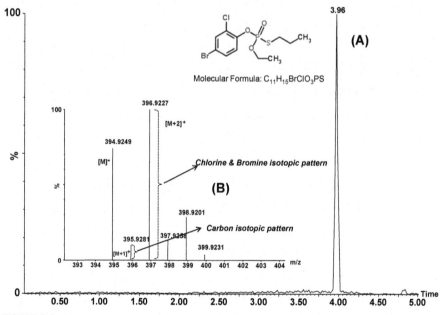

FIGURE 6.6

Evidence of carbon, chlorine, and bromine isotopic pattern of profenofos in mango matrix spiked at 25 µg/kg; (A) extracted ion chromatogram of profenofos (RT 2.48 and m/z 394.9249); (B) carbon (^{13}C m/z 395.9281), chlorine, and bromine isotopic pattern (^{37}Cl and ^{83}Br m/z 396.9227).

by HRMS (Q-TOF/MS). For example, some of the pesticide residues were described as representative pesticides in specific samples and it has been considered as real sample analysis. These samples were collected from different public distribution markets. Imidacloprid (RT 1.73; m/z 278.0439) was detected in capsicum and the observed concentration was 179 µg/kg. The extracted ion chromatogram and mass spectrum of imidacloprid is shown in Fig. 6.7. Thiamethoxam (RT 1.47; m/z 314.0096), acephate (RT 0.31; m/z 206.0024), and imidacloprid (RT 1.73; m/z 278.0419) were detected in different green chili samples; the concentrations were 236, 484, and 278 µg/kg, respectively. An automatic nontargeted screening LC-TOF method successfully detected and identified 210 pesticide residues in fruit and vegetable samples. Seventy eight samples showed residues greater than 0.01 mg/kg (Mezcua et al., 2011). Taylor et al. (2008) analyzed 100 pesticide residues in strawberry by UHPLC-TOF-MS using matrix-matched calibration. The residues were reported in the concentration range from 0.025 to 0.28 mg/kg. A total of 101 pesticides and their degradation products were estimated in fruit and vegetable samples by LC-TOF-MS. About 82% of the samples analyzed (a total of 20) showed at least two pesticides and 15% of the samples were found to have more than the MRL (Ferrer & Thurman, 2007).

Ferrer, García-Reyes, et al. (2005) estimated pesticide residues in fruits and vegetables by LC-TOF-MS. Carbendazim and methomyl were detected in apple; the

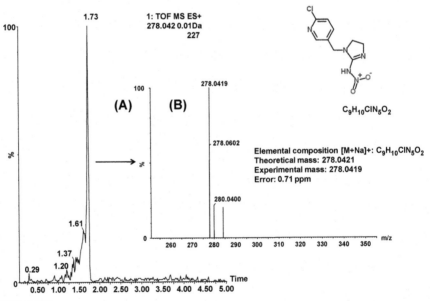

FIGURE 6.7

Imidacloprid detected in capsicum by ultrahigh-performance liquid chromatography—high-resolution mass spectrometry; (A) extracted ion chromatogram of imidacloprid (RT 1.73; m/z 278.0439); (B) accurate mass spectrum of imidacloprid.

concentration was 0.21 and 0.32 mg/kg, respectively. Carbendazim, hexaflumuro, imidaloprid, methomyl, spinosad, and azoxystrobin were found in strawberry sample and the observed concentrations were 0.25, 0.27, 0.10, 0.53, 0.13, and 0.14 mg/kg, respectively.

In strawberry samples (STR1 & STR2), qualitative and quantitative analyses were performed in 100 targeted pesticide residues by UPLC/TOFMS. The pesticide residues detected were azoxystrobin (0.038 mg/kg), pyrimethanil (0.176 mg/kg), myclobutanil (0.024 mg/kg), fenhexamid (0.075 mg/kg), and bupirimate (0.025 mg/kg) in strawberry STR1 sample. In case of strawberry STR2 sample, boscalid (0.271 mg/kg), myclobutanil (0.059 mg/kg), cyprodinil (0.045 mg/kg), pyraclostrobin (0.045 mg/kg), and fludioxinil (0.029 mg/kg) were detected (Taylor et al., 2008). Sivaperumal et al. (2015) estimated pesticide residues in fruit and vegetable samples by UHPLC-TOFMS. The result showed that acephate, propoxur, benfuracarb, carbendazim, chlorpyrifos, difenoconazole, dimethoate, ethion, monocrotophos, myclobutanil, and profenofos were found in the samples. Acephate, ethion, monocrotophos, and profenophos were found frequently, for which the detection range was 14–2774, 13–71, 8–253, and 16–164 μg/kg, respectively.

6.7 **CONCLUSION**

Trace level multiresidue analysis demands detection of more number of pesticide residues with lower LODs. Compliance to regulatory requirements by various national and international agencies to ensure food safety would remain the driving force for the development of generic multiresidue methods. As new pesticide molecules are introduced in the market regularly, the development of generic extraction and cleanup methods pose more challenges. In general, it is desirable to minimize sample handling and reduce the number of extraction steps. A wide variety of pesticide residues can be determined using UHPLC-HRMS following generic extraction and cleanup methods. From the advances and developments in chromatography and HRMS system, it is evident that HRMS offers great improvement in sensitivity and selectivity. This technique has made significant contributions for quantitative determination of pesticide residues in fruit and vegetable samples during recent years. HRMS covers a wide range of pesticide residues in different matrices and it includes targeted and nontargeted analytes. However, more attention is required for impact of MEs on detection and quantification, the presence of isobaric interference. Maintaining confidence in the assignment of identity remains the major limitation for methods employing LC-MS for the determination of chemical contaminants in complex food matrices. The increasing use of hybrid mass spectrometers, incorporating mass analyzers that are capable of high mass resolution and accurate mass measurements, reduces some of the problems associated with selectivity and identification, but further technological development is required to minimize MEs.

ACKNOWLEDGMENTS

The authors are grateful to the Director, National Institute of Occupational Health, Ahmedabad-16, for granting permission to publish the book chapter. The authors acknowledge the Indian Council of Medical Research for the financial assistance of the project (IRIS ID No. 2011-02530). A very appreciative thanks to laboratory staff and family members for their assistance in finalizing this chapter.

REFERENCES

Beltrán, E., Ibáñez, M., Portolés, T., Ripollés, C., Sancho, J., Yusà, V., et al. (2013). Development of sensitive and rapid analytical methodology for food analysis of 18 mycotoxins included in a total diet study. *Analytica Chimica Acta, 783*(6), 39—48.

Brenton, A., & Godfrey, A. (2010). Accurate mass measurement: Terminology and treatment of data. *Journal of the American Society for Mass Spectrometry, 21*(11), 1821—1835.

Cajka, T., Hajslova, J., Lacina, O., Mastovska, K., & Lehotay, S. (2008). Rapid analysis of multiple pesticide residues in fruit-based baby food using programmed temperature vaporiser injection-low-pressure gas chromatography-high-resolution time-of-flight mass spectrometry. *Journal of Chromatography A, 1186*(1—2), 281—294.

De Dominicis, E., Commissati, I., & Suman, M. (2012). Targeted screening of pesticides, veterinary drugs and mycotoxins in bakery ingredients and food commodities by liquid chromatography-high-resolution single-stage Orbitrap mass spectrometry. *Journal of Mass Spectrometry, 47*, 1232—1241.

del Mar Gómez-Ramos, M., Rajski, L., Heinzen, H., & Fernández-Alba, A. R. (2015). Liquid chromatography orbitrap mass spectrometry with simultaneous full scan and tandem MS/MS for highly selective pesticide residue analysis. *Analytical and Bioanalytical Chemistry, 407*(21), 6317—6326.

European Commission's Directorate General for Health and Consumer Protection (DG SANCO), (2015). Guidance document on analytical quality control and validation procedures for pesticide residues analysis in food and feed. Document No SANCO 11945/ 2015. (Online) Available from: http://ec.europa.eu/food/plant/docs/plant_pesticides_mrl_ guidelines_wrkdoc_11945_en.pdf.

Farré, M., Picó, Y., & Barceló, D. (2014). Application of ultra-high pressure liquid chromatography linear ion-trap orbitrap to qualitative and quantitative assessment of pesticide residues. *Journal of Chromatography A, 1328*(7), 66—79.

Ferrer, I., Fernandez-Alba, A., Zweigenbaum, J., & Thurman, E. (2006). Exact-mass library for pesticides using a molecular-feature data base. *Rapid Communications in Mass Spectrometry, 20*(24), 3659—3668.

Ferrer, I., García-Reyes, J., Mezcua, M., Thurman, E., & Fernánndez-Alba, A. (2005). Multi-residue pesticide analysis in fruits and vegetables by liquid chromatography-time-of-flight mass spectrometry. *Journal of Chromatography A, 1082*(1), 81—90.

Ferrer, C., Lozano, A., Agüera, A., Girón, A., & Fernández-Alba, A. (2011). Overcoming matrix effects using the dilution approach in multiresidue methods for fruits and vegetables. *Journal of Chromatography A, 1218*(42), 7634—7639.

Ferrer, I., & Thurman, E. (2007). Multi-residue method for the analysis of 101 pesticides and their degradates in food and water samples by liquid chromatography/time-of-flight mass spectrometry. *Journal of Chromatography A, 1175*(1), 24—37.

Ferrer, I., Thurman, E., & Fernández-Alba, A. (2005). Quantitation and accurate mass analysis of pesticides in vegetables by LC/TOF-MS. *Analytical Chemistry, 77*(9), 2818–2825.

García-Reyes, J., Ferrer, I., Thurman, E., Molina-Díaz, A., & Fernández-Alba, A. (2005). Searching for non-target chlorinated pesticides in food by liquid chromatography/time-of-flight mass spectrometry. *Rapid Communications in Mass Spectrometry, 19*(19), 2780–2788.

García-Reyes, J., Hernando, M., Ferrer, C., Molina-Díaz, A., & Fernández-Alba, A. (2007). Large scale pesticide multiresidue methods in food combining liquid chromatography–time-of-flight mass spectrometry and tandem mass spectrometry. *Analytical Chemistry, 79*(19), 7308–7323.

García-Reyes, J., Molina-Díaz, A., & Fernandez-Alba, A. (2007). Identification of pesticide transformation products in food by liquid chromatography/time-of-flight mass spectrometry via "fragmentation-degradation" relationships. *Analytical Chemistry, 79*(1), 307–321.

Geis-Asteggiante, L., Lehotay, S., Lightfield, A., Dutko, T., Ng, C., & Bluhm, L. (2012). Ruggedness testing and validation of a practical analytical method for >100 veterinary drug residues in bovine muscle by ultrahigh performance liquid chromatography-tandem mass spectrometry. *Journal of Chromatography A, 1258*(10), 43–54.

Gilbert-López, B., García-Reyes, J., Ortega-Barrales, P., Molina-Díaz, A., & Fernández-Alba, A. (2007). Analyses of pesticide residues in fruit-based baby food by liquid chromatography/ electrospray ionization time-of-flight mass spectrometry. *Rapid Communications in Mass Spectrometry, 21*(13), 2059–2071.

Grimalt, S., Pozo, O., Sancho, J., & Hernández, F. (2007). Use of liquid chromatography coupled to quadrupole time-of-flight mass spectrometry to investigate pesticide residues in fruits. *Analytical Chemistry, 79*(7), 2833–2843.

Grimalt, S., Sancho, J., Pozo, O., & Hernández, F. (2010). Quantification, confirmation and screening capability of UHPLC coupled to triple quadrupole and hybrid quadrupole time-of-flight mass spectrometry in pesticide residue analysis. *Journal of Mass Spectrometry, 45*(4), 421–436.

van der Heeft, E., Bolck, Y., Beumer, B., Nijrolder, A., Stolker, A., & Nielen, M. (2008). Full-scan accurate mass selectivity of ultra-performance liquid chromatography combined with time-of-flight and orbitrap mass spectrometry in hormone and veterinary drug residue analysis. *Journal of the American Society for Mass Spectrometry, 20*(3), 451–463.

Hernández, F., Grimalt, S., Pozo, O., & Sancho, J. (2009). Use of ultra-high-pressure liquid chromatography-quadrupole time-of-flight MS to discover the presence of pesticide metabolites in food samples. *Journal of Separation Science, 32*(13), 2245–2261.

Hernández, F., Ibáñez, M., Sancho, J., & Pozo, O. (2004). Comparison of different mass spectrometric techniques combined with liquid chromatography for confirmation of pesticides in environmental water based on the use of identification points. *Analytical Chemistry, 76*(15), 4349–4357.

Hernández, F., Pozo, O., Sancho, J., Bijlsma, L., Barreda, M., & Pitarch, E. (2006). Multiresidue liquid chromatography tandem mass spectrometry determination of 52 non gas chromatography-amenable pesticides and metabolites in different food commodities. *Journal of Chromatography A, 1109*(2), 242–252.

Hird, S. J., Lau, B. P. Y., Schuhmacher, R., & Krska, R. (2014). Liquid chromatography-mass spectrometry for the determination of chemical contaminants in food. *Trends in Analytical Chemistry, 59*(7–8), 59–72.

Kellmann, M., Muenster, H., Zomer, P., & Mol, H. (2009). Full scan MS in comprehensive qualitative and quantitative residue analysis in food and feed matrices: How much resolving power is required. *Journal of the American Society for Mass Spectrometry,* *20*(8), 1464–1476.

Kotretsou, S. I., & Koutsodimou, A. (2006). Overview of the applications of tandem mass Spectrometry (MS/MS) in food analysis of nutritionally harmful compounds. *Food Reviews International, 22*(2), 125–172.

Lacina, O., Urbanova, J., Poustka, J., & Hajslova, J. (2010). Identification/quantification of multiple pesticide residues in food plants by ultra-high-performance liquid chromatography-time-of-flight mass spectrometry. *Journal of Chromatography A, 1217*(5), 648–659.

Lambropoulou, D., & Albanis, T. (2007). Methods of sample preparation for determination of pesticide residues in food matrices by chromatography-mass spectrometry-based techniques: A review. *Analytical and Bioanalytical Chemistry, 389*(6), 1663–1683.

van Leeuwen, S. P., & de Boer, J. (2008). Advances in the gas chromatographic determination of persistent organic pollutants in the aquatic environment. *Journal of Chromatography A, 1186*(1–2), 161–182.

Lehotay, S., de Kok, A., Hiemstra, M., & Bodegraven, V. (2005). Validation of a fast and easy method for the determination of residues from 229 pesticides in fruits and vegetables using gas and liquid chromatography and mass spectrometric detection. *Journal of AOAC International, 88*(2), 595–614.

López, M., Fussell, R., Stead, S., Roberts, D., Mccullagh, M., & Rao, R. (2014). Evaluation and validation of an accurate mass screening method for the analysis of pesticides in fruits and vegetables using liquid chromatography-quadrupole-time of flight-mass spectrometry with automated detection. *Journal of Chromatography A, 1373*(12), 40–50.

Malato, O., Lozano, A., Mezcua, M., Agüera, A., & Fernandez-Alba, A. (2011). Benefits and pitfalls of the application of screening methods for the analysis of pesticide residues in fruits and vegetables. *Journal of Chromatography A, 1218*(42), 7615–7626.

Meng, C., Zweigenbaum, J., Fürst, P., & Blanke, E. (2010). Finding and confirming non targeted pesticides using GC/MS, LC/quadrupole-time-of-flight MS, and databases. *Journal of AOAC International, 93*(2), 703–711.

Mezcua, M., Ferrer, C., García-Reyes, J., Martínez-Bueno, M., Albarracín, M., Claret, M., et al. (2008). Determination of selected non-authorized insecticides in peppers by liquid chromatography time-of-flight mass spectrometry and tandem mass spectrometry. *Rapid Communications in Mass Spectrometry, 22*(9), 1384–1392.

Mezcua, M., Malato, O., García-Reyes, J., Molina-Díaz, A., & Fernández-Alba, A. (2009). Accurate-mass databases for comprehensive screening of pesticide residues in food by fast liquid chromatography time-of-flight mass spectrometry. *Analytical Chemistry, 81*(3), 913–929.

Mezcua, M., Malato, O., Martinez-Uroz, M., Lozano, A., Agüera, A., & Fernández-Alba, A. (2011). Evaluation of relevant time-of-flight-MS parameters used in HPLC/MS full-scan screening methods for pesticide residues. *Journal of AOAC International, 94*(6), 1674–1684.

Mol, H. G., Zomer, P., & de Koning, M. (2012). Qualitative aspects and validation of a screening method for pesticides in vegetables and fruits based on liquid chromatography coupled to full scan high resolution (Orbitrap) mass spectrometry. *Analytical and Bioanalytical Chemistry, 403*, 2891–2908.

Moulard, Y., Bailly-Chouriberry, L., Boyer, S., Garcia, P., Popot, M. A., & Bonnaire, Y. (2011). Use of benchtop exactive high resolution and high mass accuracy orbitrap mass spectrometer for screening in horse doping control. *Analytica Chimica Acta, 700*(1−2), 126−136.

Niessen, W., Manini, P., & Andreoli, R. (2006). Matrix effects in quantitative pesticide analysis using liquid chromatography-mass spectrometry. *Mass Spectrometry Reviews, 25*(6), 881−899.

Núñez, O., Moyano, E., & Galceran, M. T. (2004). Time-of-flight high resolution versus triple quadrupole tandem mass spectrometry for the analysis of quaternary ammonium herbicides in drinking water. *Analytica Chimica Acta, 525*(2), 183−190.

O'Mahony, J., Clarke, L., Whelan, M., O'Kennedy, R., Lehotay, S., & Danaher, M. (2013). The use of ultra-high pressure liquid chromatography with tandem mass spectrometric detection in the analysis of agrochemical residues and mycotoxins in food − Challenges and applications. *Journal of Chromatography A, 1292*(5), 83−95.

Ortelli, D., Edder, P., & Corvi, C. (2005). Pesticide residues survey in citrus fruits. *Food Additives and Contaminants, 22*(5), 423−428.

Picó, Y., Blasco, C., Farré, M., & Barceló, D. (2009). Analytical utility of quadrupole time-of-flight mass spectrometry for the determination of pesticide residues in comparison with an optimized column high-performance liquid chromatography/tandem mass spectrometry method. *Journal of AOAC International, 92*(3), 734−744.

Picó, Y., Blasco, C., & Font, G. (2003). Environmental and food applications of LC-tandem mass spectrometry in pesticide-residue analysis: An overview. *Mass Spectrometry Reviews, 23*(1), 45−85.

Picó, Y., la Farré, M., Segarra, R., & Barceló, D. (2010). Profiling of compounds and degradation products from the postharvest treatment of pears and apples by ultra-high pressure liquid chromatography quadrupole-time-of-flight mass spectrometry. *Talanta, 81*, 281−293.

Picó, Y., la Farré, M., Soler, C., & Barceló, D. (2007). Identification of unknown pesticides in fruits using ultra-performance liquid chromatography-quadrupole time-of-flight mass spectrometry. Imazalil as a case study of quantification. *Journal of Chromatography A, 1176*(1−2), 123−134.

Picó, Y., Farré, M., Soler, C., & Barceló, D. (2007). Confirmation of fenthion metabolites in oranges by IT-MS and QqTOF-MS. *Analytical Chemistry, 79*(24), 9350−9363.

Picó, Y., Farré, M., Tokman, N., & Barceló, D. (2008). Rapid and sensitive ultra-high-pressure liquid chromatography-quadrupole time-of-flight mass spectrometry for the quantification of amitraz and identification of its degradation products in fruits. *Journal of Chromatography A, 1203*(1), 36−46.

Picó, Y., Font, G., Ruiz, M., & Fernández, M. (2006). Control of pesticide residues by liquid chromatography-mass spectrometry to ensure food safety. *Mass Spectrometry Reviews, 25*(6), 917−960.

Pihlström, T., Blomkvist, G., Friman, P., Pagard, U., & Osterdahl, B. (2007). Analysis of pesticide residues in fruit and vegetables with ethyl acetate extraction using gas and liquid chromatography with tandem mass spectrometric detection. *Analytical and Bioanalytical Chemistry, 389*(6), 1773−1789.

Polgár, L., García-Reyes, J., Fodor, P., Gyepes, A., Dernovics, M., Abrankó, L., et al. (2012). Retrospective screening of relevant pesticide metabolites in food using liquid chromatography high resolution mass spectrometry and accurate-mass databases of parent molecules and diagnostic fragment ions. *Journal of Chromatography A, 1249*(8), 83−91.

Rajski, L., Gomez-Ramos Mdel, M., & Fernandez-Alba, A. R. (2014). Large pesticide multi-residue screening method by liquid chromatography-Orbitrap mass spectrometry in full scan mode applied to fruit and vegetables. *Journal of Chromatography A, 1360*, 119—127.

Rodriguez-Aller, M., Gurny, R., Veuthey, J., & Guillarme, D. (2012). Coupling ultra high-pressure liquid chromatography with mass spectrometry: Constraints and possible applications. *Journal of Chromatography A, 1292*(5), 2—18.

Sivaperumal, P., Anand, P., & Riddhi, L. (2015). Rapid determination of pesticide residues in fruits and vegetables, using ultra-high-performance liquid chromatography/time-of-flight mass spectrometry. *Food Chemistry, 168*(2), 356—365.

Soler, C., Hamilton, B., Furey, A., James, K., Mañes, J., & Picó, Y. (2007). Liquid chromatography quadrupole time-of-flight mass spectrometry analysis of carbosulfan, carbofuran, 3-hydroxycarbofuran and other metabolites in food. *Analytical Chemistry, 79*(4), 1492—1501.

Soler, C., James, K., & Picó, Y. (2007). Capabilities of different liquid chromatography tandem mass spectrometry systems in determining pesticide residues in food. Application to estimate their daily intake. *Journal of Chromatography A, 1157*(1—2), 73—84.

Taylor, M., Keenan, G., Reid, K., & Fernández, D. (2008). The utility of ultra-performance liquid chromatography/electrospray ionisation time-of-flight mass spectrometry for multi-residue determination of pesticides in strawberry. *Rapid Communications in Mass Spectrometry, 22*(17), 2731—2746.

Thurman, E., Ferrer, I., & Fernández-Alba, A. (2005). Matching unknown empirical formulas to chemical structure using LC/MS TOF accurate mass and database searching: Example of unknown pesticides on tomato skins. *Journal of Chromatography A, 1067*(3), 127—134.

Thurman, E., Ferrer, I., Malato, O., & Fernández-Alba, A. (2006). Feasibility of LC/TOFMS and elemental database searching as a spectral library for pesticides in food. *Food Additives and Contaminants, 23*(11), 1169—1178.

Valverde, A., Aguilera, A., Ferrer, C., Camacho, F., & Cammarano, A. (2010). Analysis of forchlorfenuron in vegetables by LC/TOF-MS after extraction with the buffered QuEChERS method. *Journal of Agricultural and Food Chemistry, 58*(5), 2818—2823.

Wang, J., Chow, W., & Leung, D. (2010). Applications of LC/ESI-MS/MS and UHPLC QqTOF MS for the determination of 148 pesticides in fruits and vegetables. *Analytical and Bioanalytical Chemistry, 396*(4), 1513—1538.

Wang, J., & Leung, D. (2009). Applications of ultra-performance liquid chromatography electrospray ionization quadrupole time-of-flight mass spectrometry on analysis of 138 pesticides in fruit- and vegetable-based infant foods. *Journal of Agricultural and Food Chemistry, 57*(6), 2162—2173.

Wang, J., Leung, D., & Chow, W. (2010). Applications of LC/ESI-MS/MS and UHPLC QqTOF MS for the determination of 148 pesticides in berries. *Journal of Agricultural and Food Chemistry, 58*(10), 5904—5925.

Williamson, L., & Bartlett, M. (2007). Quantitative liquid chromatography/time-of-flight mass spectrometry. *Biomedical Chromatography, 21*(6), 567—576.

Zomer, P., & Mol, H. (2015). Simultaneous quantitative determination, identification and qualitative screening of pesticides in fruits and vegetables using LC-Q-Orbitrap™-MS. *Food Additives & Contaminants. Part A: Chemistry, Analysis, Control, Exposure & Risk Assessment, 32*(10), 1628—1636.

Application of HRMS in Pesticide Residue Analysis in Food From Animal Origin

7

Roberto Romero-González, Antonia Garrido Frenich

University of Almería, Almería, Spain

7.1 INTRODUCTION

In Chapter 6, the current methods focused on the determination of pesticide residues in vegetable and fruit samples based on high-resolution mass spectrometry (HRMS) have been described in detail. However, fewer methods have been developed for the determination of pesticides in food from animal origin (Stachniuk & Fornal, 2016), being meat, fish, honey, egg, and baby food the most widely evaluated matrices. In addition, animal feed could also be contaminated with pesticides, considering that most of feeds used for animal farming are made from plant raw material, which may also contain pesticide residues used in agricultural practices (Nácher-Mestre, Ibáñez, Serrano, Pérez-Sánchez, & Hernández, 2013). In these matrices, other types of contaminants such as veterinary drugs have been studied (Bogialli & Di Corcia, 2009), although the presence of pesticide residues is also a main concern to ensure food safety, bearing in mind the increased consumers' awareness and legal—toxicological implications. Therefore, these contaminants in these types of matrices have been regulated by Europe (European Union, Council Regulation 395/2005) and the United States (United States, EPA, 2016), as well as by other international organizations such as Codex (Codex Alimentarius, 2015).

To fulfill these requirements, liquid chromatography (LC) (Souza Tette, Rocha Guidi, Abreu Glória, & Fernandes, 2016) or gas chromatography (GC) (Li, Wang, Yan, Fu, & Dai, 2013) coupled to low-resolution mass spectrometry (LRMS) has been mainly used for the determination of pesticide residues in these type of matrices, although sometimes they are not selective enough and only unit mass resolution is provided as well as a "limited" number of compounds could be simultaneously determined in one single run. Therefore, in the last 6 years, HRMS has also been utilized (Botitsi, Garbis, Economou, & Tsipi, 2011; Masiá, Blasco, & Picó, 2014), bearing in mind the advantages described throughout this book, such as sensitivity, nontarget approaches, posttarget analysis, and unlimited (theoretically) determination of compounds. Therefore, it is a suitable tool for large screening purposes, considering that, currently, the risk associated with mixtures of pesticides is a real challenge, and therefore, new analytical methods that cover a wide range of compounds are demanded by global food industry to detect potentially hazardous compounds applying nontargeted methods (Romero-González, 2015).

Therefore, because of the great variety of pesticides that could be detected in this type of matrices, HRMS is a powerful technique to cover a wide range of pesticides, reducing the cost of the analysis and allowing and improving the cost-effectiveness of the developed analytical methods.

Bearing in mind that LC coupled to HRMS is the most flexible and effective technique used for the determination of pesticide residues in food matrices, this chapter discusses the application of HRMS in pesticide residue analysis in food from animal origin and feed, describing the requirements needed by the analyzer to achieve a suitable identification of the target compounds, as well as the extraction procedures and chromatographic conditions most commonly used for this purpose.

7.2 INSTRUMENTAL REQUIREMENTS

Feed as well as food from animal origin, such as meat, fish, honey, milk, or egg, are complex matrices, and when HRMS is used, in addition to the ions associated with the target compounds, high number of peaks related to the matrix can be observed. Moreover, these are commonly much higher than the peaks corresponding to the target compounds, bearing in mind that they are present in the samples at higher concentrations. Thus, the selection of the resolution power of the analyzer can be a key parameter to perform a reliable assignment of target compounds (Kellmann, Muenster, Zomer, & Mol, 2009). This is very interesting if Orbitrap technology is used, because this parameter can be set by the analyst, and therefore, the performance of the identification process can be seriously modified. This is a great advantage in relation to conventional time-of-flight (TOF) analyzers, which usually work at a fixed and lower resolving power than Orbitrap. One question is whether the medium mass resolving power provided by conventional TOF instruments is sufficient to enable the detection and accurate mass measurement of compounds at low levels in complex matrices, or if higher resolution is needed (De Dominicis, Commissati, Gritti, Catellani, & Suman, 2015). In relation to this, Kellmann et al. (2009) observed that the couple resolving power-concentration is critical on the correct assignment of masses. Moreover, these authors performed an interesting study where the effect of mass resolution was evaluated. Thus, mass resolution levels ranging from 10,000 to 100,000 FWHM (full width at half maximum) were studied for the analysis of pesticides and other contaminants (veterinary drugs, mycotoxins, and plant toxins) in feed matrices and honey. It was observed that high resolving power (>50,000 FWHM) was needed for a reliable mass assignment (<2 ppm) at low concentrations in feed samples, whereas lower resolution (25,000 FWHM) was enough if honey matrices were utilized. As it can be observed in Table 7.1, several resolutions have been applied for a reliable detection of pesticides, and they range

Table 7.1 High-Resolution Mass Spectrometry Conditions for Pesticides Determination in Food From Animal Origin

Compounds	Matrix	Analyzer	Resolution	Identification Criteria	Working Mode	LOQ	References
Pesticides and VDs (No. >350)	Honey	Exactive-Orbitrap (+ and −)[a]	25,000 FWHM at m/z 200	RTW Mass accuracy <5 ppm Isotopic pattern	Database and quantitation purposes	<10 μg/kg (LOD)	Gómez-Pérez et al. (2012)
Pesticides and antibiotics (No. 83)	Honey	Exactive-Orbitrap (+)[b]	25,000 FWHM at m/z 200	RT tolerance: ±0.15 min Mass accuracy <5 ppm Isotopic pattern	Quantitative and multivariate approach	0.1–10 μg/kg (LOD)	Cotton et al. (2014)
Pesticides, VDs, and mycotoxins (No. 36)	Milk	Exactive-Orbitrap (+ and −)	100,000 FWHM at m/z 200	RT variation	Semiquantitative approach	<100 μg/kg	De Dominicis et al. (2012)
Pesticides, antibiotics, and mycotoxins (No. 33)	Milk	Exactive-Orbitrap (+)	50,000 FWHM at m/z 200	RTW Mass accuracy <5 ppm Peak threshold >10,000 units Isotopic pattern	Quantitative approach (retrospective analysis)[c]	<10 μg/kg	De Dominicis et al. (2015)
Contaminants including pesticides (No. 118)	Milk, muscle, and liver tissues	Exactive-Orbitrap	50,000 FWHM at m/z 200	RT: ±0.25 min Mass accuracy: ±6 mDa	Quantitative approach	Not provided	Filigenzi et al. (2011)
Pesticides and VDs (No. >350)	Meat	Exactive-Orbitrap (+ and −)	25,000 FWHM at m/z 200	RTW Mass accuracy <5 ppm Fragments	Database with fragmentation (quantitation purposes)	<2 μg/kg (>300 compounds)	Gómez-Pérez et al. (2014)

Continued

Table 7.1 High-Resolution Mass Spectrometry Conditions for Pesticides Determination in Food From Animal Origin—cont'd

Compounds	Matrix	Analyzer	Resolution	Identification Criteria	Working Mode	LOQ	References
Pesticides (No. 54)	Fish	LTQ-Orbitrap (+)	30,000 FWHM at m/z 400	RT tolerance: 2.5% Mass accuracy <5 ppm, Productions Isotopic pattern	Quantitative approach and nontarget analysis	LOD final extract (ng/mL): 0.2–2	Farré et al. (2014)
Pesticides, antibiotics, and mycotoxins (No. 82)	Fish feed and Fish	Q-TOF (+)	10,000 FWHM at m/z 556	RTW Mass accuracy <5 ppm One fragment	Screening mode	<100 µg/kg	Nácher-Mestre et al. (2013)
Pesticides, VDs, mycotoxins, and plant toxins (No. 151)	Honey and feed	Exactive-Orbitrap (+)	25,000 FWHM for honey and 50,000 FWHM for feed at m/z 200	RT variation Mass accuracy <5 ppm	Screening mode	Not established	Kellmann et al. (2009)
Pesticides and VDs (No. > 250)	Animal Feed (chicken, hen, horse, rabbit)	QqTOF (+ and −)	9,000 FWHM at m/z 556	RTW Mass accuracy <5 ppm One fragment Isotopic pattern	Quantitative approach	<50 µg/kg for 79% of the compounds	Aguilera-Luiz et al. (2013)
Pesticides and VDs (No. > 300)	Animal Feed (chicken, hen, horse, rabbit)	Q-Exactive (+ and −)	25,000 FWHM at m/z 200	RTW Mass accuracy <5 ppm One fragment Isotopic pattern	Comparison between QqTOF and Orbitrap	<12.5 (>90% of the compounds)	Gómez-Pérez et al. (2015b)

Analytes	Matrix	Instrument	Resolution	Identification criteria	Approach	LOD/LOQ	Reference
Pesticides, veterinary drugs, ergot alkaloids, plant toxins, and other substances (No. 425)	Feed	Exactive-Orbitrap (+ and −)	50,000 FWHM at m/z 200	RT tolerance: 2.5% Mass accuracy <5 ppm Isotopic pattern Ratio of the fragment ions	Quantitative and retrospective analysis	Not provided for pesticides	León et al. (2016)
Pesticides and VDs (333)	Baby food	Q-Exactive (+ and −)	70,000 FWHM at m/z 200	RTW Mass accuracy <2.5 ppm Product ions	Quantitative approach: screening and confirmation approach	0.01−9.27 µg/kg	Jia et al. (2014)
Pesticides and VDs (>300)	Baby food	Exactive-Orbitrap (+ and −)	25,000 FWHM at m/z 200	RTW Mass accuracy <5 ppm One fragment Isotopic pattern	Database with fragmentation (quantitation purposes)	10−100 µg/kg	Gómez-Pérez et al. (2015a)

FWHM, full width at half maximum; LOD, limit of detection; LOQ, limit of quantification; LTQ-Orbitrap, linear ion trap quadrupole-Orbitrap; Q-TOF, quadrupole time-of-flight analyzer; RT, retention time; RTW, retention time window; VDs, veterinary drugs.

[a] Ionization mode in brackets.
[b] For MS/MS analysis a Q-Exactive was used.
[c] Retrospective analysis performed for mycotoxins.

from 9000–10,000 FWHM when TOF is used to 100,000 FWHM when Exactive-Orbitrap is applied in a semiquantitative approach. However, the highest resolution in Exactive-Orbitrap or related analyzers is not usually applied when quantitative applications are performed, and 25,000 or 10,000 FWHM is commonly used (Gómez-Pérez, Plaza-Bolaños, Romero-González, Martínez Vidal, & Garrido-Frenich, 2012). This can be explained considering that the higher the resolution used, the lower the points per peak, and this can affect the reliability of the quantification process. This fact is accentuated if several acquisition functions or ionization modes are used, which implies higher duty cycle, and therefore, if working resolution is very high, the points per peak can be dramatically reduced, especially if sub-2-μm stationary phases were used. This effect has been carefully evaluated by De Dominicis et al. (2015), who observed that the reduction of resolution from 100,000 FWHM to 50,000 FWHM, which increases the scan rate from 1 Hz to 2 Hz, is an acceptable compromise between selectivity and scan speed, improving precision. Another solution is based on the activation of the fragmentation mode only if the ion from the parent compound is higher than the threshold value (Farré, Picó, & Barceló, 2014). This approach is only valid if precursor ion selection could be performed, as in hybrid analyzers such as Quadrupole (Q) coupled to TOF or Orbitrap, whereas in Exactive-Orbitrap this approach cannot be used. Moreover, this tandem mass spectrometry (MS/MS) experiments can be useful for the correct identification of some pesticides that can be isobaric. For instance, this approach has been used for the discrimination between acetochlor and alachlor, which have the same empirical formula and coelute, but they have different MS/MS spectra, and therefore, these structural isomers can be distinguished (Farré et al., 2014).

In addition to this parameter, the mass extraction window to obtain the extracted ion chromatograms also plays an important role when screening methods are developed. It is well known that the narrower the mass window, the higher the selectivity as it can be observed in Fig. 7.1, where the example of the detection of pirimicarb in an animal feed matrix is shown. It can be observed that the peak corresponding to the target compound is clearly defined when a mass extraction window of 2 ppm is used. In other matrices such as baby food it was observed that 2.5 ppm extraction window could be enough, working at resolving power of 70,000 FWHM (Jia, Chu, Ling, Huang, & Chang, 2014). Thus, the use of narrow mass tolerance windows is only feasible when the instrument provides sufficient resolving power to discriminate analytes from isobaric coeluting sample matrix compounds. However, this parameter could be carefully evaluated to avoid both false-positive and false-negative results (Kaklamanos, Vincent, & Von Holst, 2013). Thus if the mass window is too narrow, the compound cannot be detected increasing the number of false negatives. On the other hand, if the selected window is too broad, the number of false positives (compounds detected but subsequently rejected during the confirmation process) could enlarge, and the time needed for confirmation purposes increases dramatically.

At low concentrations, if a mass resolution of 10,000 FWHM is used, mass error of 10 ppm is considered as acceptable (Ojanperä et al., 2006), although currently, mass error values lower than 5 ppm are commonly obtained (De Dominicis et al., 2015).

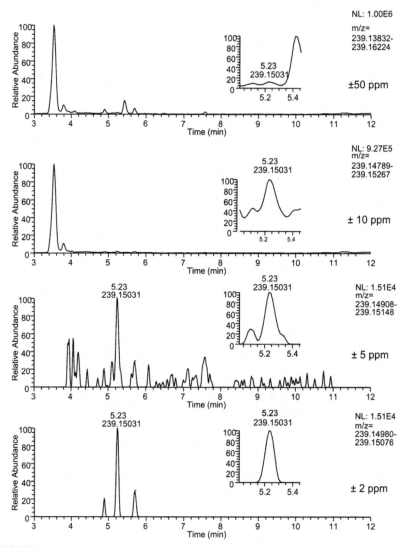

FIGURE 7.1

Effect of mass extraction window on selectivity. Extracted ion chromatograms for the pesticide pirimicarb (MH+, $C_{11}H_{19}N_4O_2$, m/z_{theo} 239.15028, retention time 5.23 min) in animal feed matrix at 10 ng/g; resolving power: 100,000 FWHM (full width at half maximum) at m/z 200.

Reprinted from Kellmann, M., Muenster, H., Zomer, P., & Mol, H. (2009). Full scan MS in comprehensive qualitative and quantitative residue analysis in food and feed matrices: How much resolving power is required? Journal of American Society of Mass Spectrometry, 20, 1464–1476, with permission from Springer behalf of American Society for Mass Spectrometry.

Moreover, this error can be affected by the coelution of isobaric compounds, and therefore if the mass tolerance window is too narrow, the probability of false negatives can be increased.

In relation to the applications developed up to now, Table 7.1 shows the main characteristics of the MS conditions used to detect and identify pesticides in food from animal origin focused on qualitative or quantitative analysis. Basically, Orbitrap analyzer (Exactive-Orbitrap, linear ion trap-quadrupole-Orbitrap, or Q-Exactive) is commonly used, although tandem quadrupole/time-of-flight (QqTOF) analyzer has also been applied. In both cases, positive and negative ionization modes are commonly used, to increase the scope of the analysis, although sometimes, two injections should be performed (Aguilera-Luiz, Romero-González, Plaza-Bolaños, Martínez Vidal, & Garrido Frenich, 2013). Nevertheless, as it has been commented previously, the resolution should also be fixed bearing in mind the complexity of the matrix.

Identification criteria should be carefully set to minimize the number of false positives and negatives during the identification process when HRMS is applied. Generally, the combination of several parameters such as retention time window, mass accuracy of the characteristic ion or fragments, and isotopic pattern are commonly used of identification purposes (see Table 7.1). In this sense, SANTE guidelines (European Union, SANTE, 2015) are usually applied, and although 5 ppm is commonly used as threshold value for mass accuracy, some authors (Jia et al., 2014) set lower limits such as 2.5 ppm, increasing the reliability of the identification process.

Conventionally, during the identification of a positive sample, two stages are used. First, the detection of the potential compound, which can be automatically performed using specific software, is carried out. In this case, only retention time and the characteristic ion are commonly used as parameters. After that, other criteria such as fragments and/or isotopic pattern are evaluated to identify the target compound (De Dominicis et al., 2015). Some authors observed that during the first stage (Gómez-Pérez et al., 2012), the number of false positives (compounds detected but not identified using the indicated criteria) is lower than 4%, indicating that this automatic screening should be improved to minimize false positives and, therefore, reduce the number of pesticides submitted to identification—quantification steps.

7.3 ANALYTICAL PROCEDURES: EXTRACTION AND CHROMATOGRAPHIC CONDITIONS

It is widely accepted that nonpolar pesticides are accumulated in fatty tissues, and that is why laboratories are mainly focused on these type of compounds when food from animal origin should be analyzed. However, modern pesticides, which are considered less harmful to mammals, can be detected in the food chain (Lichtmannegger, Fischer, Steemann, Unterluggauer, & Masselter, 2015), and therefore generic and fast methods are necessary to increase the scope of the analytical methods, and analyte losses should be avoided during this step. Nevertheless, the high proportion of lipids and proteins is an important analytical task in comparison to fruits and vegetables (Hildmann, Gottert, Frenzel, Kempe, & Speer, 2015), and that is why the development of an efficient extraction method is critical during this step.

Pesticides have been commonly extracted from feed and food from animal origin using several techniques such as Soxhlet (Chung & Chen, 2011), pressurized liquid extraction (Vázquez-Roig & Picó, 2015), accelerated solvent extraction (Sun, Ge, Lv, & Wang, 2012), matrix solid-phase dispersion (MSPD) (Hildmann et al., 2015), or conventional solid–liquid extraction (SLE) (Yáñez, Martín, Bernal, Nozal, & Bernal, 2014).

Because of the high number of available pesticides, the use of generic sample extraction procedures is advisable to increase the scope of the methods and to include as many compounds as possible in the analysis. Then, the lack of selectivity of these procedures can be compensated by the high selectivity provided by the HRMS analyzer, although efficient extraction methods are still needed to minimize matrix effect as well as the presence of interferences in the final extract, which could hamper the identification of the target compounds. For instance, most of the matrices covered in this chapter contain a large number of matrix compounds, such as lipids and proteins, and therefore a cleanup step based on solid-phase extraction (SPE) or dispersive SPE (d-SPE) is commonly used (Nácher-Mestre et al., 2013). Thus, this cleanup step also reduces matrix effect (Gómez-Pérez, Romero-González, Martínez Vidal, & Garrido Frenich, 2015a), although depending on the type of matrix, this can be avoided (Filigenzi, Ehrke, Aston, & Poppenga, 2011).

In relation to the extraction procedures, generic approaches such as "dilute and shoot" and QuEChERS (acronym of quick, easy, cheap, effective, rugged, and safe) have been widely used as it can be observed in Table 7.2, although conventional SLE using McIlvaine solution is also used (De Dominicis et al., 2015). Thus, "dilute and shoot" has been applied for the simultaneous extraction of pesticides and veterinary drugs from different samples such as honey, meat, baby food, fish, and animal feed. Basically, the solvent used is acetonitrile acidified with formic acid at 1% (see Table 7.2) although for some applications, such as the simultaneous extraction of pesticides, antibiotics, and mycotoxins, a mixture of acetonitrile:water (80:20 v/v) acidified with formic acid at 0.1% is recommended (Nácher-Mestre et al., 2013).

It is important to highlight that extraction time should be ~1 h to get suitable extraction of the target compounds. Although this could be a drawback for this methodology, it must be considered that more than 10 samples (normally 15–20 samples) could be simultaneously extracted, increasing the sample throughput.

Although some of these studies have mainly focused on screening and performed qualitative results, some of them also evaluated the recovery of the extraction procedure and it is usually ranged from 60% to 125% for most of the selected compounds, indicating that this approach could be used.

In relation to QuEChERS, European (European Union, CEN, 2008) as well as American (United States, AOAC, 2007) version has been used as it has been indicated in Table 7.2, although some modifications were also included. For instance, Farré et al. (2014), who determined 54 pesticides in fish, included a lyophilization step before the extraction procedure. On the other hand, Jia et al. (2014) modified

Table 7.2 Extraction Procedures Developed for the Determination of Pesticides in Food From Animal Origin

Compounds	Matrix	Extraction	Addition of Water	Extraction Time	Cleanup Step	Recovery (%)	References
Pesticides and VDs (No. >350)	Honey	SLE (dilute and shoot): acetonitrile (1% formic acid, v/v)	Yes: 2.5 mL to 2.5 g of sample	1 h (rotary shaker)	No	60–120 (381 compounds)	Gómez-Pérez et al. (2012)
Pesticides and antibiotics (No. 83)	Honey	QuEChERS—American version	Yes: 4 mL to 5 g of sample	1 h (rotary shaker)	No	43–114	Cotton et al. (2014)
Pesticides, VDs, and mycotoxins (No. 36)	Milk	QuEChERS—American version	No	Few min	Yes: PSA and C_{18}	Not provided	De Dominicis et al. (2012)
Pesticides, antibiotics, and mycotoxins (No. 33)	Milk	SLE (Na$_2$EDTA-McIlvain solution buffer 0.01 N pH 4)	No	5 min (mechanical shaker)	Yes: PSA and C_{18}	64–135	De Dominicis et al. (2015)
Contaminants including pesticides (No. 118)	Milk, muscle, and liver tissues	QuEChERS—American version	Yes: 10 mL of water (0.1 formic acid) to 10 mL of milk	No	Yes: PSA and C_{18} (only for liver)		Filigenzi et al. (2011)
Pesticides and VDs (No. >350)	Meat	SLE (dilute and shoot): acetonitrile (1% formic acid, v/v)	Yes: 2.5 mL to 2.5 g of sample	1 h (rotary shaker)	No	70–120 (>300 compounds)	Gómez-Pérez et al. (2014)
Pesticides (No. 54)	Fish	QuEChERS—European version (lyophilized samples)	Yes: 7.5 mL to 1 g of lyophilized sample	1 min	Yes: PSA and GCB	70–100 (82% of compounds)	Farré et al. (2014)
Pesticides, antibiotics, and mycotoxins (No. 82)	Fish feed and fish	SLE: acetonitrile/water (80:20) 0.1% formic acid	No	1 h	No	Qualitative results	Nácher-Mestre et al. (2013)

Analytes	Sample	Extraction procedure	Dilution	Extraction time	Cleanup	Recovery (%)	Reference
Pesticides, VDs, mycotoxins, and plant toxins (No. 151)	Honey and feed	SLE: acetonitrile containing 1% of formic acid	Yes: 5 mL to 2.5 g of samples	1 h (rotary shaker)	No	Theoretical work evaluating resolving power	Kellmann et al. (2009)
Pesticides and VDs (No. >250)	Animal feed (chicken, hen, horse, rabbit)	SLE (dilute and shoot): acetonitrile (1% formic acid, v/v)	Yes: 2.5 mL to 2.5 g of sample	1 h (rotary shaker)	Yes: florisil cartridge	60–125 (75% of the compounds)	Aguilera-Luiz et al. (2013)
Pesticides and VDs (No. >300)	Animal feed (chicken, hen, horse, rabbit)	SLE (dilute and shoot): acetonitrile (1% formic acid, v/v)	Yes: 2.5 mL to 2.5 g of sample	1 h (rotary shaker)	Yes: florisil cartridge	60–125 (>300 compounds)	Gómez-Pérez et al. (2015b)
Pesticides, veterinary drugs, ergot alkaloids, plant toxins, and other substances (No. 425)	Feed	QuEChERS—European version	Yes: 6 mL to 2 g of sample	30 min (head-over-head)	No	Not provided	León et al. (2016)
Pesticides and VDs (333)	Baby food	QuEChERS—American version (acetonitrile:water 84:16 v/v).	No	1 min	No	80–111	Jia et al. (2014)
Pesticides and VDs (>300)	Baby food	SLE (dilute and shoot): acetonitrile (1% formic acid, v/v).	Yes: 2.5 mL to 2.5 g of sample	1 h (rotary shaker)	Yes: florisil cartrige	70–120 (>351 compounds)	Gómez-Pérez et al. (2015a)

GCB, graphitized black carbon; PSA, primary and secondary amine; SLE, solid liquid extraction; VDs, veterinary drugs.

the composition of the extraction solvent and they used a mixture of acetonitrile:water (84:16 v/v) for the extraction of pesticides and veterinary drugs from baby food. They compared the results using different solvents such as mixtures of methanol:water, acetone:water, and acetonitrile:water, observing that the best results were obtained in terms of matrix effect and recovery, when acetonitrile:water was used. It can be observed from Table 7.2 that the extraction time is considerably shorter when QuEChERS approach is applied than when "dilute and shoot" methodology is applied. Thus, if both approaches provide similar results, QuEChERS should be used instead of "dilute and shoot," bearing in mind that sample throughput increases.

In both approaches ("dilute and shoot" and QuEChERS) a cleanup step is sometimes necessary because of the complexity of the matrix, as it has been indicated previously. Although most of the applications avoided this step, for very complex matrices such as animal feed, it is needed. Aguilera-Luiz et al. (2013) studied the effect of this step, evaluating d-SPE using florisil and C_{18} as sorbents and conventional SPE utilizing florisil cartridges. They observed that the addition of a cleanup step minimized the number of interferent compounds (cleaner chromatograms), as well as recovery and sensitivity improved when SPE using florisil cartridges were utilized. In the same way, De Dominicis et al. (2015) included the d-SPE step during the extraction of pesticides, veterinary drugs, and mycotoxins from milk. In this case, they used a mixture of PSA (300 mg) and C_{18} (100 mg), to get suitable results. In general, recoveries usually ranged between 70.0% and 100.7% for target compounds, showing the effectiveness of this approach. In this case, the extraction step is faster than the "dilute and shoot" because the 1-h agitation step could be avoided. For other matrices such as fish, other sorbents such as PSA and graphitized black carbon are satisfactorily used during this cleanup step (Farré et al., 2014).

Although in some applications, especially in vegetables, the extract is sometimes diluted before the injection into the chromatographic system (Ferrer, Lozano, Agüera, Girón, & Fernández-Alba, 2011), when food from animal origin is analyzed, this step is commonly avoided, and the addition of the commented cleanup step is preferred. However, for honey, dilution of heated honey with different solvents or with a mixture of solvents is usually the first step (Souza Tette et al., 2016).

In addition to this fact, when both methodologies are used, sometimes the addition of water is needed to improve the recovery and repeatability of the extraction process. This is well-known for the QuEChERS general approach (QuEChERS, 2011), which indicates that for matrices with low water content, certain volume of water should be added to get reproducible results, and it has also been used when "dilute and shoot" is used. For instance, for the extraction of pesticides from honey, 2.5 mL of water was added to 2.5 g of sample (Gómez-Pérez et al., 2012). The same procedure was used for other matrices such as animal feed (Aguilera-Luiz et al., 2013), meat (Gómez-Pérez et al., 2014), or baby food (Gómez-Pérez, Romero-González, Martínez Vidal, & Garrido Frenich, 2015b). For lyophilized samples, 7.5 mL of water was added to 1 g of lyophilized fish (Farré et al., 2014).

After the extraction procedure, a chromatographic stage is commonly applied for the separation of the pesticides. LC coupled to HRMS is the most effective and flexible technique that can be used to determine pesticide residues in different food matrices from animal origin as well as in feed. In this sense, and although HRMS can provide an excellent selectivity, chromatographic separation is still necessary to separate the target compounds from the interferences and, thus, improve the reliability of the identification of the target compounds. Moreover, although high resolution can provide better results during the identification of the compounds, if fast LC such as ultrahigh performance liquid chromatography (UHPLC) is used, lower resolving power, which implies faster scan rates, is necessary to get suitable peak shapes.

For the chromatographic separation, C_{18} was used as the stationary phase (see Table 7.3) with a particle size lower than 2 μm in most of the applications, reducing the running time, which usually ranges from 14 (Gómez-Pérez et al., 2012) to 28 min (De Dominicis et al., 2015), although other compounds such as mycotoxins and antibiotics are also analyzed. However, some authors used stationary phases with a higher particle size, ranging from 2.6 (Jia et al., 2014) to 4.0 μm (Kellmann et al., 2009), being the running time shorter than 15 min in both cases or longer if 5 μm was used as particle size (Cotton et al., 2014). In all the cases, the length of the column is 100 or 150 mm, although Farré et al. (2014) used a shorter column (50 mm), obtaining good separation and efficiency, as well as peak shape, with a baseline peak width of 25−40 s.

Concerning the mobile phase, methanol has been commonly used as the organic solvent utilized for the separation of the pesticides (see Table 7.3), although there are some differences in the modifiers added as well as in the composition of the aqueous solution. Thus, acetic or formic acid is commonly added to both phases, although in the same studies, volatile salts such as ammonium formate or acetate are also added to improve the ionization step when both positive and negative ion modes are used.

Conventional flow rate is commonly set (0.2 or 0.3 mL/min), whereas injection volume ranges from 5 to 20 μL, as it can be observed in Table 7.3.

Although chromatographic separation is usually considered less important when it is coupled to HRMS, because of the high resolution provided by current analyzers, special attention should be paid to this step when isobaric compounds must be determined. Thus, this step is important in the case of isobaric species with the same characteristic ions such as simazine and desethylterbuthylazine (Fig. 7.2). In addition to this, matrix effect can be minimized if a suitable chromatographic method is developed, reducing the presence of interferent compounds that coelute at the same retention time than the target compound.

7.4 QUANTITATIVE AND QUALITATIVE APPLICATIONS

HRMS is an efficient alternative to conventional MS/MS using triple quadrupole (QqQ) for large screening purposes. Most of the cited applications using LC−HRMS have mainly focused on qualitative purposes, being applied for the elucidation of the structure of pesticides and pesticide transformation products

Table 7.3 Chromatographic Conditions for Pesticides Determination in Food From Animal Origin

Compounds	Matrix	Stationary Phase	Mobile Phase	Running Time	Flow (mL/min)	Injection Volume	References
Pesticides and VDs (No. >350)	Honey	Hypersil GOLD aQ C$_{18}$ column (100 mm × 2.1 mm, 1.7 µm particle size)	Eluent A: 0.1% (v/v) formic acid and ammonium formate 4 mM in water Eluent B: 0.1% (v/v) formic acid and ammonium formate 4 mM in methanol	14.0 min	0.3	10 µL	Gómez-Pérez et al. (2012)
Pesticides and antibiotics (No. 83)	Honey	XTerra C$_{18}$ (150 mm × 2.1 mm, 5 µm particle size)	Eluent A: water (formic acid 0.1%) Eluent B: acetonitrile (formic acid 0.1%)	30 min	0.3	10 µL	Cotton et al. (2014)
Pesticides, VDs, and mycotoxins (No. 36)	Milk	Acquity UPLC BEH C$_{18}$ (100 mm × 2.1 mm, 1.7 µm particle size)	Eluent A: water (1 mM ammonium acetate with 0.05% acetic acid) Eluent B: methanol (0.05% acetic acid)	22.0 min	0.2	10 µL	De Dominicis et al. (2012)
Pesticides, antibiotics, and mycotoxins (No. 33)	Milk	Eclipse Plus C$_{18}$ (50 mm × 2.1 mm, 1.8 µm particle size)	Eluent A: Water (1 mM ammonium acetate with 0.1% acetic acid) Eluent B: Methanol (1 mM ammonium acetate with 0.1% acetic acid)	28.0 min	0.2	10 µL	De Dominicis et al. (2015)

Contaminants including pesticides (No. 118)	Milk, muscle, and liver tissues	Hypersil GOLD aQ column (50 mm × 2.1 mm, 1.9 μm particle size)	Eluent A: water (0.1% formic acid) Eluent B: methanol (0.1% formic acid)	12.0 min	0.38	20 μL	Filigenzi et al. (2011)
Pesticides and VDs (No. >350)	Meat	Hypersil GOLD aQ C_{18} column (100 × 2.1 mm, 1.7 μm particle size)	Eluent A: 0.1% (v/v) formic acid and ammonium formate 4 mM in water Eluent B: 0.1% (v/v) formic acid and ammonium formate 4 mM in methanol	14.0 min	0.3	10 μL	Gómez-Pérez et al. (2014)
Pesticides (No. 54)	Fish	Kinetex XB-C_{18} (50 mm × 2.10 mm, 1.7 μm particle size)	Eluent A: water (0.1% formic acid) Eluent B: methanol (0.1% formic acid)	15.0 min	0.3	5 μL	Farré et al. (2014)
Pesticides, antibiotics, and mycotoxins (No. 82)	Fish feed and Fish	Acquity UHPLC BEH C_{18} (100 mm × 2.1 mm, 1.7 μm particle size)	Eluent A: water containing 0.01% formic acid and 0.1 mM ammonium acetate Eluent B: methanol 0.01% formic acid and 0.1 mM ammonium acetate	18.0 min	0.3	20 μL	Nácher-Mestre et al. (2013)
Pesticides, VDs, mycotoxins, and plant toxins (No. 151)	Honey and feed	Synergi C_{18} (100 mm × 2.0 mm, 4 μm particle size)	Eluent A: water/5% methanol (20 μL formic acid and 2 mM ammonium formate) Eluente B: methanol/5% water (20 μL formic acid and 2 mM ammonium formate)	24.0 min	0.3	5 μL	Kellmann et al. (2009)

Continued

Table 7.3 Chromatographic Conditions for Pesticides Determination in Food From Animal Origin—cont'd

Compounds	Matrix	Stationary Phase	Mobile Phase	Running Time	Flow (mL/min)	Injection Volume	References
Pesticides and VDs (No. >250)	Animal Feed (chicken, hen, horse, rabbit)	Acquity UPLC BEH C$_{18}$ column (100 mm × 2.1 mm, 1.7 μm particle size)	Eluent A: 0.1% (v/v) formic acid and ammonium formate 4 mM in water Eluent B: 0.1% (v/v) formic acid and ammonium formate 4 mM in methanol	14.0 min	0.3	5 μL	Aguilera-Luiz et al. (2013)
Pesticides and VDs (No. >300)	Animal feed (chicken, hen, horse, rabbit)	Hypersil GOLD aQ C$_{18}$ column (100 mm × 2.1 mm, 1.7 μm particle size)	Eluent A: 0.1% (v/v) formic acid and ammonium formate 4 mM in water Eluent B: 0.1% (v/v) formic acid and ammonium formate 4 mM in methanol	14.0 min	0.3	10 μL	Gómez-Pérez et al. (2015b)
Pesticides, veterinary drugs, ergot alkaloids, plant toxins, and other substances (No. 425)	Feed	Hypersil GOLD aQ C$_{18}$ column (100 mm × 2.1 mm, 1.9 μm particle size)	Eluent A: water (0.1% formic acid) Eluent B: methanol: acetonitrile (90:10 v/v) containing formic acid (0.1%)	14.5 min	0.4	10 μL	León et al. (2016)

Pesticides and VDs (333)	Baby food	Thermo Accucore C$_{18}$ aQ (100 mm × 2.1 mm, i.d., 2.6 µm particle size)	Eluent A: 0.1% (v/v) formic acid and ammonium formate 4 mM in water Eluent B: 0.1% (v/v) formic acid and ammonium formate 4 mM in methanol	15.0 min	Not indicated	10 µL	Jia et al. (2014)
Pesticides and VDs (>300)	Baby food	Hypersil GOLD aQ C$_{18}$ column (100 × 2.1 mm, 1.7 µm particle size)	Eluent A: 0.1% (v/v) formic acid and ammonium formate 4 mM in water Eluent B: 0.1% (v/v) formic acid and ammonium formate 4 mM in methanol	14.0 min	0.3	10 µL	Gómez-Pérez et al. (2015a)

VDs, Veterinary drugs.

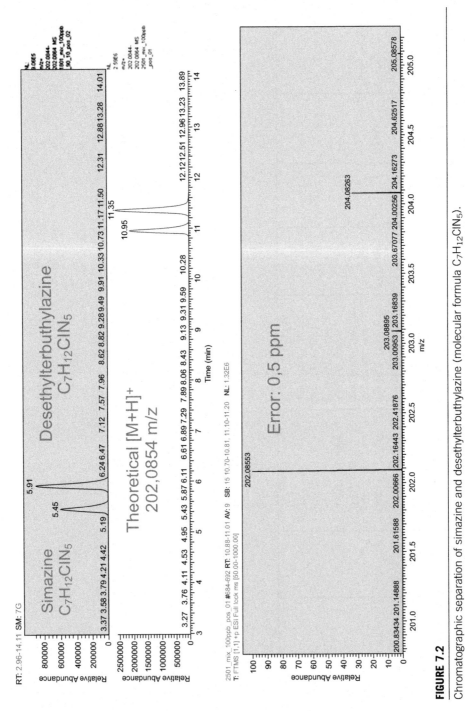

FIGURE 7.2

Chromatographic separation of simazine and desethylterbuthylazine (molecular formula $C_7H_{12}ClN_5$).

Reprinted from De Dominicis, E., Commissati, I., & Suman, M. (2012). Targeted screening of pesticides, veterinary drugs and mycotoxins in bakery ingredients and food commodities by liquid chromatography-high-resolution single-stage Orbitrap mass spectrometry. Journal of Mass Spectrometry, 47, 1232–1241, with permission from Wiley.

(Portolés, Ibáñez, Sancho, López, & Hernández, 2009), and only few studies have been applied for quantitative determination in multiresidue analysis (León, Pastor, & Yusà, 2016). Moreover, this approach can also be used in routine analysis of diagnosis of poisonings by unknown agents (Filigenzi et al., 2011). In relation to samples, honey, meat, baby food, fish, animal feed, and milk have been evaluated, showing an overview of the positive samples in Table 7.4.

One interesting type of food is baby food, which covers a wide range of different matrices such as cereal-, fish-, meat-, and vegetable-based food (Gómez-Pérez et al., 2015a; Jia et al., 2014), bearing in mind that some pesticides cannot be applied in agricultural production used for the production of baby food (European Union. Commission Directive 2003/13/EC). Jia et al. (2014) analyzed 93 commercial baby food samples, and coumaphos, azoxystrobin, and thiacloprid were detected in only three samples (cereal-, fruit-, and vegetable-based food) but at very low concentrations (up to 2.13 µg/kg). The analyzer used to perform these determinations is a Q-Orbitrap, which allows the determination of these compounds at very low concentrations. Moreover, Gómez-Pérez et al. (2015a) analyzed 46 baby food but no positive samples were identified.

Table 7.4 Overview of the Positive Results Identified in the Analyzed Samples

Matrix	No. of Analyzed Samples	No. of Detected Pesticides	Concentration Range	References
Baby food	93 (3)[a]	3	1.92–2.31 µg/kg	Jia et al. (2014)
Baby food	46 (-)	–	–	Gómez-Pérez et al. (2015a)
Honey	26 (4)	4	Traces to 5.1 µg/kg	Gómez-Pérez et al. (2012)
Honey	76 (70)	35	Not indicated	Cotton et al. (2014)
Meat (chicken, beef, pork)	18 (-)	–	–	Gómez-Pérez et al. (2014)
Fish and Feed	11 (2) fish fillet and 10 (8) feed	1	<20 µg/kg	Nácher-Mestre et al. (2013)
Fish	4 (3)	3	16.9–765.9 µg/kg[b]	Farré et al. (2014)
Animal feed	18 (4)	1	18–193 µg/kg	Aguilera-Luiz et al. (2013)

[a] Number of positive samples in brackets.
[b] Dry weight.

Honey is also a complex matrix, which theoretically should be free of toxic compounds, and it is commonly included in monitoring programs to ensure that no residues and contaminants are detected. For instance, Gómez-Pérez et al. (2012) analyzed 26 honey samples, including organic ones, and only four compounds were detected (azoxystrobin, coumaphos, dimethoate, and thiacloprid) at low levels (higher value 5.1 μg/kg). An interesting study in honey was performed by Cotton et al. (2014), who analyzed 55 pesticides and 28 antibiotics in 76 honey samples. They observed that 35 of the targeted compounds were detected in the analyzed samples at concentrations below the regulatory limits, and 74 out of 76 honeys were contaminated. Among these compounds, carbendazim, amitraz, and chlorfenvinphos were the analytes most frequently detected (70, 59, and 35 times, respectively). Moreover, they applied nontarget approach for the identification of other chlorine molecules, due to the characteristic isotopic pattern of this atom, and they identified thiacloprid, imidacloprid, cyproconazole, and tebuconazole, applying additional criteria for identification purposes such as relative abundance of ^{37}Cl, retention time, and chromatographic area higher than 10^6 arbitrary units.

Meat and meat-related products are one of the most important sources of protein worldwide, and the presence of pesticide residues in these matrices should be evaluated. Several types of meats, such as chicken, pork, and beef samples, were analyzed using Exactive-Orbitrap (Gómez-Pérez et al., 2014), but no pesticides were detected above the established limits of quantification (LOQs) of the method.

The presence of pesticide residues in fish muscle is a problem related to environment and food security, and therefore, fish and feed are also matrices evaluated by HRMS. Nácher-Mestre et al. (2013) analyzed 11 fish fillets and 10 feed samples, but only pirimiphos methyl was detected in the analyzed samples. Although this compound was only detected in 2 samples of fish fillet, 8 out of 10 feed samples were contaminated with this compound. However, the concentration detected was below the threshold value set by the authors, who indicated that to confirm the presence of the detected compounds LC—QqQ-MS/MS should be used to reanalyze the samples and confirm the detected compounds. Moreover, Farré et al. (2014) analyzed four samples of fish and azinphos ethyl, chlorpyrifos, and diazinon were detected in the analyzed samples, chlorpyrifos being the compound most widely detected (three out of four samples). However, the same samples were also reanalyzed using QqQ, and imidacloprid, which was not detected using linear ion trap-Orbitrap, was quantified in one sample at 1.041 μg/kg.

The applicability of the QqTOF for the quantification of pesticides in complex matrices such as animal feed has been performed by Aguilera-Luiz et al. (2013), who evaluated the presence of pesticide residues in different types of feed such as rabbit, chicken, hen, pig, lamb, and horse feed. Only one compound, chlorpyrifos, was detected in four samples at concentrations ranging from 18 to 193 μg/kg. Considering the maximum residue limit (MRL) set by Codex Alimentarius for some primary animal feed commodities such as wheat straw and alfalfa fodder (5 mg/kg) (Codex Alimentarius, 2015), the concentrations found were below this value. The presence of this compound can be explained because chlorpyrifos has been defined by the

United States Environmental Protection Agency (EPA) as "one of the most widely used organophosphate insecticide" (United States, EPA, 2013) and feed is mainly composed by agricultural products such as cereals, oleaginous, and leguminous plants. As an example, Fig. 7.3 shows a positive rabbit feed sample containing chlorpyrifos, monitoring two fragment ions in addition to the characteristic ion when Exactive-Orbitrap is used (Fig. 7.3A) or only one fragment ion when QqTOF analyzer is utilized (Fig. 7.3B), confirming the presence of this compound in the sample.

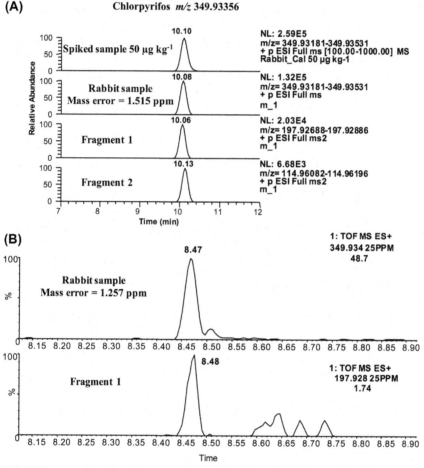

FIGURE 7.3

Chromatograms of the characteristic ion and fragments of chlorpyrifos detected in positive rabbit sample (75 µg/kg) using Exactive-Orbitrap (A) or QqTOF (B) as analyzers.

Reprinted from Gómez-Pérez, M.L., Romero-González, R., Martínez Vidal, J.L., & Garrido Frenich, A. (2015b). Analysis of veterinary drug and pesticide residues in animal feed by high-resolution mass spectrometry: Comparison between time-of-flight and Orbitrap. Food Additives & Contaminants: Part A, 32(10), 1637–1646 *with permission from Taylor and Francis.*

Regarding the type of compounds evaluated, although pesticides can only be the target compounds (Farré et al., 2014), most of the studies included more contaminants such as veterinary drugs (Gómez-Pérez et al., 2015b) and/or mycotoxins (De Dominicis, Commissati, & Suman, 2012).

Furthermore, retrospective analysis has also been performed to search for additional compounds, avoiding the performance of additional analyses. Thus, Nácher-Mestre et al. (2013) applied the posttarget approach to detect other compounds not included in the preliminary list. For instance, ethoxyquin (which is used as a pesticide in agriculture and a preservative in food) was identified in 17 samples, showing in Fig. 7.4 the identification of this compound in a posttarget way, and mass error values lower than 2.3 ppm were obtained for the characteristic ion and two fragments. Moreover, León et al. (2016) tentatively identified pirimiphos methyl in feed samples when retrospective analysis was performed, indicating that standards

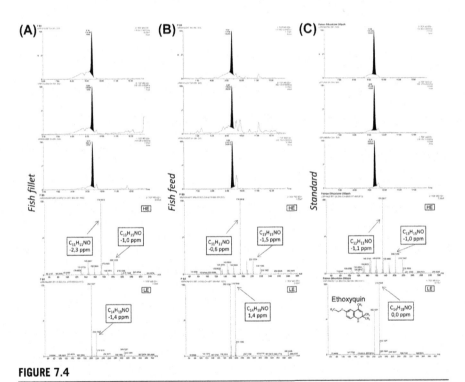

FIGURE 7.4

Narrow window-extracted ion chromatograms for protonated and fragment ions and accurate mass spectra [both low (LE) and high energy (HE)] for ethoxyquin in (A) fish fillet, (B) fish feed, and (C) standard (200 ng/mL), respectively.

should be used to achieve definitive conclusions of the presence of this compound in the detected samples.

In addition to the determination of parent compounds, HRMS is also a powerful tool for the determination of transformation products, which sometimes are more toxic than parent compounds (Martínez Vidal, Plaza-Bolaños, Romero-González, & Garrido Frenich, 2009) and they are included in MRL definitions (European Union, Council Regulation, 2005). Thus, HRMS allows the identification of transformation products, working at posttarget or unknown analysis, as it has been shown in Fig. 7.5, where several coumaphos metabolites were identified in honeybees using HRMS. Thus, Gómez-Pérez, Romero-González, Martínez Vidal, and Garrido Frenich (2015c) detected the presence of 3,5,6-trichloro-2-pyridinol, a transformation product of chlorpyrifos (organophosphate insecticide used to control pest insects), in animal feed and Cotton et al. (2014) identified 2,6-dichlorobenzamide, a metabolite of

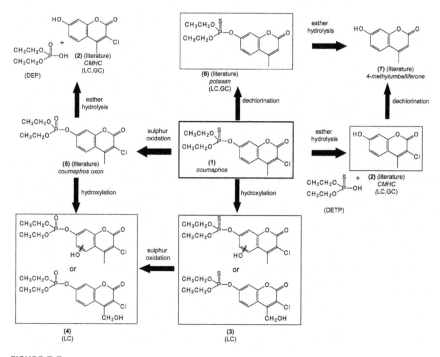

FIGURE 7.5

Coumaphos metabolites identified in honeybees, showing proposed degradation. "LC, GC" indicates that the metabolite was identified in the honeybee sample. *GC*, gas chromatography; *LC*, liquid chromatography.

Reprinted from Portolés, T., Ibáñez, M., Sancho, J.V., López, F.J., & Hernández, F. (2009). Combined use of GC-TOF MS and UHPLC-(Q)-TOF MS to investigate the presence of nontarget pollutants and their metabolites in a case of honeybee poisoning. Journal of Agricultural and Food Chemistry, 57, 4079–4090 with permission from American Chemical Society.

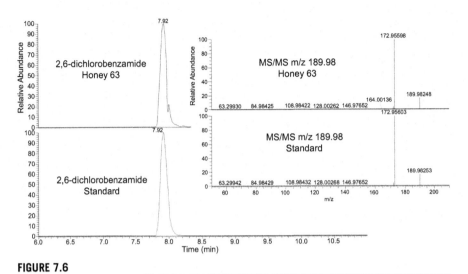

FIGURE 7.6

Identification of 2,6-dichlorobenzamide in honey ([M+H]$^+$ = 189.9821, m/z ± 8 ppm).

Reprinted from Cotton, J., Leroux, F., Broudin, S., Marie, M., Corman, B., Tabet, J.C., et al. (2014).
High-resolution mass spectrometry associated with data mining tools for the detection of pollutants and
chemical characterization of honey samples. Journal of Agricultural and Food Chemistry, 62, 11335–11345
with permission from American Chemical Society.

dichlobenil in honey samples, which was banned in France since 2010. Fig. 7.6 shows that the compound detected in the sample matches the retention time, accurate mass, and isotopic pattern of the standard, ensuring the presence of the compound in the positive samples.

7.5 DIFFERENCES BETWEEN LOW-RESOLUTION MASS SPECTROMETRY AND HIGH-RESOLUTION MASS SPECTROMETRY ANALYTICAL METHODS

Despite the advantages of HRMS, nowadays, LRMS is still most widely used for the determination of pesticide residues in food samples (Stachniuk & Fornal, 2016). Moreover, QqQ is the analyzer commonly coupled to LC because of its robustness and sensitivity, and it fulfills the requirements of regulatory agencies that impose even lower limits. This is one of the main differences when LRMS is compared with HRMS. The sensitivity provided by HRMS analyzers is lower than LRMS and sometimes QqQ is needed because a reliable identification by HRMS cannot be achieved because of the low sensitivity of fragments if Exactive-Orbitrap is used (Gómez-Pérez et al., 2014). Nevertheless, this issue can be overcome using hybrid analyzers such as QqTOF or Q-Exactive, where precursor ion can be selected.

In published methods, LOQs are usually lower than 5 µg/kg and even lower than 1 µg/kg in fish (Lazartigues et al., 2011) and fish tissue (Gan et al., 2016). However, when HRMS analyzers are used, if complex matrices are analyzed, limits of identification could be higher than 20 µg/kg (Nácher-Mestre et al., 2013).

However, one of the main drawbacks of the QqQ is its scan speed, which could limit the number of compounds that can be acquired simultaneously (Bogialli & Di Corcia, 2009). Therefore, the number of pesticides analyzed by LRMS is commonly lower than that analyzed by HRMS. Thus, whereas more than 50 or 100 pesticides are commonly analyzed by HRMS, there are methods based on QqQ where only 8 (Tanner & Czerwenka, 2011) or 13 (Lazartigues et al., 2011) compounds are determined, and most of them are based on the determination of a single class of compounds such as organofluorine pesticides in fish (Gan et al., 2016) or neonicotinoid insecticides in honey (Gbylik-Sikorska, Sniegocki, & Posyniak, 2015). However, in monitoring studies more compounds have been included. For instance, Kasiotis, Anagnostopoulos, Anastasiadou, and Machera (2014) monitored 115 pesticides in honey and bee pollen. In addition, it should be emphasized that retrospective analysis cannot be performed using QqQ, which is also a main drawback in comparison with HRMS.

Although the number of pesticides is lower using LRMS, the chromatographic running time is similar and it is ~ 20 min (Lazartigues et al., 2011), although faster methods have also been used (Xie et al., 2015).

In relation to the extraction procedures, although, currently, generic extraction procedures such as QuEChERS are being used during the determination of pesticide residues in fish (Lazartigues et al., 2011), egg (Choi, Kim, Shin, Kim, & Kim, 2015), meat (Wei et al., 2015), and baby food (Gilbert-López, García-Reyes, & Molina-Díaz, 2012), complex procedures have been used. Thus, Xie et al. (2015) used a low-temperature cleanup during the extraction of pesticides and other compounds from milk and yogurt. Hildmann et al. (2015) employed MSPD for the determination of pesticide residues in eggs, adding a cleanup step based on SPE.

Some of the proposed methods based on QqQ have developed approaches for the determination of LC- and GC-amenable pesticides, applying common extraction

Table 7.5 Comparison Between High-Resolution Mass Spectrometry (HRMS) and Low-Resolution Mass Spectrometry (LRMS) Procedures for the Determination of Pesticide Residues in Food From Animal Origin

Characteristic	HRMS	LRMS
Number of analytes	Higher number of target compounds	Single-class analysis
Sensitivity	Worst sensitivity	Lower limits (<1 µg/kg)
Extraction procedure	Generic approaches	Specific methods
Retrospective analysis	Yes	No

procedures although two analyses should be performed to cover the compounds included in the target list (Hildmann et al., 2015; Lichtmannegger et al., 2015).

Finally, Table 7.5 shows an overview of the main differences between HRMS- and LRMS-based methods developed for pesticide residues determination of food from animal origin.

7.6 OVERVIEW AND FUTURE PERSPECTIVES

LC coupled to HRMS is a powerful tool for the reliable identification and quantification of pesticide residues in food from animal origin. UHPLC provides several advantages that favor the use of HRMS when complex matrices are analyzed. Thus, high separation power, narrow peaks, and short analysis time could be linked to HRMS to provide efficient analytical methods, which allow the determination of "unlimited" number of compounds. For that purpose, generic extraction, chromatographic, and MS conditions should be applied, and currently, the main efforts have been devoted to the development of efficient, fast, and clean extraction procedures that allow the simultaneous extraction of a wide range of compounds. Although this procedure has been widely settled for pesticide residues in fruits and vegetables, it is not well-implemented in food from animal origin.

In relation to spectrometric detection and identification of target compounds, MS parameters such as extraction mass window and resolving power are important to avoid false positives and negatives, and a careful study of them should be performed to improve the reliability of the detection process. Moreover, it must be considered that high resolution is not the only factor that should be considered for correct mass assignment, because the concentration of the compound, the ratio of the analyte/interferent concentrations, and sample preparation also play an important role as it has been discussed in previous sections.

Although most of the applications have mainly focused on qualitative purposes (detection and identification of target and nontarget compounds), HRMS can also be applied for quantitative purposes, improving cost-effectiveness, because the number of analytes determined by a single procedure is maximized. However, more efforts are needed in this field to simplify the quantification step, considering the high number of compounds that can be simultaneously analyzed. Sometimes, this step is tedious and very time-consuming and it is one of the main drawbacks for the implementation of these methodologies in routine laboratories. To overcome this step, the combination of qualitative (retention time window, mass accuracy, detection of fragments) and quantitative (linearity, suitable recoveries, etc.) parameters should be evaluated to produce a single response, which could provide a first-sight decision by the analyzer to minimize data treatment in routine analysis.

Related to this, when HRMS is used during screening methods, software plays an important role because the data processing time is important to minimize data treatment stage, as well as if the software is connected with a suitable database, identification of new compounds could be easily extended without needing reference substances.

Therefore, once the HRMS instrumentation has been well established, software tools needed for a fast and reliable detection, identification, and quantification are demanding by researchers and routine laboratories, to minimize data treatment and provide fast and efficient methodologies.

REFERENCES

Aguilera-Luiz, M. M., Romero-González, R., Plaza-Bolaños, P., Martínez Vidal, J. L., & Garrido Frenich, A. (2013). Wide-scope analysis of veterinary drug and pesticide residues in animal feed by liquid chromatography coupled to quadrupole-time-of-flight mass spectrometry. *Analytical and Bioanalytical Chemistry, 405*, 6543–6553.

Bogialli, S., & Di Corcia, A. (2009). Recent applications of liquid chromatography-mass spectrometry to residue analysis of antimicrobials in food of animal origin. *Analytical and Bioanalytical Chemistry, 395*, 947–966.

Botitsi, H. V., Garbis, S. D., Economou, A., & Tsipi, D. F. (2011). Current mass spectrometry strategies for the analysis of pesticides and their metabolites in food and water matrices. *Mass Spectrometry Reviews, 30*, 907–939.

Choi, S., Kim, S., Shin, J. Y., Kim, M., & Kim, J.-H. (2015). Development and verification for analysis of pesticides in eggs and egg products using QuEChERS and LC-MS/MS. *Food Chemistry, 173*, 1236–1242.

Chung, S. W. C., & Chen, B. L. S. (2011). Determination of organochlorine pesticide residues in fatty foods: A critical review on the analytical methods and their testing capabilities. *Journal of Chromatography A, 1218*, 5555–5567.

Codex Alimentarius. (2015). *Codex Alimentarius pesticide residues in food and feed.* (Online). Available from http://www.codexalimentarius.net/pestres/data/pesticides/details.html?id=17.

Cotton, J., Leroux, F., Broudin, S., Marie, M., Corman, B., Tabet, J. C., et al. (2014). High-resolution mass spectrometry associated with data mining tools for the detection of pollutants and chemical characterization of honey samples. *Journal of Agricultural and Food Chemistry, 62*, 11335–11345.

De Dominicis, E., Commissati, I., Gritti, E., Catellani, D., & Suman, M. (2015). Quantitative targeted and retrospective data analysis of relevant pesticides, antibiotics and mycotoxins in bakery products by liquid chromatography-single-stage Orbitrap mass spectrometry. *Food Additives and Contaminants: Part A, 32*(10), 1617–1627.

De Dominicis, E., Commissati, I., & Suman, M. (2012). Targeted screening of pesticides, veterinary drugs and mycotoxins in bakery ingredients and food commodities by liquid chromatography-high-resolution single-stage Orbitrap mass spectrometry. *Journal of Mass Spectrometry, 47*, 1232–1241.

European Union. (2005). Council Regulation No 396/2005 of 23 February 2005 on maximum residue levels of pesticides in or on food and feed of plant and animal origin and amending directive no 91/414. *Official Journal of the European Communities. L70*(1) (16.03.05). Available on http://ec.europa.eu/food/plant/pesticides/eu-pesticides-database-redirect/index_en.htm.

European Union. (2003). Commission Directive 2003/13/EC of 10 February 2003 amending directive 96/5/EC on processed cereal-based foods and baby foods for infants and young children. *Official Journal of European Union. L41*(33) (14.02.03). Available from http://eur-lex.europa.eu/LexUriServ/LexUriServ.do?uri=OJ:L:2003:041:0033:0036:EN:PDF.

European Union, CEN. (2008). *CEN Standard Method EN 15662: Food of plant origin determination of pesticide residues using GC-MS and/or LC-MS/MS following acetonitrile extraction/portioning and clean-up by dispersive SPE QuECHERS method.* Available from http://www.cen.eu.

European Union, SANTE. (2015). *SANTE/11945/2015. Guidance document on analytical quality control and method validation procedures for pesticide residues analysis in food and feed.* Available from http://ec.europa.eu/food/plant/docs/plant_pesticides_mrl_guidelines_wrkdoc_11945_en.pdf.

Farré, M., Picó, Y., & Barceló, D. (2014). Application of ultra-high pressure liquid chromatography linear ion-trap to qualitative and quantitative assessment of pesticide residues. *Journal of Chromatography A, 1328*, 66−79.

Ferrer, C., Lozano, A., Agüera, A., Girón, A. J., & Fernández-Alba, A. R. (2011). Overcoming matrix effects using the dilution approach in multiresidue methods for fruit and vegetables. *Journal of Chromatography A, 1218*, 7634−7639.

Filigenzi, M. S., Ehrke, N., Aston, L. S., & Poppenga, R. H. (2011). Evaluation of a rapid screening method for chemical contaminants of concern in four food-related matrices using QuEChERS extraction, UHPLC and high resolution mass spectrometry. *Food Additives and Contaminants: Part A, 28*(10), 1324−1339.

Gan, J., Lv, L., Peng, J., Li, J., Xiong, Z., Chen, D., et al. (2016). Multi-residue method for the determination of organofluorine pesticides in fish tissue by liquid chromatography triple quadrupole tandem mass spectrometry. *Food Chemistry, 207*, 195−204.

Gbylik-Sikorska, M., Sniegocki, T., & Posyniak, A. (2015). Determination of neonicotinoid insecticides and their metabolites in honey bee and honey by liquid chromatography tandem mass spectrometry. *Journal of Chromatography B, 990*, 132−140.

Gilbert-López, B., García-Reyes, J. F., & Molina-Díaz, A. (2012). Determination of fungicide residues in baby food by liquid chromatography-ion trap tandem mass spectrometry. *Food Chemistry, 135*, 780−786.

Gómez-Pérez, M. L., Plaza-Bolaños, P., Romero-González, R., Martínez Vidal, J. L., & Garrido-Frenich, A. (2012). Comprehensive qualitative and quantitative determination of pesticides and veterinary drugs in honey using liquid chromatography-Orbitrap high resolution mass spectrometry. *Journal of Chromatography, 1248*, 130−138.

Gómez-Pérez, M. L., Romero-González, R., Martínez Vidal, J. L., & Garrido Frenich, A. (2015a). Analysis of pesticide and veterinary drug residues in baby food by liquid chromatography coupled to Orbitrap high resolution mass spectrometry. *Talanta, 131*, 1−7.

Gómez-Pérez, M. L., Romero-González, R., Martínez Vidal, J. L., & Garrido Frenich, A. (2015b). Analysis of veterinary drug and pesticide residues in animal feed by high-resolution mass spectrometry: Comparison between time-of-flight and Orbitrap. *Food Additives & Contaminants: Part A, 32*(10), 1637−1646.

Gómez-Pérez, M. L., Romero-González, R., Martínez Vidal, J. L., & Garrido Frenich, A. (2015c). Identification of transformation products of pesticides and veterinary drugs in food and related matrices: Use of retrospective analysis. *Journal of Chromatography A, 1389*, 133−138.

Gómez-Pérez, M. L., Romero-González, R., Plaza-Bolaños, P., Génin, E., Martínez Vidal, J. L., & Garrido Frenich, A. (2014). Wide-scope analysis of pesticide and veterinary drug residues in meat matrices by high resolution MS: Detection and identification using Exactive-Orbitrap. *Journal of Mass Spectrometry, 49*, 27−36.

Hildmann, F., Gottert, C., Frenzel, T., Kempe, G., & Speer, K. (2015). Pesticide residues in chicken eggs — A sample preparation methodology for analysis by gas and liquid chromatography/tandem mass spectrometry. *Journal of Chromatography A, 1403*, 1—20.

Jia, W., Chu, X., Ling, Y., Huang, J., & Chang, J. (2014). High-throughput screening of pesticide and veterinary drug residues in baby food by liquid chromatography coupled to quadrupole Orbitrap mass spectrometry. *Journal of Chromatography A, 1347*, 122—128.

Kaklamanos, G., Vincent, U., & Von Holst, C. (2013). Multi-residue method for the detection of veterinary drugs in distillers grains by liquid chromatography-Orbitrap high resolution mass spectrometry. *Journal of Chromatography A, 1322*, 38—48.

Kasiotis, K. M., Anagnostopoulos, C., Anastasiadou, P., & Machera, K. (2014). Pesticide residues in honeybees, honey and bee pollen by LC-MS/MS screening: Reported death incidents in honeybees. *Science of the Total Environment, 485—486*, 633—642.

Kellmann, M., Muenster, H., Zomer, P., & Mol, H. (2009). Full scan MS in comprehensive qualitative and quantitative residue analysis in food and feed matrices: How much resolving power is required? *Journal of American Society of Mass Spectrometry, 20*, 1464—1476.

Lazartigues, A., Wiest, L., Baudot, R., Thomas, M., Feidt, C., & Cren-Olivé. (2011). Multi-residue method to quantify pesticides in fish muscle by QuEChERS-based extraction and LC-MS/MS. *Analytical and Bioanalytical Chemistry, 400*, 2185—2193.

León, N., Pastor, A., & Yusà, V. (2016). Target analysis and retrospective screening of veterinary drugs, ergot alkaloids, plant toxins and other undesirable substances in feed using liquid chromatography-high resolution mass spectrometry. *Talanta, 149*, 43—52.

Li, Y., Wang, M., Yan, H., Fu, S., & Dai, H. (2013). Simultaneous determination of multiresidual phenyl acetanilide pesticides in different food commodities by solid-phase cleanup and gas chromatography-mass spectrometry. *Journal of Separation Science, 36*(6), 1061—1069.

Lichtmannegger, K., Fischer, R., Steemann, F. X., Unterluggauer, H., & Masselter, S. (2015). Alternative QuEChERS-based modular approach for pesticide residue analysis in food of animal origin. *Analytical and Bioanalytical Chemistry, 407*, 3727—3742.

Martínez Vidal, J. L., Plaza-Bolaños, P., Romero-González, R., & Garrido Frenich, A. (2009). Determination of pesticide transformation products: A review of extraction and detection methods. *Journal of Chromatography A, 1216*, 6767—6788.

Masiá, A., Blasco, C., & Picó, Y. (2014). Last trends in pesticide residue determination by liquid chromatography-mass spectrometry. *Trends in Environmental Analytical Chemistry, 2*, 11—24.

Nácher-Mestre, J., Ibáñez, M., Serrano, R., Pérez-Sánchez, J., & Hernández, F. (2013). Qualitative screening of undesirable compounds from feeds to fish by liquid chromatography coupled to mass spectrometry. *Journal of Agricultural and Food Chemistry, 61*, 2077—2087.

Ojanperä, S., Pelander, A., Pelzing, M., Krebs, I., Vuori, E., & Ojanperä, I. (2006). Isotopic pattern and accurate mass determination in urine drug screening by liquid chromatography/time-of-flight mass spectrometry. *Rapid Communications in Mass Spectrometry, 20*(7), 1161—1167.

Portolés, T., Ibáñez, M., Sancho, J. V., López, F. J., & Hernández, F. (2009). Combined use of GC-TOF MS and UHPLC-(Q)-TOF MS to investigate the presence of nontarget pollutants and their metabolites in a case of honeybee poisoning. *Journal of Agricultural and Food Chemistry, 57*, 4079—4090.

QuEChERS. (2011). (Online) Available from http://quechers.cvua-stuttgart.de.

Romero-González, R. (2015). Food safety: How analytical chemists ensure it. *Analytical Methods, 7*, 7193—7201.

Souza Tette, P. A., Rocha Guidi, L., Abreu Glória, M. B., & Fernandes, C. (2016). Pesticides in honey: A review on chromatographic analytical methods. *Talanta, 149*, 124—141.

Stachniuk, A., & Fornal, E. (2016). Liquid chromatography-mass spectrometry in the analysis of pesticide residues in food. *Food Analytical Methods, 9*(6), 1654—1665.

Sun, H., Ge, X., Lv, Y., & Wang, A. (2012). Application of accelerated solvent extraction in the analysis of organic contaminants, bioactive and nutritional compounds in food and feed. *Journal of Chromatography A, 1237*, 1—23.

Tanner, G., & Czerwenka, C. (2011). LC-MS/MS analysis of neonicotinoid insecticides in honey: Methodology and residue findings in Austrian honeys. *Journal of Agricultural and Food Chemistry, 59*, 12271—12277.

United States, AOAC. (2007). *AOAC official method 2007.01. Pesticide residues in foods by acetonitrile extraction and partitioning with magnesium sulphate.* Available from http://www.eoma.aoac.org/methods/info.asp?ID=48938.

United States, EPA. (2013). *Interim reregistration eligibility decision for chlorpyrifos, United States Environmental Protection Agency, prevention, pesticides and toxic substances, EPA 738-R-01—007.* Available from http://www.epa.gov/oppsrrd1/REDs/chlorpyrifos_ired.pdf.

United States, EPA. (2016). *Tolerances and exemptions for pesticide chemical residues in food.* Available from http://www.ecfr.gov/cgi-bin/text-idx?SID=652f6661f1c740545053c400d-fe56616&node=pt40.24.180&rgn=div5.

Vázquez-Roig, P., & Picó, Y. (2015). Pressurized liquid extraction of organic contaminants in environmental and food samples. *Trends in Analytical Chemistry, 71*, 55—64.

Wei, H., Tao, Y., Chen, D., Xie, S., Pan, Y., Liu, Z., et al. (2015). Development and validation of a multi-residue screening method for veterinary drugs, their metabolites and pesticide meat using liquid chromatography-tandem mass spectrometry. *Food Additives and Contaminants. Part A, 32*(5), 686—701.

Xie, J., Peng, T., Zhu, A., He, J., Chang, Q., Hu, X., et al. (2015). Multi-residue analysis of veterinary drugs, pesticides and mycotoxins in dairy products by liquid chromatography-tandem mass spectrometry using low-temperature cleanup and solid phase extraction. *Journal of Chromatography B, 1002*, 19—29.

Yáñez, K. P., Martín, M. T., Bernal, J. L., Nozal, M. J., & Bernal, J. (2014). Determination of spinosad at trace levels in bee pollen and beeswax with solid-liquid extraction and LC-ESI-MS. *Journal of Separation Science, 37*(3), 204—210.

Recent Advances in HRMS Analysis of Pesticide Residues Using Atmospheric Pressure Gas Chromatography and Ion Mobility

Lauren Mullin, Gareth Cleland, Jennifer A. Burgess

Waters Corporation, Milford, MA, United States

Abbreviations

APCI Atmospheric pressure chemical ionization
APGC Atmospheric pressure gas chromatography
CI Chemical ionization
CID Collision-induced dissociation
Da Daltons
DDD Dichlorodiphenyl dichloroethane
DDE 1,1-Dichloro-2,2-bis(*p*-chlorophenyl) ethylene
DDT Dichlorodiphenyl trichloroethane
DTIMS Drift time ion mobility spectrometry
EC Electron capture detection
ECNI Electron capture negative ionization
EI Electron ionization
ESI Electrospray ionization
FAIMS Field-asymmetric waveform ion mobility spectrometry
FI Flame ionization
FWHM Full width at half-maximum height
GC Gas chromatography
HCB Hexachlorobenzene
HRMS High-resolution mass spectrometry
IM Ion mobility
IMS Ion mobility separation
LC Liquid chromatography
MALDI Matrix-assisted laser desorption/ionization
MRM Multiple reaction monitoring
MS Mass spectrometry
MS/MS Tandem mass spectrometry

Applications in High Resolution Mass Spectrometry. http://dx.doi.org/10.1016/B978-0-12-809464-8.00008-7

m/z Mass-to-charge ratio
NCI Negative ion chemical ionization
QuEChERS Quick, easy, cheap, effective, rugged, and safe
RPLC Reversed-phase liquid chromatography
SRM Single reaction monitoring
TOF MS Time-of-flight mass spectrometry
TQ MS Tandem quadrupole mass spectrometry
TW-IMS Traveling wave ion mobility spectrometry
UPLC Ultra performance liquid chromatography

8.1 INTRODUCTION

Chemicals have been employed as insecticides for thousands of years. However, it is the relatively recent expansion of synthetic pesticide use since the 1940s that has resulted in a list of 1700+ compounds that are categorized as pesticides (Wood, 2016). Many of these compounds are not currently registered for use and some have trade restrictions (Rotterdam Convention, 2016) or are prohibited from production (Stockholm Convention, 2016; World Health Organization, 2016). The vast compound diversity and array of chemical properties exhibited by these compounds, and their metabolites, pose a great challenge to the analytical chemist.

In the 1970s, gas chromatography (GC) analysis with electron capture (EC) or flame ionization detection was typically deployed for the analysis of known target pesticides. With the introduction of GC—mass spectrometry (MS), selectivity of the detection method had improved and this became the method of choice (Alder, Greulich, Kempe, & Vieth, 2006). To ionize the pesticides for detection by MS, three main mechanisms were used: electron ionization (EI), negative chemical ionization, or positive chemical ionization. For the vast majority of analyses, EI was deployed, with chemical ionization (CI) techniques usually being reserved for challenging and labile compounds or where additional sensitivity was required. EI is a highly energetic process, which results in many fragments of the compounds being created in the source of the MS system and these spectra have been used to create spectral libraries. The fragmentation induced by EI, however, may pose a challenge for the use of tandem mass spectrometry (MS/MS), which relies on the selection and subsequent fragmentation of a molecular ion of the pesticide of interest (Alder et al., 2006).

With the introduction of the atmospheric ionization techniques, such as electrospray ionization (ESI) and atmospheric pressure chemical ionization (APCI) in the 1980s, the use of liquid chromatography (LC)—MS and LC—MS/MS techniques for pesticide residue analysis has rapidly expanded. Today, there are more LC-amenable pesticides than GC-amenable pesticides and many of the multiresidue methods available today reflect this (Alder et al., 2006; Hird, Lau, Schuhmacher, & Krska, 2014). The soft ionization that is achieved using these atmospheric techniques lends itself to MS/MS. The ability to select the molecular ion and subsequently fragment it to monitor specific fragments provides the selectivity required for multiresidue analysis in complex matrices. The use of multiple reaction monitoring (MRM) provides a high level of selectivity and has become the method of choice for multiresidue pesticide methods. Recent advances in the sensitivity, simplicity, and ruggedness of high-resolution MS techniques, such as time-of-flight MS, mean that accurate mass full spectrum data are available for pesticide residue chemists and new ways to deploy these technologies are being investigated and even incorporated into recent method guidelines (EU Directorate-General for Health and Food Safety, 2015). TOF MS is especially suited to the acquisition of rapid full spectrum data whereas scanning instruments (such as quadrupoles and ion traps) are most well suited to targeted analysis of specific compounds (MRM or SRM modes).

The selectivity of accurate-mass, high-resolution mass spectrometers is achieved through a combination of mass accuracy (which is in the low ppm range) and their ability to distinguish between closely related masses (the mass resolution) (Hird et al., 2014). Ions entering the pusher region of the time-of-flight system are accelerated into the flight tube simultaneously. There they separated based on their mass-to-charge ratio (m/z) such that those with higher m/z values take longer to travel through the flight tube to reach the detector. The separation occurs in the microsecond time frame, so multiple measurements can be made within the time frame of a chromatographic peak, even for the narrow peaks that are achieved with GC and ultra performance liquid chromatography (UPLC).

Recent advances in MS technology have seen an increase in sensitivity of TOF MS that is comparable to that on MS/MS quadrupole systems operated in MRM mode. The advantage of TOF MS is that all ions are analyzed, rather than just the targeted compounds included in the MRM method. This advantage has resulted in a growing interest in the technology for nontargeted contaminant screening (Hird et al., 2014; Kauffman, 2012).

Other improvements with TOF MS include the coupling of an additional dimension of separation using ion mobility (IM). By coupling chromatographic separations, ion mobility separations (IMSs), and high-resolution MS, the overall peak capacity of the system is increased and has the potential to separate compounds that would otherwise be detected as a single component in a non-IMS system. UPLC-IM-MS has been demonstrated to facilitate the identification of pesticides in QuEChERS (quick, easy, cheap, effective, rugged, and safe) extracts of fruits and vegetables (Goscinny, Joly, De Pauw, Hanot, & Eppe, 2015), and interest is growing for its use in this field.

The rapid increase in atmospheric techniques for interfacing LC to MS also renewed interest in atmospheric pressure ionization techniques for interfacing GC to MS, giving rise to the technique known as atmospheric pressure gas chromatography (APGC)−MS. The APGC source can be used directly with MS systems designed for the integration of an LC system using APCI or ESI. The source can therefore be changed between LC and GC, without breaking the vacuum of the MS system. Owing to the wide range of physiochemical properties of pesticides and the need for both GC and LC, the ability to perform both techniques with the same acquisition method on the same MS system is advantageous. The application of nontargeted screening approaches relies on comprehensive sample coverage, generic sample preparation, and full spectrum acquisition. GC and LC coverage is also an important consideration for full compound coverage.

The challenge of pesticide residue analysis in such a diverse range of commodities and products relies on the ability to extract and detect many compounds within a method. Techniques that increase the peak capacity of the system, through either high-resolution chromatography or orthogonal techniques such as IM, help to separate compounds of interest from coextracted matrix components and from compounds with similar properties (e.g., isomeric pesticides). In combination with the rapid, full-spectrum analysis provided by TOF MS, comprehensive information can be interpreted by modern informatics to enhance compound detection, elucidation, and identification.

8.2 ATMOSPHERIC PRESSURE GAS CHROMATOGRAPHY
8.2.1 INTRODUCTION

APCI can be used as an alternative means of interfacing GC to MS platforms. In GC–APCI, also commonly referred to as APGC, the GC effluent is swept into an atmospheric pressure MS source where ionization is induced by means of a corona discharge. Compared to the 70 eV energy typically imparted during EI (McLafferty & Tureček, 1993, p. 6), APGC is a "softer" technique resulting in less fragmentation (Li, Gan, Bronja, & Schmitz, 2015). Reduced fragmentation of the precursor ion within the source allows for controlled collision cell fragmentation. Sensitivity using APGC–MS is increased over EI MS given the reduced fragmentation in the source. Moreover, the atmospheric pressure source design affords the switching between LC and GC inlets on a single MS system, increasing analyte coverage (Li et al., 2015). This is advantageous for pesticide residue analysis as both LC–MS and GC–MS analyzers are required to cover the full suite of pesticides. The following sections explore the development and application of APGC coupled with HRMS for pesticide analysis.

8.2.2 BACKGROUND

Development of an atmospheric pressure source coupled to a GC (APGC) began in the early 1970s, by Horning et al. in the interest of achieving picogram (pg) level detection of metabolites and hormones in biological samples (Horning, Horning, Carroll, Dzidic, & Stillwell, 1973). Initial studies used [63]Ni decay for ionization but later work published by Carroll et al. compared this approach with corona discharge (Carroll, Dzidic, Stillwell, Haegele, & Horning, 1975; Li et al., 2015). Compared to typical CI sources the APGC system requires much lower gas input to generate suitable ionization conditions (Carroll et al., 1975). Negative polarity ionization was also assessed for the analysis of halogenated pesticides and other environmental contaminants in these early experiments. The flow from the GC system was split between both electron capture detection (ECD) and the corona discharge API MS source. Ion molecule reactions with gaseous superoxide $(O_2)^-$ as a reagent resulted in relatively simple mass spectra, which provided additional information not obtained using ECD (Horning et al., 1977).

Implementation of APGC−MS using the negative mode ionization approach continued in the 1980s to detect low-levels of tetrachlorodibenzo-p-dioxins, aminonitropyrene, nitronapthalene, and mononitrofluoranthenes (Korfmacher, Mitchum, Hileman, & Mazer, 1983; Korfmacher et al., 1984; McEwen & McKay, 2005; Mitchum, Moler, & Korfmacher, 1980). Widespread usage of APGC−MS was not embraced, however, because of the lack of commercially available instrumentation (Li et al., 2015). The development of a modified LC/MS source by McEwen and McKay (2005) resulted in the commercialization of the APGC source in 2008. The source was developed on a TOF MS system, to fully harness HRMS capabilities and the system could also be used as an LC interface with the ESI probe, which could be blocked using a metal flange when the LC was not in use (McEwen & McKay, 2005). Important findings from this prototype included the observation of moisture in the source resulting in protonation, and the improvement of ionization for nonpolar analytes by evacuating the source of moisture with N_2 overnight and subsequent ionization via charge transfer between N_2^+/N_4^+ ions and analyte molecules (McEwen & McKay, 2005). It was also reported that source contaminants coming from the GC capillary column bleed and polyamide coating could be reduced by conditioning the column and system at elevated temperatures before analysis (McEwen & McKay, 2005). Based on the analysis of semivolatile organic compounds using this prototype, it was observed that the sensitivity of APGC−MS as compared to traditional CI was good or better (when in negative polarity mode) and that gas load into the source could be increased as compared to EI. This added benefit afforded the ability to use shorter, wide-bore capillary GC columns to elute less volatile analytes in a shorter period of time (McEwen & McKay, 2005). As a consequence of these promising results, commercial scale development of this source was implemented by Waters Corporation shortly thereafter.

8.2.3 CURRENT DESIGN

The current commercial design of the APGC−MS interface and source (referred to hereafter as APGC) is shown in Fig. 8.1. The corona discharge is delivered from the pin as in the McEwen et al.'s design. The heated transfer line interfaces the GC analytical column to the MS source. An extended ionization chamber was introduced to stabilize and focus the ionization process close to the MS sample cone aperture. Nitrogen is introduced into the ionization chamber from the sample cone gas, auxiliary gas, and the heated transfer line from the GC interface. Gas flows for the heated transfer line are typically optimized between 150 and 350 mL/min.

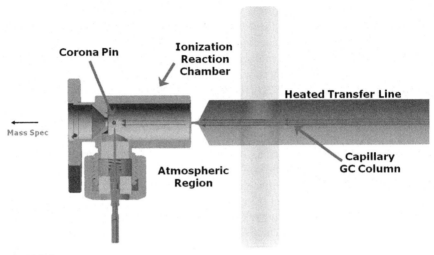

FIGURE 8.1

Commercially available Waters Corporation atmospheric pressure gas chromatography source schematic.

Ionization occurs through reactions between the charged nitrogen plasma (containing N_2^+ and N_4^+ cations) and analyte molecules. The resulting ions are primarily the $M^{+\bullet}$ ion for nonpolar analytes, as described in the following sections. Alternatively, introduction of a protic solvent (e.g., water) into the source will allow for protonation reactions generating the $[M+H]^+$ ion for more polar molecules.

8.2.4 IONIZATION MECHANISMS

When the source is sufficiently dry by removing the solvent from the source (achieved by utilizing moisture traps on gas lines and increasing the cone gas flow), nitrogen cations generated from the corona discharge and analyte molecules undergo a charge transfer reaction, as shown in the following schema (Li et al., 2015):

$$N_2 + e^- \rightarrow N_2^+ + 2e$$

$$N_2^+ + 2N_2 \rightarrow N_4^{+\bullet} + N_2$$

$$N_4^{+\bullet} + M \rightarrow M^{+\bullet} + 2N_2$$

When moisture or another source of protons is present, protonation of the analyte occurs as shown in the following schema (Horning et al., 1973):

$$N_4^{+\bullet} + H_2O \rightarrow H_2O^{+\bullet} + 2N_2$$

$$H_3O^+ + H_2O + N_2 \rightarrow H^+(H_2O)_2 + N_2$$

$$H^+(H_2O)_{n-1} + H_2O + N_2 \rightarrow H^+(H_2O)_n + N_2$$

$$H^+(H_2O)_n + M \rightarrow [M+H]^+ + (H_2O)_n$$

Under negative conditions where oxygen is present in the source, the following reactions and ions will occur for halogenated (X) compounds (Fernando et al., 2015; Horning et al., 1977):

$$O_2 + e^- \rightarrow O_2^-$$
$$O_2^- + M \rightarrow [MO_2]^-$$
$$[MO_2]^- \rightleftharpoons O_2 + M^- \leftarrow M + e^- \quad \text{or}$$
$$[MO_2]^- \rightarrow [M - X + O]^- + XO^{\bullet}$$

As described previously, conditions that most strongly affect ionization are purity of gas supply and gas flows in the source. Ion formation preference will depend on chemical properties of the analyte and are discussed in the following section.

8.2.5 IONIZATION TRENDS FOR PESTICIDES

To establish the best possible conditions for pesticide analysis using APGC, ionization preference has proven to be the most critical parameter. Of the three ionization mechanisms highlighted earlier, charge transfer and protonation have been observed in the positive ion mode for pesticide analysis (Cheng et al., 2016; Cherta et al., 2015; Nacher-Mestre et al., 2014; Portolés, Mol, Sancho, López, & Hernández, 2014; Portolés, Sancho, Hernández, Newton, & Hancock, 2010). Negative ion analysis has not yet been performed on pesticides using APGC Q-TOF MS to the authors' awareness, although work with coplanar compounds such as dibenzo-p-dioxins indicates promise regarding this approach (Fernando et al., 2015). Based on polarity of the molecules, the best sensitivity is achieved using one of two source conditions. For the majority of the pesticides analyzed in published studies, the best sensitivity has been achieved by generating the protonated $[M+H]^+$ ion, which is achieved through supplying water or methanol in the heated source door. For highly nonpolar pesticides, such as the organochlorine and legacy pesticides, charge transfer mechanisms are the preferred means of ionization and require a dry source. Ionization is not always exclusive to either $[M+H]^+$ or $M^{+\bullet}$, and if moisture is present in the source both ions can be seen for some pesticides. This is true particularly for pesticides where heteroatoms are present, for example, pentachloronitrobenzene and endrin (Cheng et al., 2016). Source gas flow rates, specifically the gas flow applied to the MS sample cone interface, will also impact the ionization formation.

To achieve the best sensitivity, identifying ionization preference is recommended. Table 8.1 summarizes the ionization preferences observed in various pesticide screening experiments.

8.2.6 PESTICIDE SCREENING USING ATMOSPHERIC PRESSURE GAS CHROMATOGRAPHY WITH HIGH-RESOLUTION MASS SPECTROMETRY

Increasing interest in using APGC coupled to quadrupole (Q)-TOF HRMS systems has developed in pesticide screening laboratories in recent years. The main reasons

Table 8.1 Ionization Preferences of 144 Pesticides Surveyed

Pesticide	Pesticide Class	Ionization Mode
2-Phenylphenol	Phenol	Protonation[b,c]
4-4'-Dichlorobenzophenone	Breakdown product	Protonation[b,c]
Alachlor[a]	Chloroacetanilide	Protonation[b,c]
Aldrin	Cyclodiene	Charge transfer[d], protonation[b,c]
Alpha-chlordane	Cyclodiene	Charge transfer[d]
Alpha-endosulfan	Cyclodiene	Charge transfer[d], protonation[b,c]
Alpha-HCH[a]	Cyclodiene	Charge transfer[b,c]
Atrazine	Chlorotriazine	Protonation[b,c]
Atrazine desethyl	Metabolite	Protonation[b,c]
Atrazine desisopropyl	Metabolite	Protonation[b,c]
Azinphos methyl	Organothiophosphate	Protonation[b,c]
Azoxytrobin	Methoxyacrylate strobilurin	Protonation[b,c]
Beta-chlordane	Cyclodiene	Charge transfer[d]
Beta-endosulfan	Cyclodiene	Protonation[b,c]
Beta-HCH[a]	Cyclodiene	Charge transfer[b,c]
Bifenthrin	Pyrethroid	Protonation[b,c]
Bromophos	Organothiophosphate	Protonation[b,c]
Bromophos ethyl	Organothiophosphate	Protonation[b,c]
Bromopropylate	Bridged diphenyl	Protonation[b]
Buprofezin	Chitin synthesis inhibitor	Protonation[b,c]
Cadusafos	Organothiophosphate	Protonation[b,c]
Captafol	Pthalamide	Protonation[b,c]
Captan	Pthalamide	Protonation[b,c]
Carbaryl[a]	Carbamate	Protonation[b,c]
Carbofuran	Carbamate	Protonation[b,c]
Carbophenothion	Organothiophosphate	Protonation[b,c]
Carfentrazone ethyl	Triazolone	Protonation[b,c]
Chinomethionat	Quinoxaline	Protonation[b,c]
Chlorfenapyr	Pyrrole	Protonation[b,c]
Chlorfenson	Bridged diphenyl	Protonation[b,c]
Chlorfenvinphos	Organophosphate	Protonation[b,c]
Chloropyrifos ethyl	Organothiophosphate	Protonation[b,c]
Chloropyrifos methyl	Organothiophosphate	Protonation[b,c]
Chlorothalonil	Aromatic	Charge transfer[d]
Chlorothalonil	Aromatic	Protonation[b,c]
Chlorpropham[a]	Carbanilate	Protonation[b,c]
Coumaphos	Organothiophosphate	Protonation[b,c]

Table 8.1 Ionization Preferences of 144 Pesticides Surveyed—cont'd

Pesticide	Pesticide Class	Ionization Mode
Cyanazine	Chlorotriazine	Protonation[b,c]
Cyanophos	Organothiophosphate	Protonation[b,c]
Cyfluthrin	Pyrethroid	Protonation[b,c]
Cypermethrin	Pyrethroid	Protonation[b,c]
Cyprodinil	Anilinopyrimidine	Protonation[b,c]
Delta-HCH[a]	Cyclodiene	Charge transfer[b,c]
Deltamethrin	Pyrethroid	Protonation[b,c]
Diflufenican	Anilide	Protonation[b,c]
Diazinon	Organothiophosphate	Protonation[b,c]
Dichlofenthion	Organothiophosphate	Protonation[b,c]
Dichloran	Aromatic	Protonation[b,c]
Dichlorvos	Organophosphate	Protonation[b,c]
Dieldrin	Cyclodiene	Protonation[b,c]
Dimethoate	Organothiophosphate	Protonation[b,c]
Dioxathion[a]	Organothiophosphate	Protonation[b,c]
Diphenylamine	Bridged diphenyl	Protonation[b,c]
Endosulfan ether	Metabolite	Protonation[b,c]
Endosulfan sulfate	Metabolite	Protonation[b,c]
Endrin	Cyclodiene	Protonation[b,c,d]
Endrin	Cyclodiene	Charge transfer[d]
EPN	Phenylphosphonothioate	Protonation[b,c]
Ethalfluralin	Dinitroaniline	Protonation[b,c]
Ethion	Organothiophosphate	Protonation[b,c]
Ethoxyquin	Quinoline	Protonation[b,c]
Etofenprox[a]	Pyrethroid	Protonation[b,c]
Famphur	Organothiophosphate	Protonation[b,c]
Fenamiphos	Organophosphate	Protonation[b,c]
Fenarimol	Pyrimidine	Protonation[b,c]
Fenhexamid	Anilide	Protonation[b,c]
Fenitrothion	Organothiophosphate	Protonation[b,c]
Fenoxycarb	Carbamate	Protonation[b,c]
Fenthion	Organothiophosphate	Protonation[b,c]
Fenvalerate	Pyrethroid	Protonation[b,c]
Fipronil	Phenylpyrazole	Protonation[b,c]
Flucythrinate[a]	Pyrethroid	Protonation[b,c]
Fludioxonil	Pyrrole	Charge transfer[c]
Folpet	Pthalimide	Protonation[c]
Gamma-HCH[a]	Cyclodiene	Charge transfer[b,c,d]
HCB	Aromatic	Charge transfer[b,c]
Heptachlor[a]	Cyclodiene	Charge transfer[b,c]
Heptachlor epoxide A	Organochlorine	Protonation[b,c]

Continued

Table 8.1 Ionization Preferences of 144 Pesticides Surveyed—cont'd

Pesticide	Pesticide Class	Ionization Mode
Heptachlor epoxide B	Organochlorine	Protonation[b,c]
Hexachlorobutadiene	Chlorinated diene	Charge transfer[b,c]
Imazalil	Conazole	Protonation[b,c]
Iprodione	Imidazole	Protonation[b,c]
Isodrin	Cyclodiene	Protonation[b,c]
Lambda-cyhalothrin	Pyrethroid	Protonation[b,c]
Leptophos	Phenylphosphonothioate	Protonation[b,c]
Malathion	Organothiophosphate	Protonation[b,c]
Metalaxyl	Anilide	Protonation[b,c]
Methamidophos	Phosphoroamidothionate	Protonation[b,c]
Methidathion	Organothiophosphate	Protonation[b,c]
Methiocarb[a]	Carbamate	Protonation[b,c]
Methoxychlor[a]	Organochlorine	Protonation[b,c]
Metolachor	Chloroacetanilide	Protonation[b,c]
Metribuzin	Triazinone	Protonation[b,c]
Mirex[a]	Cyclodiene	Charge transfer[b,c,d]
Molinate	Thiocarbamate	Protonation[b,c]
o,p′-DDT	Organochlorine	Charge transfer[d]
o,p-DDD	Metabolite	Charge transfer[d]
o,p′-DDE	Metabolite	Charge transfer[d]
Omethoate	Organothiophosphate	Protonation[b]
Oxadixyl	Anilide	Protonation[b,c]
Oxyfluorfen	Nitrophenyl	Protonation[b,c]
p,p′-DDT	Organochlorine	Charge transfer[b,c,d]
p,p′-DDD[a]	Metabolite	Charge transfer[b,c,d]
p,p′-DDE	Metabolite	Charge transfer[b,c,d]
Parathion ethyl	Organothiophosphate	Protonation[b,c]
Parathion methyl	Organothiophosphate	Protonation[b,c]
PCNB	Aromatic	Protonation[d]
PCNB	Aromatic	Charge transfer[d]
Pendimethalin	Dinitroaniline	Protonation[b,c]
Pentachlorobenzene	Organochlorine	Charge transfer[b,c]
Permethrin[a]	Pyrethroid	Protonation[b,c]
Phorate	Organothiophosphate	Protonation[b,c]
Phosmet	Organothiophosphate	Protonation[b,c]
Phosphamidon	Organophosphate	Protonation[b]
Pirimicarb	Dimethylcarbamate	Protonation[b,c]
Pirimiphos methyl	Organothiophosphate	Protonation[b,c]
Procymidone	Dichlorophenyl dicarboximide	Protonation[b,c]

Table 8.1 Ionization Preferences of 144 Pesticides Surveyed—cont'd

Pesticide	Pesticide Class	Ionization Mode
Propachlor	Chloroacetanilide	Protonation[b]
Propetamphos	Phosphoroamidothionate	Protonation[b,c]
Propham[a]	Carbanilate	Protonation[b,c]
Propiconazole	Conazole	Protonation[b,c]
Propoxur	Carbamate	Protonation[b,c]
Propyzamide	Amide	Protonation[b,c]
Pyriproxyfen	Juvenile hormone mimic	Protonation[b,c]
Quinalphos	Organothiophosphate	Protonation[b,c]
Resmethrin	Pyrethroid	Protonation[b,c]
Simazine	Chlorotriazine	Protonation[b,c]
Tau-fluvalinate	Pyrethroid	Protonation[b,c]
Tefluthrin	Pyrethroid	Protonation[b,c]
Terbacil[a]	Uracil	Protonation[b,c]
Terbumeton	Methoxytriazine	Protonation[b,c]
Terbumeton desethyl	Metabolite	Protonation[b,c]
Terbuthylazine	Chlorotriazine	Protonation[b,c]
Terbuthylazine desethyl	Metabolite	Protonation[b,c]
Terbutryn	Methylthiotriazine	Protonation[b,c]
Tetradifon	Bridged diphenyl	Protonation[b,c]
Thiabendazole	Benzimidazole	Protonation[b,c]
Toclofos methyl	Organophosphorus	Protonation[b,c]
Tolyfluanid[a]	Phenylsulfamide	Protonation[b,c]
Trans-chlordane	Cyclodiene	Charge transfer[b,c]
Triadimefon	Conazole	Protonation[b,c]
Triflumizole	Conazole	Protonation[b,c]
Trifluralin	Dinitroaniline	Protonation[b,c]
Vinclozolin	Dichlorophenyl dicarboximide	Protonation[b,c]

Ionization via protonation is generally implemented with the exception of some cyclodiene pesticides, which are nonpolar in nature and ionize by charge transfer in most cases.
[a] *Product ion monitored.*
[b] *Portolés et al. (2014).*
[c] *Nacher-Mestre et al. (2014).*
[d] *Cheng et al. (2016).*

cited are that the softer ionization enables preservation of the precursor ions compared to EI, resulting in conservation of the molecular ion and reduced fragmentation. Better stability in response and broader analyte coverage (compared to traditional CI) are also observed (Cheng et al., 2016; Cherta et al., 2015; Nacher-Mestre et al., 2014; Portolés et al., 2014, 2010). Fig. 8.2A and B shows the comparison between an EI and APGC spectrum for the pesticide chloroneb. In Fig. 8.2B, the

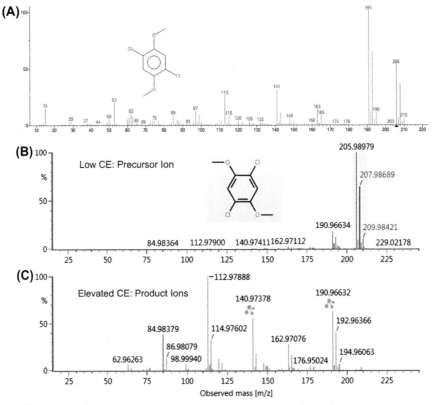

FIGURE 8.2

(A) Electron ionization (EI) spectra for chloroneb (precursor mass of 206 Da), from NIST Library (NIST, 2011). (B and C) Atmospheric pressure gas chromatography quadrupole time-of-flight mass spectrometry spectra for chloroneb. The precursor ion is well conserved in the low collision energy (CE) spectra, while product ions for confirmatory identification purposes are generated using an elevated CE state. Also, isotopic ratio conservation is observed for the double-chlorinated species in the presented spectra, as is seen in the EI spectrum as well.

(A) Reprinted courtesy of the National Institute of Standards and Technology, U.S. Department of Commerce. Not copyrightable in the United States.

spectrum shows the APGC Q-TOF MS spectrum of chloroneb acquired with low collision energy and the preservation of the precursor ion is apparent, with very little fragmentation from the ionization process compared to the EI spectrum in Fig. 8.2A. By increasing the energy on the collision cell, the ions undergo collision-induced dissociation to form fragment ions. This acquisition mode is referred to as MS^E and is described later in this chapter. The spectrum in Fig. 8.2C displays the resulting high-energy spectrum for chloroneb. As can be seen from Fig. 8.2B and C, Q-TOF

MS provides full-spectrum data with accurate mass measurement of all the precursor (B) and fragment (C) ions. Furthermore, the isotope pattern of the ions is maintained for both precursors and fragment ions. The ability of Q-TOF MS to perform full-spectrum, accurate mass measurements for both precursor and product ions in a single injection, at acquisition rates that can easily provide sufficient data points across a GC peak is another reason for interest in its use for pesticide screening. Given that full-spectrum data are acquired for the entire analytical run, full chemical profiling and historical data review (mining the data set after acquisition for newly identified masses of interest) are also possible.

8.2.7 IMPROVED SELECTIVITY AND SENSITIVITY WITH ATMOSPHERIC PRESSURE GAS CHROMATOGRAPHY

The most prominent characteristics of APGC-generated data are the improved selectivity that can be obtained compared to traditional EI as a result of retaining the precursor. To capture as many compounds as possible and harness the power of full spectral acquisitions, nonspecific sample preparation approaches are widely implemented for pesticide screening experiments (Cheng et al., 2016; Cherta et al., 2015; Goscinny et al., 2015; Nacher-Mestre et al., 2014; Portolés et al., 2014, 2010). The use of nonselective sample preparation results in the coextraction of more matrix interferences. Selectivity of the detection system is critical to be able to distinguish analytes of interest from the thousands of components that may be present in these complex extracts. With the high-resolution capabilities of Q-TOF MS, the exact mass measurement of the compounds provides a high degree of selectivity. Using a narrow window of 0.01 Da (10 mDa) to generate an extracted ion chromatogram for targeted compounds, several studies have demonstrated this capability with pesticide identifications in fruit, vegetable, and fish matrices (Cheng et al., 2016; Cherta et al., 2015; Nacher-Mestre et al., 2014; Portolés et al., 2014, 2010).

To confirm identifications in pesticide screening experiments, product ion formation is also required according to European guidance SANTE/11945/2015. Product ion formation during Q-TOF MS operation is generally executed in one of two ways. The traditional approach is MS/MS, where the precursor ion is selected in the quadrupole and fragmented in the collision cell, followed by full spectral acquisition of the product ions. A more widely taken approach within the context of pesticide screening experiments is to operate the quadrupole in radio frequency (RF)-only mode, so that all ions pass through without mass selection (Cheng et al., 2016; Cherta et al., 2015; Nacher-Mestre et al., 2014; Portolés et al., 2014, 2010). The collision cell is then switched between a low collision energy setting and an elevated collision energy setting. The spectra from the low- and high-energy acquisitions are recorded in separate channels. This acquisition mode is referred to as MS^E. Product ions from the elevated energy channel are chromatographically aligned with precursors from the low collision energy channel (Waters, 2011). For the aforementioned studies using APGC Q-TOF MS, MS^E was used to detect pesticides and other food contaminants using both precursor and product ions (Cheng et al., 2016; Cherta et al., 2015; Nacher-Mestre et al., 2014; Portolés et al., 2014, 2010).

Using the SANTE/11945/2015 criteria for identification in a validation study conducted for 132 pesticides in 10 matrices (apple, cucumber, lettuce, tomato, pepper, carrot, grape, orange, strawberry, and cauliflower), Portolés et al. (2014) found that 70% of the compounds could be identified at the 0.01 mg/kg level, 87% at the 0.05 mg/kg level, and 93% at the 0.20 mg/kg level. A study by Nacher-Mestre et al. (2014) performed in fish and feed established the screening detection limit (SDL, defined as the lowest concentration at which a compound could be detected in 95% of the samples analyzed) for 133 pesticides. With detection requiring a mass accuracy within 5 ppm mass error and a diagnostic product ion, 76% of pesticides had an SDL of 0.01 mg/kg, which increased to 91% of the pesticides at the 0.05 mg/kg level. In the years since these studies, further improvements in ion optics have been implemented and the sensitivity of the systems available today has increased. These improvements will result in a lowering of the SDLs and an increase in the percentage of pesticides that can be detected at these levels.

8.2.8 CARRIER GAS FLOW RATE INCREASE

Another advantage of using an atmospheric source GC/MS interface highlighted in recent research is the ability to increase the flow rate of the GC carrier gas beyond normal levels used in EI sources (Tienstra, Portolés, Hernández, & Mol, 2015). The motivation to take an approach where the carrier gas flow rate is increased is described by Tienstra et al. (2015) as reducing the analytical run time of a GC method similar to or less than that of a modern UPLC system. To develop a fast GC method, a 15 m GC analytical column was utilized (rather than the typically employed 30 m for pesticide analyses), with a flow rate of 6 mL/min He and rapid temperature program starting at 120°C and increasing to 330°C at a rate of 30°C/min (Tienstra et al., 2015). Injection conditions were optimized to consider the requirement of a higher oven starting temperature in this method and a hot split injection with a split of 1:10 with a split flow of 60 mL/min was used (Tienstra et al., 2015). The split injection also allowed for the use of acetonitrile as a solvent, which is not ideal for GC splitless injections because of its high vapor volume and lack of compatibility with the nonpolar stationary phase. The compatibility of acetonitrile is of importance, however, because it is the final solvent of a QuEChERS extract (Tienstra et al., 2015). Source conditions were set to promote charge transfer reactions, as the compounds analyzed in this work were composed of polychlorinated biphenyls and nonpolar pesticides, which are traditionally analyzed by GC.

A total of 49 pesticides were analyzed using the method and all eluted at or before 5.25 min, with the total run time of 10 min including the time for cooling down (Tienstra et al., 2015). The goal of reducing the run time to typical UPLC analysis times was therefore achieved. Reducing the amount of time spent at high temperatures of the GC program is beneficial for thermolabile analytes, and, in general, peak shape was improved compared to the longer 30 m column method used for reference (Tienstra et al., 2015). Peak resolution was slightly compromised, but still deemed acceptable within the existing criteria for the analysis of the pesticides in this study (Tienstra et al., 2015). A validation study was performed in wheat, beet pulp, sunflower seed meal, wheat yeast concentrate, and maize using this method to assess linearity, selectivity, matrix effects, trueness, and precision. With the exception of several pesticides, data were acceptable for the quantification of pesticides (Tienstra et al., 2015). Results also adhered to the criteria in SANTE/11945/2015 for identification (Tienstra et al., 2015). By diminishing residence time on the analytical column under high temperatures, the peak shape for labile compounds can be improved, which contributes to improved signal-to-noise. For this reason and the possibility of greatly increasing speed of analysis, fast GC has proven to be an advantageous approach that should be more widely investigated in future experiments with APGC.

8.3 TIME-OF-FLIGHT MASS SPECTROMETRY

TOF MS systems have multiple configurations but (Hoffmann & Stroobant, 2007, p. 126; Watson & Sparkman, 2007) all use the measurement of the time it takes for an ion to travel through a flight tube to determine the *m/z*. In TOF MS, ions are accelerated and have the same kinetic energy but separate from one another during their travel under high vacuum based on their mass (Watson & Sparkman, 2007). Ions of the same mass then arrive in packets at the detector (Watson & Sparkman, 2007). Unlike the tandem quadrupole (TQ) MS systems, exact mass is able to be determined using modern TOF instrumentation (Watson & Sparkman, 2007), typically to four decimal places. Exact mass allows for determination of the molecular formula of an unknown without much ambiguity and without relying on fragmentation information as in a quadrupole MS system.

The full mass range is generally acquired in TOF MS analysis, rather than specified nominal masses as in MRM experiments. Although the acquisition of a wide mass range (referred to as full scan or full spectrum) is not unique to TOF MS systems, TOF systems are able to provide full spectrum analysis without compromising acquisition speed. In contrast to quadrupole and ion trap systems, which require scanning through the mass range to obtain a full spectrum analysis, TOF systems obtain the full spectrum with a single "push" of the ions into the flight tube. Full spectrum analysis is therefore more sensitive on TOF MS systems than full spectrum analysis on TQ MS systems.

In TOF MS systems, the ions are accelerated into the flight tube using a pusher, which is located prior to the flight tube. The ions are pushed orthogonally to the ion source (orthogonal acceleration), which reduces the spread of ions of the same mass but different kinetic energies. A voltage is applied to a plate above the ions, and they are propelled in a direction orthogonal to their original trajectory

(Guilhaus, Selby, & Mlynski, 2000; Hoffmann & Stroobant, 2007, p. 141). The frequency of activation of the pusher is dependent on the mass range specified for the experiment and is repeated multiple times throughout an acquisition (Hoffmann & Stroobant, 2007, p. 129). The ions then traverse through the flight tube, which, in most modern TOF MS systems, utilize some form of a reflectron (Hoffmann & Stroobant, 2007, p. 131). The reflectron, which is an electric field that reflects the arriving ions, acts to further compensate for differences in kinetic energy of ions with the same m/z and also to increase the flight path (Hoffmann & Stroobant, 2007, p. 132; Watson & Sparkman, 2007). Both of these are important in regard to the resolution that can be achieved with the TOF MS system. The use of ion mirrors in TOF instruments can increase the path distance without lengthening the actual flight tube (Watson & Sparkman, 2007). Other features of TOF MS systems include the introduction of a quadrupole prior to the flight tube, as well as collision cells, such that MS/MS experiments can be performed (Watson & Sparkman, 2007).

Advances in ion optics and detector design and the use of analog-to-digital converter (ADC) replacing time-to-digital converter (TDC) detectors have increased both the sensitivity and overall performance of TOF MS systems. The use of ADC detectors results in faster signal processing and recording than the older TDC designs (Waters, 2013). With an ADC, up to 25 simultaneous ion arrivals are possible during a single TOF ion event, resulting in an increase in the reported signal, as well as improving the linear dynamic range of the system (Waters, 2013). Ion optics have been improved through the application of components that narrow the ion band as it passes through the series of lenses on its way to the pusher region (Waters, 2014). An example of ion optics improvements for the quadrupole time-of-flight hybrid mass spectrometer involves the use of a segmented quadrupole for the collision cell to improve ion transmission. In this system an RF field is applied between the opposing rods of the segmented quadrupole, which confines the ions to a focused central beam. Simultaneously, a DC gradient is applied to the segmented quadrupole, and ions are rapidly transmitted into the pusher region (Waters, 2014). Future improvements to MS designs are likely to continue to be focused on improving sensitivity, based in part on increasingly lower legislative requirements in contaminant analysis. The desire for increased sensitivity also stems from the ability to reduce matrix effects through sample dilution. With increased sensitivity, a higher dilution factor can be used as part of the sample preparation methodology.

TOF MS for the analysis of organic pollutants in complex matrices has been used in several documented studies (Eljarrat & Barceló, 2002; Hajšlová, Pulkrabová, Poustka, Čajka, & Randák, 2007) and holds much promise with regard to identification of pesticides. The rapid, sensitive full spectrum analysis that can be obtained with TOF MS systems along with the accurate mass data is ideal for nontargeted pesticide screening. The full mass range can be acquired simultaneously, allowing for a potentially unlimited number of compound identifications.

8.4 ION MOBILITY SEPARATION
8.4.1 BACKGROUND AND THEORY

IMS has seen revived interest in recent years, especially with advances that enable it to be coupled with MS. IMS was originally termed plasma chromatography (Collins & Lee, 2002), although the name has since changed to more accurately reflect its function. IMS separations are obtained by the application of an electrical potential to a gas-filled drift cell. Ions travel through the drift cell and arrive at a detector. The differing velocities of ions as they pass through the drift cell determine the drift time (measured in milliseconds) and are a result of the number of collisions an ion has with the neutral gas molecules, such as N_2, which populate the drift cell (Collins & Lee, 2002; Stach & Baumbach, 2002). Average ion velocity (v_d) can be described by the following equation (Collins & Lee, 2002; Stach & Baumbach, 2002):

$$v_d = KE$$

where K is the IM constant ($cm^2/V\ s$) and E represents the applied electrical field strength (V/cm). The determination of the value for K is dependent on the molecule in question (Stach & Baumbach, 2002). The following equation describes this (Collins & Lee, 2002; Stach & Baumbach, 2002):

$$K = 3/16 \left[ze/N_0 (2\pi/\mu kT)^{1/2} (1/\Omega_D) \right]$$

In this equation, z is the charge of an ion, e is the charge of an electron (1.602^{-19} Coulombs), N_0 is the density of the drift gas (molecules/cm), k is Boltzmann's constant, T is the temperature (K) of the drift cell, and Ω_D is the average cross-sectional area of the ion based on three-dimensional movements through the drift gas. The term μ, which represents the colliding ion mass (m) and neutral drift-gas mass (M) pair (Collins & Lee, 2002), also deserves an explanation, which is described below (Collins & Lee, 2002; Stach & Baumbach, 2002):

$$\mu = mM/(m + M)$$

Different methods of IM include drift time ion mobility spectrometry (DTIMS), field-asymmetric waveform ion mobility spectrometry (FAIMS), and the more recently introduced traveling wave ion mobility spectrometry (TW-IMS) (Kanu, Dwivedi, Tam, Matz, & Hill, 2008). The use of DTIMS is often not differentiated from IMS, as it was the initial form (Kanu et al., 2008), and utilizes simply the movement of an ion through a gas-filled drift cell at a single applied electrical field, as described previously (Kanu et al., 2008). In FAIMS, an increased strength electrical field is applied between two electrodes (Collins & Lee, 2002; Kanu et al., 2008). The ion migration is perpendicular to the drift gas direction and difference between the two electrodes. The result is divergent mobility depending on the electrode the ion is moving toward (Kanu et al., 2008). In this way, mobility-based separation of ions occurs. TW-IMS utilizes a high electrical field that is moved through the IM cell segments, in this way causing IM (Kanu et al., 2008). As a result, ions come in pulses through the IM cell through interactions with the waves of electrical field and mobility gas (Kanu et al., 2008). TW-IMS utilizes lower pressures in the IM cell than the aforementioned techniques. This allows for more efficient transfer of ions to the detector (Kanu et al., 2008).

There are varied ways in which ions can be produced prior to IMS separations. Earlier systems employed radiation sources such as ^{63}Ni to produce β-radiation, similar to those found in EC detectors, while other ionization techniques such as ESI, CI, photo, and laser induced (such as matrix-assisted laser desorption/ionization, or MALDI) have been implemented (Stach & Baumbach, 2002). For postseparation detection, IMS coupled to MS has shown to be quite effective (Collins & Lee, 2002; Kanu et al., 2008; Stach & Baumbach, 2002). Previously employed IMS systems utilized a Faraday plate, which, although easy to use and inexpensive, did not provide mass specific information (Collins & Lee, 2002). Fig. 8.3 displays a schematic of a modern IMS-MS system.

An ion's behavior in the IM cell is dependent on its shape, charge, and size (Stach & Baumbach, 2002) and is defined by the term Ω_D (collision cross section, CCS). A simplified and nonabsolute measurement of the various ions traveling through the IM cell is the drift time [measured in milliseconds (ms)], mentioned previously. Drift gas composition is also an important factor affecting the nature of an ion's mobility,

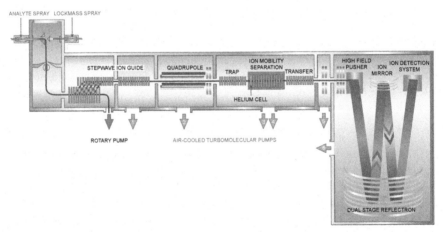

FIGURE 8.3

Schematic of a traveling wave ion mobility spectrometry (TW-IMS) quadrupole TOF MS system (SYNAPT G2-Si from Waters Corporation). Before and after the ion mobility separation cell reside trap and transfer triwave regions, where collision-induced dissociation fragmentation can be selectively induced. Prior to entering the drift cell, ions are cooled by He gas, as indicated by the highlighted yellow helium cell label. Upon entering the drift cell, in packets, the ions are separated from one another based on their mobility through N_2, He, or CO_2 gas.

and various types have been employed, including N_2, SF_6, Ar, Ar/methane mixtures, and CO_2. The impact that the gas choice will have on the separation of ions is dependent on its density and the polarity of gas molecules (Collins & Lee, 2002; Stach & Baumbach, 2002). For the majority of work done on pesticides between the mass range of 50–800 Da, N_2 has shown to be the most suitable drift gas choice based on resolution and mobility profile (Goscinny et al., 2015; McCullagh et al., 2014). As can be seen by the equation for K, the temperature of the gas in the IM region is a contributing factor for mobility values of a particular ion. Work by Bush, Campuzano, and Robinson (2012); Bush et al. (2010) using TW-IMS to standardize the determination of Ω_D for peptides has afforded for the use of an IM calibration procedure using DL-polyalanine. Using such a calibration procedure, relative recorded drift times of an ion can be adjusted to an absolute Ω_D value for that ion, irrespective of laboratory variations that will occur. These values can thus be utilized as an additional comparative value for identification (Bush et al., 2012, 2010). Although this work was specific to biomolecules, the application potential to pesticides and other small molecule analyzer is described in the following section.

8.4.2 APPLICATION OF TRAVELING WAVE ION MOBILITY SPECTROMETRY TO PESTICIDE SCREENING

Drift time and Ω_D, or collision cross section as it will be hereafter referred to, indicate physical properties of an ion and thus provide valuable identification information with regard to pesticide screening. The EU Directorate-General for Health and Food Safety guidance document SANTE/11945/2015 states that the SDL of the qualitative screening method is the lowest level at which an analyte has been detected in at least 95% of the samples [i.e., an acceptable false-negative rate of 5% (EU Directorate-General for Health and Food Safety, 2015)]. In a study by Goscinny et al. (2015) using Q-TOF IMS the use of drift time as additional identification point for the screening methodology described in SANTE/11495/2015 was assessed. Following optimization of conditions to enhance both ion transmission and mobility separations for the full mass range of pesticides analyzed in this study, 100 pesticide drift times were obtained. In some cases, these drift times were for prominent Na^+ adducts and dominant product ions, which were present in the low collision energy channel. Six complex fruit and vegetable matrices (onion, leek, mandarin, pepper, grape, and apple) were spiked with the pesticides and the robustness of their drift time is assessed. Drift times are generally seen to increase with increasing mass of the analyte, as was apparent in this study. One trend in the data was that for analytes with similar retention times, a separation in drift occurred (Goscinny et al., 2015). Conservation of the drift times over six analytical batches injected at the 10 ng/mL level in both the standards and matrices was observed. Statistical analyzes during this study revealed the following key findings: (1) no significant variation in drift time was observed between matrix and standards for the pesticides studied and (2) a 2% variation tolerance from standard obtained drift time was appropriate when

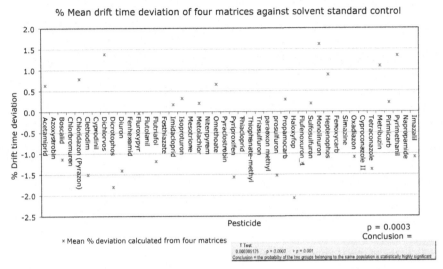

FIGURE 8.4

McCullagh et al. (2014) drift time analysis for 50 pesticides, where mean % drift time deviation from solvent standard measurement is assessed across four matrices (leek, ginger, pear, and mandarin).

using drift time as a criterion for pesticide detection. Fig. 8.4 highlights results from a similar study for 50 pesticides in a mandarin extract from a previous European Reference Laboratory proficiency test (McCullagh et al., 2014). Matching against the library value generated from pesticide standards was found to aid in pesticide detection. Drift time was particularly useful in one case where mass error for quinalphos exceeded the SANTE guidance tolerance, and drift time could be used with retention time to detect that pesticide (Goscinny et al., 2015).

Calibrating the drift time measurement to obtain CCS values, measured in units of angstroms squared (Å^2), enables comparisons across different platforms, laboratories, and users. Fig. 8.5 shows the percentage error on the CCS values for 40 pesticides in the mandarin matrix compared to solvent standard values. The study was carried out in multiple laboratories. All systems had been mass and mobility calibrated and CCS measurements were well maintained across all the systems (McCullagh et al., 2014). In contrast, the retention time variation was much higher, especially in regard to the more complex matrices such as leek and ginger and because of interlab variations such as mobile phase composition, LC system setup, and human error (McCullagh et al., 2014).

As already discussed, the comprehensive analysis of pesticides requires both GC and LC and there are a number of pesticides that can be analyzed using either technique. A study performed to assess both chromatographic techniques with the collection of CCS highlighted the conservation of these values, regardless of the

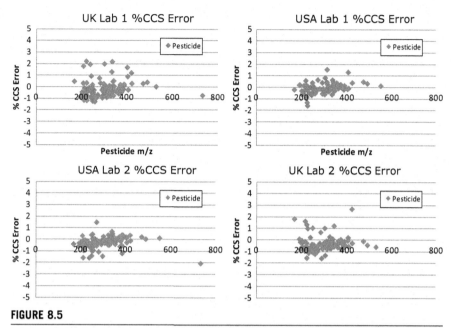

FIGURE 8.5

Interlab study showing reproducibility of collision cross section (CCS) measurements for 40 pesticides across four laboratories. Percentage deviations less than ±2.0 are considered acceptable criteria for identification, for which the vast majority of measurements fall under.

chromatographic method used (Mullin, Cleland, & McCullagh, 2016). In this work, APGC and UPLC were interfaced separately with a Q-TOF TW-IMS and ionization was performed such that $[M+H]^+$ ions were generated for a suite of 73 pesticides. As would be expected, a comparison of the GC and LC retention time for each pesticide shows no correlation, as shown in Fig. 8.6. In contrast, CCS values for both analyzers were highly correlated, with a regression analysis giving an R^2 value of 0.998, as seen in Fig. 8.7.

8.4.3 SPECTRAL SELECTIVITY ENHANCEMENT

Differentiation of analytes of interest from matrix components is challenging for broad scope pesticide screening experiments. Sample preparation approaches such as QuEChERS are popular because of their extensive coverage and nonselective nature. QuEChERS extractions often lead to a complicated spectral view, even when using HRMS, because of coeluting matrix components (Goscinny et al., 2015). Goscinny et al. (2015) selected matrices specifically to introduce complexity to the analysis due to known constituents in those matrices, for example, sulfur-containing compounds, pigments, and high levels of sugar. The samples were blended in methanol, filtered to remove solid particles, and then diluted

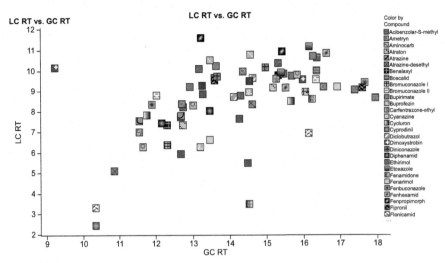

FIGURE 8.6

Retention times (RTs) (min) for 73 pesticides injected by liquid chromatography (LC) and gas chromatography (GC), represented as averages of five injections under each chromatographic condition.

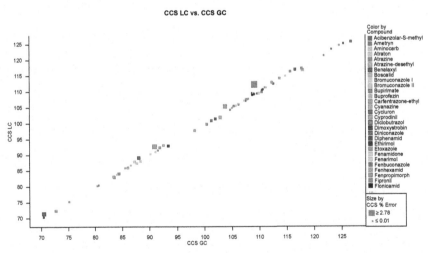

FIGURE 8.7

Collision cross section (CCS) values across both chromatographic platforms (same injections as previous figure), showing high correlation of CCS measurements between the two. The sizes of the boxes indicate the %RSD (relative standard deviation) for the 10 averaged measurements. These data show the robustness of CCS values as a specific property of an ion, which is independent of the pathway used to generate the ion.

(Goscinny et al., 2015; Hanot, Goscinny, & Deridder, 2015). They were purposefully not carried through any additional cleanup procedures. In the case of the leek matrix, it was observed that there was a large matrix component that strongly interfered with the spectral peak for indoxacarb in the sample when IM was not deployed. Through the combination of IM with HRMS, any spectral peaks that were not present in the drift time of indoxacarb were removed and a cleaner spectrum was obtained.

First, alignment of both precursor and product ion peaks occurs through the IMS of the molecules in an injected sample. Compounds in the samples, which may share the same retention time or nominal mass, are separated by drift time. Following this, CID occurs in a collision cell, resulting in product ions, which share the same drift time as the precursor ion. This process is illustrated in Fig. 8.8. Fig. 8.9A and B from McCullagh et al. (2014) shows spectra from thiabendazole in a mandarin matrix. Fig. 8.9A shows the precursor and product ion spectra when IM is not deployed. When drift time is used to align only the fragment ions that have the same drift time as the thiabendazole precursor, the spectra are much cleaner, as shown in Fig. 8.9B. This capability vastly enhances the selectivity for every component in the analysis.

8.4.4 PROTOMER VISUALIZATION

IMS also possesses the unique ability of being able to separate ions of the same molecule but with different sites of protonation (or deprotonation) (Lalli et al., 2012).

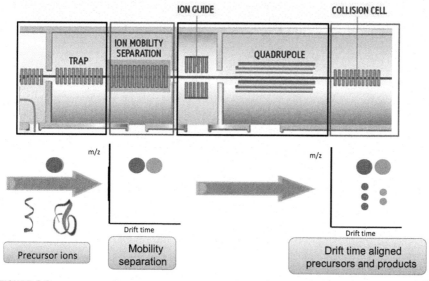

FIGURE 8.8

Schematic of product ion formation following drift separation of precursor ions in an ion mobility separation mass spectrometry system.

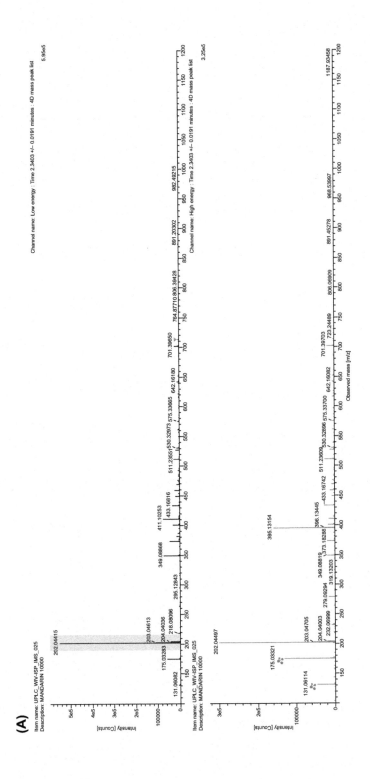

(A)

(B)

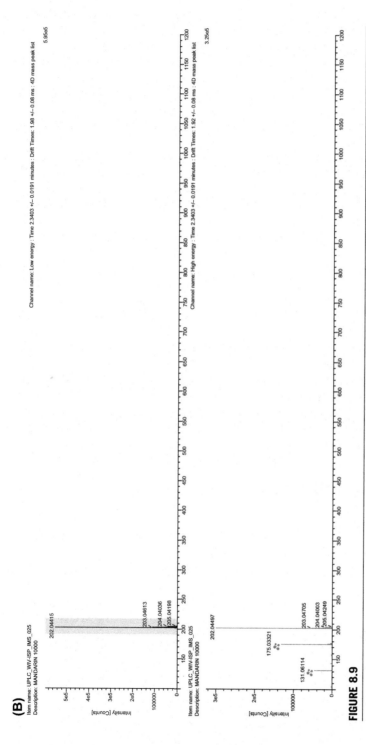

FIGURE 8.9

Conventional low and elevated collision energy spectra for thiabendazole in mandarin extract without ion mobility separation (A). Conventional low and elevated collision energy spectra for thiabendazole in mandarin extract with ion mobility separation (B). When all spectral peaks that do not share a drift time within ±0.06 ms are removed (B), the spectral specificity is significantly improved for both low and high collision energy spectra.

Molecules that have these multiple sites of ionization are referred to as "protomers," and arise out of multiple basic or acidic sites on the molecule (Lalli et al., 2012). Site-specific ionization of these molecules is dependent on the pH of the solution they are in, and the proportions of different protomers can therefore vary (Lalli et al., 2012). There is no chromatographic separation of these protomers, as they are formed through the ionization process and, because they have the same mass, they are not discernible via MS. Additionally, because of differential charge distribution of the molecule based on the site of ionization, varied fragmentation patterns and relative intensities can arise between the different protomers (Lalli et al., 2012). This leads to confusion if the specific protomers present in the gas phase prior to CID is not known or characterized and has been suggested as giving rise to false negatives in methods using MRM mode on tandem quadrupole MS systems (Kaufmann et al., 2009). IMS (specifically TW-IMS) has been shown to separate protomers of aniline, porphyrins (Lalli et al., 2012), beta-agonists (Beucher, Dervilly-Pinel, Prévost, Monteau, & Le Bizec, 2015), and fluoroquinolone veterinary drugs (McCullagh, Stead, Williams, de Keizer, & Bergwerff, 2013) based on their differential drift times. Based on IMSs, the respective spectra for each protomer can be obtained and fragmentation patterns are described.

The implications of differential fragmentation patterns have been highlighted in reports of high variability of tandem quadrupole data in interlab studies for the analysis of veterinary drugs in meat samples (Kaufmann et al., 2009). Protomer proportions are subject to liquid and gas phase conditions (Kaufmann et al., 2009), so during ionization varying electronic and chemical conditions in the source will impact product ion abundances. Separation of the protomer species, which is afforded exclusively by IMS, enables the generation of a product ion spectrum for each respective site of protonation/deprotonation. Thus using IMS-MS, abundance and product ion assignment is possible for all the protomers of a compound, as is shown in Figs. 8.10A,B and 8.11. Protomers have been observed in various pesticide screening analyzers (McCullagh et al., 2014), and because of the variety of diprotic formulations, more are likely to exist.

8.5 SUMMARY AND CONCLUSION

Because of the vast array of pesticides used today and legacy pesticides still present in the environment, methods that increase the scope of pesticide screening are highly sought after. Sample preparation that is less complex, saves time, and has maximum compound coverage is desirable. This has resulted in approaches such as QuEChERS being widely adopted for pesticide screening. The sample extracts are therefore often more complex, pushing the burden on to the analytical system. These factors combined with the huge variety of matrices that require testing mean that methods that increase the peak capacity of the system are highly advantageous. Ensuring compound coverage of both GC- and LC-amenable pesticides is critical to the success of broader scope pesticide screening methods.

(A)

Channel name: Indoxacarb [+H] : (27.8 PPM) 528.0797
Description: restek mix 100 ppb

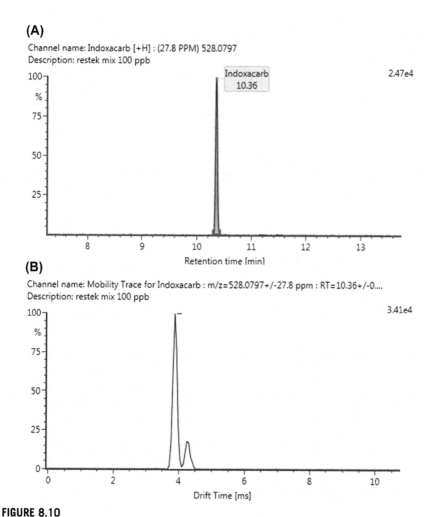

(B)

Channel name: Mobility Trace for Indoxacarb : m/z=528.0797+/-27.8 ppm : RT=10.36+/-0....
Description: restek mix 100 ppb

FIGURE 8.10

Indoxacarb extracted ion chromatogram of precursor [M+H]$^+$ ion (A), and mobility trace of the same exact mass precursor *m/z* (B). Separation in IM is based on differential charge distribution and spatial conformation of the ion when protonation occurs at different locations of the molecule. The trace indicates the relative abundance of each protomer present in the system.

APGC, which can be interfaced on the same HRMS Q-TOF system as UPLC, is a recent technology that shows promise for pesticide screening methods. APGC offers the advantage of reduced fragmentation during ionization, improving the selectivity by conserving the molecular ion. This is especially useful for compounds that fragment readily in traditional EI. The majority of the ion current is conserved to the molecular ion, which makes APGC a highly sensitive technique compared to EI where

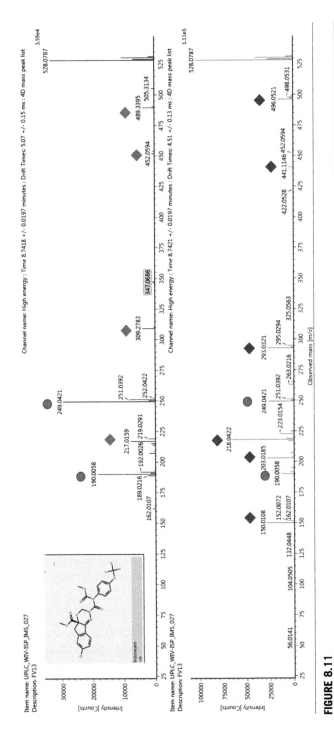

FIGURE 8.11

Product ion pattern for indoxacarb protomers, separated by drift times, from McCullagh et al. (2014). *Circles* represent common product ions between the protomers, while *diamonds* are unique to either.

the ion current is divided between multiple fragment ions. A new area of investigation into the use of APGC is in the application of higher gas flow rates for the GC separation. In contrast to traditional vacuum GC—MS techniques (such as EI), APGC operates at atmospheric pressure and is therefore not affected by higher GC flow rates. The ability to use higher flow rates in GC to reduce the long analytical run times has the potential to greatly improve laboratory efficiency.

The combination of IMS with Q-TOF MS affords an additional dimension of separation within an MS analysis. With chromatographic separation, IMS, and high-resolution accurate mass, the peak capacity of the system is increased. This is a key factor when addressing complex matrices and low-level detection of pesticides. In addition to IMS, the deployment of CCS measurements for pesticide screening provides further utility of these technologies in pesticide analysis and contaminant screening in general. Although advances in these technologies evolved separately, the amalgamation of APGC and IMS with Q-TOF MS technologies has resulted in unique analytical capabilities.

Li et al. (2015) concluded a recent review on APGC systems that in addition to several advantages of the APGC approach, the use of CCS libraries obtained from IMS experiments would assist in database searching capabilities. This is noteworthy as the focus was not on IMS; however, the benefit of IM in combination with HRMS was recognized regardless. In addition to improving selectivity and identification confidence by using accurate mass and CCS, the ability to couple GC separations with the continuously improving sensitivity of HRMS Q-TOF systems is a powerful tool for pesticide analysis. These capabilities have been shown to meet the most recent legislative guidance in regard to pesticide screening in food and to provide additional insight into contaminant properties, behavior, and analysis.

ACKNOWLEDGMENTS

The authors would like to kindly acknowledge Adam Ladak, Rhys Jones, and Jody Dunstan of Waters Corporation for diligent review of the document. Mike McCullagh of Waters Corporation is respectfully acknowledged for generation of ion mobility data and development of techniques for its use in pesticide screening.

REFERENCES

Alder, L., Greulich, K., Kempe, G., & Vieth, B. (2006). Residue analysis of 500 high priority pesticides: Better by GC-MS or LC-MS/MS? *Mass Spectrometry Reviews, 25*(6), 838–865.

Beucher, L., Dervilly-Pinel, G., Prévost, S., Monteau, F., & Le Bizec, B. (2015). Determination of a large set of β-adrenergic agonists in animal matrices based on ion mobility and mass separations. *Analytical Chemistry, 87*(18), 9234–9242.

Bush, M. F., Campuzano, I. D., & Robinson, C. V. (2012). Ion mobility mass spectrometry of peptide ions: Effects of drift gas and calibration strategies. *Analytical Chemistry, 84*(16), 7124–7130.

Bush, M. F., Hall, Z., Giles, K., Hoyes, J., Robinson, C. V., & Ruotolo, B. T. (2010). Collision cross sections of proteins and their complexes: A calibration framework and database for gas-phase structural biology. *Analytical Chemistry, 82*(22), 9557–9565.

Carroll, D. I., Dzidic, I., Stillwell, R. N., Haegele, K. D., & Horning, E. C. (1975). Atmospheric pressure ionization mass spectrometry. Corona discharge ion source for use in a liquid chromatograph-mass spectrometer-computer analytical system. *Analytical Chemistry, 47*(14), 2369–2373.

Cheng, Z., Dong, F., Xu, J., Liu, X., Wu, X., Chen, Z., et al. (2016). Atmospheric pressure gas chromatography quadrupole-time-of-flight mass spectrometry for simultaneous determination of fifteen organochlorine pesticides in soil and water. *Journal of Chromatography A, 1435*, 115–124.

Cherta, L., Portolés, T., Pitarch, E., Beltran, J., López, F. J., Calatayud, C., et al. (2015). Analytical strategy based on the combination of gas chromatography coupled to time-of-flight and hybrid quadrupole time-of-flight mass analyzers for non-target analysis in food packaging. *Food Chemistry, 188*, 301–308.

Collins, D., & Lee, M. (2002). Developments in ion mobility spectrometry–mass spectrometry. *Analytical and Bioanalytical Chemistry, 372*(1), 66–73.

Eljarrat, E., & Barceló, D. (2002). Congener-specific determination of dioxins and related compounds by gas chromatography coupled to LRMS, HRMS, MS/MS and TOFMS. *Journal of Mass Spectrometry, 37*(11), 1105–1117.

EU Directorate-General for Health and Food Safety. (2015). *Method validation & quality control procedures for pesticide residues analysis in food & feed*. Available from http://www.eurl-pesticides.eu/docs/public/tmplt_article.asp?CntID=727.

Fernando, S., Green, K., Organitini, K., Dorman, F., Jones, R., Reiner, E. J., et al. (2015). Differentiation of (mixed) halogenated dibenzo-p-dioxins by negative ion atmospheric pressure chemical ionization. *Analytical Chemistry, 88*(10), 5205–5211.

Goscinny, S., Joly, L., De Pauw, E., Hanot, V., & Eppe, G. (2015). Travelling-wave ion mobility time-of-flight mass spectrometry as an alternative strategy for screening of multi-class pesticides in fruits and vegetables. *Journal of Chromatography A, 1405*, 85–93.

Guilhaus, M., Selby, D., & Mlynski, V. (2000). Orthogonal acceleration time-of-flight mass spectrometry. *Mass Spectrometry Reviews, 19*(2), 65–107.

Hajšlová, J., Pulkrabová, J., Poustka, J., Čajka, T., & Randák, T. (2007). Brominated flame retardants and related chlorinated persistent organic pollutants in fish from river Elbe and its main tributary Vltava. *Chemosphere, 69*(8), 1195–1203.

Hanot, V., Goscinny, S., & Deridder, M. (2015). A simple multi-residue method for the determination of pesticides in fruits and vegetables using a methanolic extraction and ultra-high-performance liquid chromatography-tandem mass spectrometry: optimization and extension of scope. *Journal of Chromatography A, 1384*, 53–66.

Hird, S. J., Lau, B. P. Y., Schuhmacher, R., & Krska, R. (2014). Liquid chromatography-mass spectrometry for the determination of chemical contaminants in food. *TrAC Trends in Analytical Chemistry, 59*, 59–72.

Hoffmann, E., & Stroobant, V. (2007). *Mass spectrometry: Principles and applications* (3rd ed.). John Wiley & Sons, Inc.

Horning, E. C., Carroll, D. I., Dzidic, I., Lin, S. N., Stillwell, R. N., & Thenot, J. P. (1977). Atmospheric pressure ionization mass spectrometry: Studies of negative ion formation for detection and quantification purposes. *Journal of Chromatography A, 142*, 481–495.

Horning, E. C., Horning, M. G., Carroll, D. I., Dzidic, I., & Stillwell, R. N. (1973). New picogram detection system based on a mass spectrometer with an external ionization source at atmospheric pressure. *Analytical Chemistry, 45*(6), 936–943.

Kanu, A. B., Dwivedi, P., Tam, M., Matz, L., & Hill, H. H. (2008). Ion mobility–mass spectrometry. *Journal of Mass Spectrometry, 43*(1), 1–22.

Kaufmann, A. (2012). The current role of high-resolution mass spectrometry in food analysis. *Analytical and Bioanalytical Chemistry, 403*(5), 1233–1249.

Kaufmann, A., Butcher, P., Maden, K., Widmer, M., Giles, K., & Uría, D. (2009). Are liquid chromatography/electrospray tandem quadrupole fragmentation ratios unequivocal confirmation criteria? *Rapid Communications in Mass Spectrometry, 23*(7), 985–998.

Korfmacher, W. A., Mitchum, R. K., Hileman, F. D., & Mazer, T. (1983). Analysis of 2, 3, 7, 8-tetrachlorodibenzofuran by fused silica GC combined with atmospheric pressure ionization MS. *Chemosphere, 12*(9), 1243–1249.

Korfmacher, W. A., Rowland, K. R., Mitchum, R. K., Daly, J. J., McDaniel, R. C., & Plummer, M. V. (1984). Analysis of snake tissue and snake eggs for 2, 3, 7, 8-tetrachlorodibenzo-*p*-dioxin via fused silica GC combined with atmospheric pressure ionization MS. *Chemosphere, 13*(11), 1229–1233.

Lalli, P. M., Iglesias, B. A., Toma, H. E., Sa, G. F., Daroda, R. J., Silva Filho, J. C., et al. (2012). Protomers: Formation, separation and characterization via travelling wave ion mobility mass spectrometry. *Journal of Mass Spectrometry, 47*(6), 712–719.

Li, D. X., Gan, L., Bronja, A., & Schmitz, O. J. (2015). Gas chromatography coupled to atmospheric pressure ionization mass spectrometry (GC-API-MS): Review. *Analytica Chimica Acta, 891*, 43–61.

McCullagh, M., Cleland, G., Hanot, V., Stead, S., Williams, J., Goscinny, S., et al. (2014). *Collision cross section: A new identification point for a "Catch all" non-targeted screening approach.* Milford, MA, U.S.A.: Waters Corporation.

McCullagh, M., Stead, S., Williams, J., de Keizer, W., & Bergwerff, A. (2013). *Identification of multiple sites of intra-molecular protonation and different fragmentation patterns within the fluoroquinolone class of antibiotics in porcine muscle extracts using travelling wave ion mobility mass spectrometry.* Milford, MA, U.S.A.: Waters Corporation.

McEwen, C. N., & McKay, R. G. (2005). A combination atmospheric pressure LC/MS: GC/MS ion source: Advantages of dual AP-LC/MS: GC/MS instrumentation. *Journal of the American Society for Mass Spectrometry, 16*(11), 1730–1738.

McLafferty, F. W., & Tureček, F. (1993). *Interpretation of mass spectra.* Sausalito, CA: University Science Books.

Mitchum, R. K., Moler, G. F., & Korfmacher, W. A. (1980). Combined capillary gas chromatography/atmospheric pressure negative chemical ionization/mass spectrometry for the determination of 2, 3, 7, 8-tetrachlorodibenzo-*p*-dioxin in tissue. *Analytical Chemistry, 52*(14), 2278–2282.

Mullin, L., Cleland, G., & McCullagh, M. (2016). *Demonstration of collision cross section (CCS) value conservation across LC and GC analyses.* Milford, MA, U.S.A.: Waters Corporation.

Nacher-Mestre, J., Serrano, R., Portoles, T., Berntssen, M. H., Perez-Sanchez, J., & Hernandez, F. (2014). Screening of pesticides and polycyclic aromatic hydrocarbons in feeds and fish tissues by gas chromatography coupled to high-resolution mass spectrometry using atmospheric pressure chemical ionization. *Journal of Agricultural and Food Chemistry, 62*(10), 2165–2174.

NIST. (2011). *The NIST mass spectral search program for the NIST/EPA/NIH mass spectral library version 2.0.*

Portolés, T., Mol, J. G. J., Sancho, J. V., López, F. J., & Hernández, F. (2014). Validation of a qualitative screening method for pesticides in fruits and vegetables by gas chromatography quadrupole-time of flight mass spectrometry with atmospheric pressure chemical ionization. *Analytica Chimica Acta, 838,* 76—85.

Portolés, T., Sancho, J. V., Hernández, F., Newton, A., & Hancock, P. (2010). Potential of atmospheric pressure chemical ionization source in GC-QTOF MS for pesticide residue analysis. *Journal of Mass Spectrometry, 45,* 926—936.

Rotterdam Convention. (2016). (Online) Available from http://www.pic.int/.

Stach, J., & Baumbach, J. I. (2002). Ion mobility spectrometry-basic elements and applications. *International Journal for Ion Mobility Spectrometry, 5,* 1—21.

Stockholm Convention. (2016). (Online) Available from http://chm.pops.int/default.aspx.

Tienstra, M., Portolés, T., Hernández, F., & Mol, J. G. J. (2015). Fast gas chromatographic residue analysis in animal feed using split injection and atmospheric pressure chemical ionisation tandem mass spectrometry. *Journal of Chromatography A, 1422,* 289—298.

Waters Corporation. (2011). *An overview of the principles of MSE, the engine that drives ms performance.* Milford, MA, U.S.A.: Waters Corporation.

Waters Corporation. (2013). *QUANTOF: High-resolution, accurate mass, quantitative time-of-flight MS technology.* Milford, MA, U.S.A.: Waters Corporation.

Waters Corporation. (2014). *The XS collision cell from waters: Increased sensitivity and resolution for time-of-flight mass spectrometry.* Milford, MA, U.S.A.: Waters Corporation.

Watson, J. T., & Sparkman, O. D. (2007). *Introduction to mass spectrometry: Instrumentation, applications, and strategies for data interpretation.* England: John Wiley & Sons.

Wood, A. (2016). *Compendium of pesticide common names.* (Online) Available from http://alanwood.net/pesticides/index.html.

World Health Organization. (2016). *The WHO recommended classification of pesticides by hazard.* (Online) Available from http://www.who.int/ipcs/publications/pesticides_hazard/en/.

Direct Analysis of Pesticides by Stand-Alone Mass Spectrometry: Flow Injection and Ambient Ionization

9

E. Moyano, M.T. Galceran
University of Barcelona, Barcelona, Spain

9.1 INTRODUCTION

Although the use of pesticides in agricultural practice has resulted in a substantial increase in crop yield, pesticides and their metabolites can affect human health, remain as residues in foodstuffs, pollute natural waters, and disturb the equilibrium of the ecosystem. For these reasons stringent legislation and measurements have been promoted by national and international authorities. Acceptable daily intakes have been established, and maximum residue limits (MRLs) for pesticide residues in food commodities have been set by the European Commission (EU, 2016) and United States Environmental Protection Agency (USEPA, 2016) to ensure food safety for consumers and facilitate international trade. Regarding pesticide control policy, official laboratories play an important role in both development and application of appropriate analytical methods. However, this is not an easy task since a large number of active substances against pests belonging to more than 100 chemical classes and with a wide range of physicochemical properties are used worldwide. An additional difficulty is the diversity of matrices that must be monitored. As a result, a generic and universal method for the determination of pesticides is not applicable and work is going on in developing multiresidue methods with characteristics such as selectivity/sensitivity, speed, low cost, and high sample throughput.

Nowadays, methods used in pesticide residues laboratories for the analysis of pesticides include the extraction of the analytes from the matrix, the cleanup of the raw extracts, a concentration step, the subsequent separation by chromatographic techniques, and finally the identification and quantitation by mass spectrometry (MS), a technique that provides a high degree of selectivity and sensitivity. To guarantee the quality of the analytical results, guidelines have been produced as, for instance, those of the SANTE/11945/2015 (EU-SANTE, 2015, pp. 1–42). The changes and evolution of the methods used in the analysis of pesticides is strongly related to the development of chromatography and MS during the last 25 years.

Applications in High Resolution Mass Spectrometry. http://dx.doi.org/10.1016/B978-0-12-809464-8.00009-9

265

Multiresidue methods based on gas chromatography (GC) coupled to MS were already used in the 1980s. Nevertheless, the technological advances in chromatographic instrumentation as well as the coupling to MS (quadrupole (Q) and ion trap analyzers and more recently triple quadrupoles (QqQ)) explain that GC—MS continues to play an important role in the field of pesticide residue analysis. However, the continuous introduction of new agrochemicals not amenable by GC has led to replace this technique by liquid chromatography (LC). The exponential increase of the use of LC-MS for pesticide residue analysis in the last 10 years is also related to advances in both chromatographic and mass spectrometric techniques. The expanded use of ultrahigh performance LC has allowed increasing both efficiency and flow rates achieving fast separations with high chromatographic resolution. As regards MS analyzers, quadrupoles and ion traps have been used, but nowadays, QqQ working in multiple reaction monitoring (MRM) mode, are generally the first choice for laboratories carrying out pesticide residue analysis (Mol et al., 2015). The ion-filtering capability of the quadrupoles and improvement in fast ion scanning result in good performance in residue detection and quantitation for complex food matrices at very low levels with adequate concentration range and reproducibility. However, this approach does have limitations when the number of compounds to be analyzed in a single run is high since at least two transitions must be monitored according to the criteria described in the European guideline (EU-SANTE, 2015). Moreover, its application requires a previous selection of the compounds to be analyzed to set up the MRM acquisition method and so it is restricted to target analytes. The introduction in the pesticide field of the use of high-resolution mass analyzers, time of flight (TOF), and more recently Orbitrap, has permitted the use of full-scan mode with exact mass measurements as an alternative for the QqQ-based methods (Kaufmann, 2012). The main advantages are it is no longer necessary to decide which compounds should be tested for, the relative easiness for increasing the number of compounds in a method, and the possibility of performing retrospective analysis of the samples as the full-scan spectra information is saved. Nowadays, the use of hybrid instruments with quadrupoles or linear ion traps coupled to high-resolution MS (HRMS) analyzers (TOF and Orbitrap) has allowed reducing false positives by adding information about product ion spectra with exact mass measurements. One of the major advantages of using modern HRMS instrumentation is their ability to monitor an unlimited number of compounds, which allows the development of screening strategies based on accurate mass databases for the identification of suspect and nontarget (unknown) pesticides potentially present in the samples.

As commented earlier, over the years, the methods used for the analysis of organic compounds and particularly pesticides, have depended on the techniques available at the time. Advances in instrumentation have increased the sensitivity and selectivity of MS analyzers and analytical chemists have taken advantage of this progress by using modern mass spectrometers to simplify the overall design of analytical procedures reducing the complexity of sample treatment and separation steps. Today, relatively fast sample treatments such as QuEChERS (Quick, Easy,

Cheap, Effective, Rugged, and Safe) have achieved high popularity for the analysis of pesticides and are used in combination with fast liquid chromatographic methods. However, pesticide analysis conducted on foods is often highly time-consuming, labor intensive, and requires consistent amounts of solvents and reagents. Taking into account that questions related to food safety are sometimes better answered by using rapid screening methods and that there are pressures in modern pesticide laboratories in terms of work load, turnaround time, and low cost, it is not surprising that in the last years new methodologies based on stand-alone MS had been developed. Two approaches, flow injection analysis coupled to mass spectrometry (FIA-MS) and ambient MS (AMS) techniques, able to reduce overall cost and increase laboratory throughput have emerged (Domin & Cody, 2015; Nanita & Kaldon, 2016). In FIA, a technique specifically designed for the analysis of liquids, no chromatographic separation is applied and the liquid sample or the extract is directly injected into the MS ion source. In AMS techniques, in addition to the elimination of the chromatographic separation, sample treatment is omitted or at least reduced.

Both FIA and AMS techniques can be coupled to most types of mass spectrometers without modifying the ion optics and the vacuum system. In FIA-MS, standard atmospheric-pressure ionization sources mainly electrospray ionization (ESI) are used, whereas in AMS, what is different is the way by which desorption/ionization mechanisms are implemented. AMS comprise a large number of techniques that can be roughly grouped into those based on desorption by solid–liquid extraction followed by ESI such as desorption electrospray ionization (DESI) and those based on plasma generation such as direct analysis in real time (DART). In both FIA-MS and AMS, the quality of the results that can be obtained depends on the information that the mass spectrometer can provide. Initially these techniques were coupled to low-resolution mass analyzers (quadrupoles and ion traps) with the objective of demonstrating their applicability for the analysis of target compounds in complex samples such as food. Later, the complexity of the mass spectra, since they are generated from complex mixtures without any previous separation of the compounds, required the use of instruments able to perform tandem MS experiments or acquire data at high-resolution to extract maximum information. Today, highly sensitive and selective instruments, such as modern QqQ mass spectrometers, HRMS analyzers (TOF and Orbitrap), as well as hybrid instruments (quadrupole-TOF, quadrupole-Orbitrap, or linear ion trap-Orbitrap) are able to provide the performance needed to design reliable methods for the trace-level analysis of pesticides in complex food matrices with minimal sample preparation and high throughput.

This chapter aims to provide a general picture of the use of FIA-MS and AMS techniques for the analysis of pesticides in food samples. Brief descriptions of the fundamentals of FIA and the more common desorption/ionization techniques used in AMS are included. Moreover, special attention is devoted to the critical parameters that should be considered in each technique and to the procedures that have to be followed for developing rugged analytical methods. Capabilities of the different techniques for the analysis of target compounds, the importance of matrix effects mainly in quantitative analysis, and the applicability of the techniques for the rapid screening of target, suspected, and nontarget (unknown) compounds in food using HRMS are discussed. Advantages and drawbacks of FIA-MS and AMS techniques for the analysis of food samples are compared and critically assessed and some examples of application are included.

9.2 FLOW INJECTION ANALYSIS

FIA was developed in the mid-1970s by Ružička and Hansen (1975) as a highly efficient technique for the automated analyses of liquid samples. FIA is one example of a continuous-flow analyzer, in which samples are sequentially introduced at regular intervals into a liquid carrier stream that transports them to the detector. As a result of growing demands of high throughput, fast sample rate, and minimal sample consumption in chemical analysis, FIA has become a popular sample introduction and online treatment technology that allows for the rapid and sequential analysis of a large number of samples.

FIA is a versatile and highly efficient sample introduction technique for coupling to MS and it has been used with inductively coupled plasma for the determination of trace elements in environmental samples and with atmospheric pressure ionization (API) sources, mainly ESI, in different applications such as diffusion measurements, titrations, or high-throughput screening in pharmaceutical analysis and clinical chemistry. An overview of FIA-MS applications can be found in the Schug's LCGC blog about FIA (Schug, 2013). Despite the advantages of FIA-MS (high sampling rate, simplicity, low cost, low reagent consumption, and ease of design), there are few applications to the analysis of pesticides, probably because till recently, it has not been easy to fit the large number of compounds that must be analyzed to the velocity of the MS analyzers for scanning and interrogate the data resulting from complex mixtures. A summary of the methods using FIA-MS for the analysis of pesticides is given in Table 9.1 where the number of compounds analyzed, the matrix, the coupling conditions, MS instrumentation, and limits of quantification (LOQs) are included.

Currently, FIA-MS coupling is easily performed by using the injectors of the LC instruments simply by bypassing the column and injecting the sample directly in the ion source. The modern injectors and autosamplers of the LC instruments allow injecting low sample volumes of liquids (liquid samples or extracts) with high reliability and repeatability, which permits achieving high accuracy and precision. Multiprobe autosamplers and multiple injection loops hold great promise for an even higher throughput, although broad application of this technology remains to be applied. Nowadays, for the analysis of pesticides, standard devices are used and for this coupling several operational parameters have to be optimized.

FIA is based on the injection of a sample solution into a carrier that is continuously moving at a constant flow rate. The carrier stream transports the analyzed species from the injector to the ion source. As the sample zone moves downstream, diffusion and mixing with the carrier occurs. As a result, a Gaussian-like distribution of the sample zone profile that contains all the compounds appeared at the detector.

Table 9.1 Summary of FIA-MS Methods for the Analysis of Pesticides

Compounds	Sample	Sample Treatment	MS Analyzer/ Coupling Conditions and Carrier	Sensitivity/ Quantitation Mode	References
3 Fungicides (imazalil, thiabendazole, o-phenylphenol)	Citrus fruits	Ethyl acetate (anh. Na_2SO_4/Na_2HPO_4), direct injection	ESI (+/−), QqQ (MRM) 100 cm × 0.5 mm id PEEK capillary, methanol/water, 90:10, v/v	1–5 mg/kg Isotope dilution	Ito et al. (2003)
7 N-methyl carbamates (aldicarb, fenobucarb carbaryl, methomyl, diethiofencarb, methiocarb, pirimicarb)	Citrus fruits	Cyclohexane (anh. Na_2SO_4), evaporation, GPC (acetone/ cyclohexane), evaporation to dryness, redissolve in methanol	ESI (+), QqQ (MRM) 100 cm × 0.5 mm id PEEK capillary, methanol/0.1% formic acid 90:10, v/v	1.1 mg/kg Internal standard	Goto et al. (2003)
4 Pesticides (dichlorvos, malathion, carbaryl, 2, 4-dichlorophenoxy acetic acid)	Citrus fruits	n-Hexane (anh. Na_2SO_4), direct injection	ESI (+/−), QqQ (MRM) 100 cm × 0.5 mm id PEEK capillary, methanol/water, 90/10, v/v	0.005 mg/kg (estimated)	Nakazawa et al. (2004)
4 Sulfonylurea herbicides 2 Carbamate insecticides	Tap water, corn, lemon, pecan	Acetonitrile/phosphate buffer pH:7 75/25, v/v, cleanup with n-hexane (discarded), purification by SPE (Bond–Elut ENV)	ESI (+), QqQ (MRM) 100 cm × 0.13 mm id PEEK capillary, MeOH	1.1 mg/kg Matrix matched	Nanita et al. (2009)
Dimethoate in poisoned animals	Porcine urine, blood plasma	Acetonitrile (protein precipitation) Dilution with acetonitrile/ water 80/20,v/v (1:11,250 for plasma and 1/40,000 for urine)	ESI (+), q-LIT (MRM) ACN/0.1% formic acid 60/40,v/v	Plasma: 0.24 µg/mL Urine: 1.56 µg/mL External calibration	John et al. (2010)

Continued

Table 9.1 Summary of FIA-MS Methods for the Analysis of Pesticides—cont'd

Compounds	Sample	Sample Treatment	MS Analyzer/ Coupling Conditions and Carrier	Sensitivity/ Quantitation Mode	References
6 Sulfonylurea herbicides 2 Carbamate insecticides 2 Pyrimidine carboxylic acids 2 Anthranilic diamide insecticides	Lemon, pecan	Methanol 10- or 50-fold dilution with methanol/conc. NH$_4$OH, 98.5/1.5, v/v	ESI (+), QqQ (MRM) 100 cm × 0.13 mm id PEEK capillary, MeOH	0.05 mg/kg Matrix matched	Nanita (2011)
6 Sulfonylurea herbicides 2 Carbamate insecticides 2 Pyrimidine carboxylic acids 2 Anthranilic diamide insecticides	Lemon, pecan, soybean oil, corn meal	Methanol Dilution with methanol/ conc. NH$_4$OH, 98.5/1.5, v/v High water: 10-fold Low water: 50-fold; fat freeze	ESI (+), QqQ (MRM) 100 cm × 0.13 mm id PEEK capillary, methanol	0.01−0.05 mg/kg Matrix matched	Nanita et al. (2011)
6 Sulfonylurea herbicides 2 Carbamate insecticides 2 Pyrimidine carboxylic acids 2 Anthranilic diamide insecticides 2 Triazine herbicides	Fruits, cereals, seeds, beef, eggs, milk, urine, and blood plasma	Salting out: acetonitrile + NH$_4$Cl aq. sat. sol. 1/1 v/v. Filtration and fivefold dilution with methanol/con. NH$_4$OH, 8.5/1.5, v/v	ESI (+), QqQ (MRM) 100 cm × 0.13 mm id PEEK capillary, methanol	0.001−0.01 mg/kg matrix matched	Nanita and Padivitage (2013)
1 Sulfonylurea herbicide	Fruits, cereals, seeds, beef,	Salting out: ACN/34%w/w aqueous ammonium	ESI (+), QqQ (MRM) 150 cm × 0.13 mm id	1.1 mg/kg External standard	Nanita (2013)

Analytes	Matrix	Sample preparation	Instrument/conditions	Calibration/LOD	Reference
1 Triazinone herbicide 1 Anthranilic diamide insecticide	eggs, milk, urine, and blood plasma	formate, 1/1 v/v. Top layer 10-fold dilution with acetonitrile	PEEK capillary, carrier matching diluted reagent blank composition	calibration. Quantitation at online infinite dilution	Oellig and Schwack (2014)
7 Pesticides (azoxystrobin, fenarimol, mepanipyrim, chlorpyrifos, pirimicarb, acetamid, penconazole)	Fruits and vegetables	QuEChERS HTpSPE (silica gel NH_2), elution with acetonitrile/10 mM ammonium formate (1/1, v/v)	Nano-ESI TOF (full scan) µL-FIA, 0.5 µm frit and 100 µm id PEEK capillary Acetonitrile/water 0.05% formic acid 1/1, v/v)	0.5 mg/kg External standard calibration	Mol and van Dam (2014)
Highly polar pesticides Group 1: amitrole, cyromazine, daminozide Group 2: QUATs Group 3: glyphosate, AMPA, ethephon, fosethyl, glufosinate, maleic hydrazide	Fruits, vegetables, cereals, animal products	Water. For PQ/DQ in barley methanol/0.1M HCl (1/1, v/v) 10 times dilution with: methanol/0.1% formic for QUATs, methanol/0.1% NH_4OH for group 1, acetonitrile/0.1% NH_4OH for group 3	ESI (+,-), QqQ (MRM) 100 cm × 0.13 mm id PEEK capillary, solvents used for dilution	0.05 mg/kg Matrix matched	
6 Sulfonylurea herbicides 2 Carbamate insecticides 2 Pyrimidine carboxylic acids 2 Anthranilic diamide insecticides 2 Triazine herbicides	Water (ground, surface, and drinking)	Salting out: acetonitrile + solid NH_4Cl. Filtration and fivefold dilution with acetonitrile	ESI (+), QqQ (MRM) 100 cm × 0.13 mm id PEEK capillary, methanol	0.1–0.3 µg/L Matrix matched	Nanita et al. (2015)

AMPA, aminomethylphosphonic acid; Anh., anhydrous; DQ, diquat; ESI, electrospray ionization; FIA, flow injection analysis; GPC, gel permeation chromatography; HTpSPE, high-throughput planar solid phase extraction; LIT, linear ion trap; MRM, multiple reaction monitoring; PEEK, polyether ether ketone; PQ, paraquat; QqQ, triple quadrupole; QUATs, quaternary ammonium herbicides; QuEChERS, Quick, Easy, Cheap, Effective, Rugged, and Safe; SPE, solid phase extraction; TOF, time-of-flight.

The flow rate, the channel volume, and the channel geometry control both the diffusion and the extent of mixing of the sample plug with the carrier solvent. The importance of mixing between the sample and the carrier in FIA was clearly demonstrated by Ito et al. (2003) in one of the first papers published related to the use of FIA-MS for the analysis of pesticides. In the analysis of fungicides, they observed two peaks on the MRM profile instead of one. This curious behavior was explained taking into account that the compounds were dissolved in ethyl acetate and injected in methanol/water, 90:10 (v/v), as carrier (Table 9.1). In ethyl acetate the fungicides under study are not ionized or are difficult to be ionized under ESI conditions. The efficient ionization of the compounds requires that the sample mixes completely with the carrier and in this case, this happens only at both edges of the sample plug. In FIA-MS, to improve the performance three parameters must be controlled: sample volume (as low as possible, although it depends on sensitivity), capillary dimensions, and flow rate. When increasing the carrier flow rate, band diffusion and residence time decrease, whereas both increase at higher inner capillary volumes. So, these parameters must be optimized to achieve adequate peak shape, high signal intensity, fast analysis speed, and a peak width wide enough to allow multiple MS transitions and/or mass spectra to be recorded. Fig. 9.1 shows as an example the results obtained

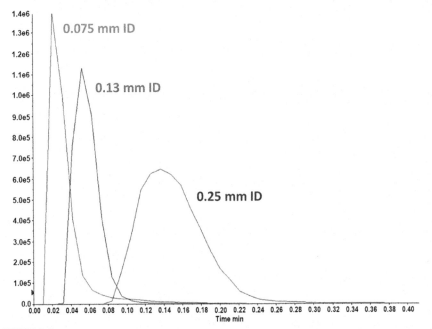

FIGURE 9.1

Effect of capillary internal diameter on peak profile obtained after flow injection of 1 μL of a solvent standard (10 ng/mL) of chlormequat. Flow rate, 400 μL/min.

Reprinted from Mol, H.G.J., & van Dam, R.C.J. (2014) Rapid detection of pesticides not amenable to multi-residue methods by flow injection-tandem mass spectrometry, Analytical and Bioanalytical Chemistry 406 (27), 6817–6825. with permission from Springer, Copyright 2014.

in the optimization of the inner capillary diameter in an FIA-MS method for the analysis of polar pesticides where it is clearly observed that the signal intensity increases at lower diameters because of the reduction of the band diffusion (Mol & van Dam, 2014). However, sensitivity is not the only parameter that must be taken into account since enough data points must be collected across the peak (at least 15) for accurate quantitation and this depends on the scanning rate of the mass analyzer. In general and as can be observed in Table 9.1, the first applications of FIA-MS to the analysis of pesticides (Ito et al., 2003; Nakazawa et al., 2004) used relatively long (1 m$-$100 cm) and wide internal diameter (id) polyether ether ketone tubing (0.5 mm). Nowadays, 100 cm $\times$ 0.13 mm id tubing is used in most of the reported methods (Table 9.1). This capillary size offers enough sensitivity and a peak width wide enough to allow multiple MS measurements, full scan, and/or precursor-to-product ion transitions, when using MS analyzers of last generation.

9.2.1 FIA METHOD DEVELOPMENT
9.2.1.1 Ionization Efficiency
At the beginning of the 1990s several API sources such as thermospray were tested for coupling FIA to MS (Geerdink, Kienhuis, & Brinkman, 1993). Nowadays, ESI sources are mainly used for the analysis of pesticides by FIA-MS (Table 9.1). In this context, a key parameter to optimize MS response is the solvent used in the coupling. One of the advantages of FIA in front of LC is that in LC-MS the solvent is largely determined by the chromatographic separation or by a compromise between good chromatographic separation and MS response, whereas in FIA only MS sensitivity must be taken into account. Consequently, most of the papers include a thorough study about solvents and additives that provide the highest response. Flexibility of FIA-MS in using carriers and sample solvents allows improving ionization efficiency, resulting in a better sensitivity that can compensate, at least partially, matrix suppression. In general, positive ESI mode is used and conditions to promote protonation and to reduce the presence of other ions such as $[M+Na]^+$ are employed, whereas negative ion mode is used only for phenolic and acidic compounds. To improve ionization, solvents currently recommended for ESI such as acetonitrile, methanol, and water are used. Among them, water is the most effective but since a high organic solvent content in the carrier liquid is needed to reduce surface tension of the sprayed charge droplets, small amounts of water (1%$-$5%) are currently added to the carrier solvent. Since methanol promotes ionization and reduces surface tension favoring ion evaporation and increasing the efficiency of the ESI source, this solvent is the most frequently used as can be observed in Table 9.1. The use of additives to favor ionization is also quite common; in some cases they are added to the carrier solvent, for instance, formic acid is added to improve protonation in ESI (Goto et al., 2003), whereas in other cases, additives are added to the sample extract. In general, an array of additives and solvents are usually evaluated to select those that provide the best sensitivities. Methanol, acetonitrile, and water at different proportions and with additives such as formic acid, acetic acid, and ammonium

hydroxide at different concentrations (0.1%—1.0%) are usually used. Conditions that provide better results depend on the analyte, for instance, the addition of ammonium hydroxide was recommended by Nanita et al. (2011) for the analysis of 12 pesticides (sulfonylurea herbicides, carbamate insecticides, pyrimidine carboxylic acid herbicides, and anthranilic diamide insecticides) providing responses threefold higher than in methanol alone. However, in other cases the use of additives is not recommended if both positive and negative ion modes must be monitored to analyze all the selected compounds, or if the additives do not produce any remarkable improvement. As an example, the same authors (Nanita, Kaldon, & Bailey, 2015) found that the use of additives (formic acid and ammonium hydroxide) did not improve ionization of chlorsulfuron, methomyl, and oxamyl, the least responsive analytes studied in their work, and proposed using acetonitrile as carrier solvent. The importance of the nature of the analyte in the selection of the solvent carrier is clearly demonstrated in the optimization of an FIA-MS method for the analysis of polar pesticides (Mol & van Dam, 2014). In this case, dramatic differences in the responses of the compounds were observed when using methanol/water and acetonitrile/water with different amounts of formic acid and ammonium hydroxide. For some of the compounds analyzed in positive mode (amitrol, cyromazine, and diazinone) the best results were found when using a carrier with high amounts of acetonitrile under alkaline conditions (0.1% ammonia). In contrast, for quaternary ammonium cations (paraquat and diquat) acidic methanol gave higher responses, whereas those compounds analyzed in negative mode were more sensitively detected using methanol with 0.1% ammonia. One advantage of using FIA-MS is that the fast analysis achieved makes possible using independent injections for the analysis of compounds that require different conditions and in this particular case, the authors recommended three injections (30 s each) to analyze all the compounds. Moreover, although some pesticides show certain instability at high pH, the addition of ammonium hydroxide is frequently used (Table 9.1) since in FIA-MS the short contact time between the sample solution and the basic carrier minimizes degradation allowing accurate results. Nevertheless, the stability of the analytes of interest at high pH must be evaluated to guaranty reproducibility and ruggedness of the method.

9.2.1.2 MS Working Conditions

One of the limitations of FIA-MS compared with chromatography-MS techniques is its lower selectivity. Chromatography contributes to selectivity by the separation of sample components in time and this resolving dimension must be compensated in FIA by increasing selectivity in the MS measurement. To be able to separate the neutral molecules in the FIA solvent the mixture of ions generated in the MS source must reflect specifically the original compounds in the mixture. To achieve this goal, the ionization must be soft as happens with atmospheric ionization sources that yield predominantly only ions of the intact molecular species (protonated and deprotonated molecules, adducts ions, etc.). However, problems can arise because of mass spectral interferences that cannot be differentiated by MS. Ions with the same *m/z* (isobaric) can be generated from different analytes or matrix components potentially present in the samples. For compounds giving molecular clusters with several isotopic ions, for instance, with several chlorine atoms, these problems may be even higher. As an example, Ito et al. (2003) found that one of the ions of the isotopic cluster of imazalil $[M+H+5]^+$ presented the same *m/z* of the $[M+H]^+$ of imazalil-d_5 used as internal standard. In this case, the use of tandem mass spectrometry (MS/MS) that generated product ions of different *m/z* (since they maintain the corresponding chlorine isotope pattern) could solve the problem. Fragment ions with *m/z* similar to that those of the analytes can also be generated through in-source fragmentation of molecular or pseudomolecular (protonated, deprotonated, or ammoniated) ions of other compounds giving as a result, false-positive identifications. This problem can be solved using HRMS if a resolution higher than 15,000 full width half maximum (FWHM) is achieved. As can be seen in Fig. 9.2, where simulated mass spectra of $[M+H]^+$ for both imazalil and

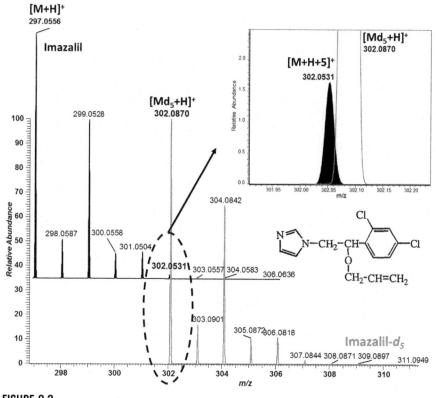

FIGURE 9.2

Simulated mass spectra of $[M+H]^+$ for imazalil and imazalil-d_5 at mass resolution of 17,000 full width half maximum.

imazalil-d_5 are shown, a mass resolution of 17,000 FWHM provides a separation of both ions at ∼50% valley.

To improve selectivity two options are available: tandem MS that reduces the number of potential interferences by increasing the number of precursor-to-product ion transitions monitored and HRMS analyzers that enhance selectivity by separating ions with m/z in close proximity. In FIA-MS, the instruments most commonly used are QqQ mass spectrometers probably because they are also the most popular when coupling LC-MS (Botitsi et al., 2011; Mol et al., 2015). The advantages of using HRMS for the analysis of pesticides by FIA-MS have not been evaluated till now and there is only one publication (Oellig & Schwack, 2014) where a TOF analyzer has been used for the multiresidue screening of pesticides (Table 9.1). As regards tandem MS, QqQ instruments working in MRM mode are commonly used, although the number of transitions monitored (usually two) reduces the selectivity. The limitation of the number of ion transitions to be monitored simultaneously in QqQ analyzers is an important drawback in FIA-MS and in most of the papers this limitation is discussed and conditions regarding the dwell time, the number of transitions, and the sample band width required to obtain reliable identification and quantitation are stabilized. As an example, in the method proposed by Nanita, Pentz, & Bramble (2009), six pesticides were analyzed by FIA-MS/MS using only one transition for each compound, a dwell time of 20 ms, and a band width of 4 s. At these conditions, they obtained 20 data points across the band, a value considered acceptable for a reliable quantitation. When the number of the compounds to be analyzed increases and two transitions are used for each compound to improve identification guarantees, a reduction of the dwell time and an enlargement of the bandwidth, generally by injecting a higher sample volume, are performed obtaining bands of 8−12 s and sampling rates of 10−15 data points per peak. The use of QqQ instruments with shorter dwell times down to 2 ms for each transition monitored, will improve the results but no applications have been published until now. In this context, the use of HRMS analyzers that allow fast full MS scanning providing both accurate mass measurements and isotopic distribution will solve most of the problems found with QqQ instruments. In addition, hybrid HRMS instruments that provide high-resolution product ion scan spectra will make possible the confirmation and characterization of the analytes. Their use in the future will enlarge the applicability of FIA-MS coupling.

To reduce potential interferences due to in-source fragmentation, presence of compounds giving isobaric ions, and MS/MS cross talk when using tandem MS, fragmentation pathways of the compounds should be known. Common ion transitions such as neutral loss of water and ammonia must be avoided and only highly specific fragments have to be selected for quantitation and confirmation to prevent overlapping of ions that could lead to serious errors. This means that those transitions used in LC-MS coupling would not always be adequate for FIA-MS. Moreover, selectivity tests with control samples of matrices of interest must also be performed since the absence of interferences from the matrix must be confirmed.

9.2.2 SAMPLE PREPARATION FOR FIA-MS-BASED METHODOLOGIES

High-throughput analysis obtained by the elimination of the separation step is the main advantage of FIA in front chromatography but it is counterbalanced by increasing the number of compounds arriving to the MS ion source that can lead to high signal suppression in the ESI source. So the usage of sample preparation procedures able to reduce the presence of matrix components seems to be mandatory. However, if one of the objectives of using FIA-MS is to reduce the analysis time, long and tedious sample procedures are not convenient. As a consequence, high-throughput sample preparation methods must be applied to obtain extracts clean enough to prevent potential interferences or, at least, to decrease matrix effect. For this purpose, solvent extraction is currently used as the first step and two approaches, the classical cleanup procedures applied in multiresidue methods of pesticide analysis by chromatographic techniques and the dilution of the sample, have been proposed.

FIA is a technique specifically designed for the analysis of liquid samples, so for solid samples the first required step is to dissolve them. Extraction using solvents of varying polarity such as ethyl acetate, acetonitrile, methanol, or cyclohexane is used for effective extraction of pesticides from the matrix. When direct injection of the extracts is performed, nonpolar solvents are frequently selected taking advantage of their ability to extract less polar analytes reducing the extraction of sample matrix that can exhibit high ion suppression in MS. This procedure is particularly convenient for samples that contain a high percentage of water and low fat content. For instance, Nakazawa et al. (2004) proposed the use of n-hexane and the addition of anhydrous sodium sulfate to extract several pesticides in citrus fruits before the FIA-MS analysis, since among the solvents tested (n-hexane, ethyl acetate, acetone, acetonitrile, and methanol) n-hexane showed the best results, high recoveries with no ion suppression, and little peak tailing. However, in most of the FIA-MS applications to the analysis of pesticides, solvents of higher polarity such as methanol or acetonitrile are used for the extraction (Table 9.1). In these cases, additional treatments are recommended to reduce matrix effects. In some cases, a cleanup step such as liquid−liquid partition or solid-phase extraction (SPE) is used. For instance, in one of the first papers published by Nanita et al. (2009) about the applicability of FIA-MS for pesticide analysis and where a systematic evaluation of this technique for quantitation of pesticide residues is performed, the authors proposed the extraction of the compounds with acetonitrile/phosphate buffer pH 7 and the subsequent purification of the extracts by n-hexane partition and SPE. However, this approach seems not to be the best option since the sample extraction and cleanup used is time-consuming reducing the benefits of the fast instrumental analysis achieved with FIA-MS/MS. A comparison between both instrumental analysis procedures, FIA-MS/MS and LC-MS/MS, without considering sample treatment, is included in this paper indicating that the instrumental analysis of 100 samples can be performed in 1.8 h by FIA-MS, whereas it requires 25−50 h by LC-MS. Therefore it is clear that high-throughput sample treatment methods are needed to maximize FIA-MS advantages. In this context, one easy option to reduce signal suppression due to the matrix is to dilute the extract. Of course, the dilution also sacrifices signal intensity but nowadays the sensitivity of MS analyzers permits this approach. This procedure known as "dilute and shoot" has been used by several authors in different applications (John et al., 2010; Nanita, 2011; Nanita et al., 2011) but the water content of the matrix must be taken into account and the solvent chosen must help to improve ionization efficiency. So, depending on the analyte, acidic (formic acid) or basic (ammonium hydroxide) diluent solvents are used. As regards the water content in the samples, it must be mentioned that the water of the matrix is released during the extraction increasing the total final volume and giving consequently low recoveries. In quantitative analysis of pesticides in high-water-content crops this problem is solved by measuring the final volume and applying the corresponding volume correction for concentration calculations (Nanita & Padivitage, 2013) or if possible, by minimizing the effect of the sample water by reducing the sample size (Nanita et al., 2011).

One problem that can be found in the analysis by FIA-MS is the formation of undesired adducts such as $[M+Na]^+$, because their high stability prevents fragmentation under tandem MS conditions. These sodium adducts can be generated in the ion source due to the presence of residual amounts of nonvolatile salts in the extracts, as sodium salts are frequently added during the extraction to dry water-immiscible solvents or to induce phase separation when water soluble ones (acetonitrile) are used. In this context, ammonium salts have demonstrated to be a good option for the salting-out cleanup of the extracts since they decomposed at the typical ion source temperatures favoring the formation of $[M+H]^+$ ions in ESI by proton transfer reactions between the analyte and ammonium adducts. Moreover, the decomposition of ammonium salts in the ion source prevents the presence of residual salts in the source thus reducing the need of instrument cleaning and maintenance. This method is known as salting-out assisted liquid−liquid extraction and several salts, such as ammonium chloride, ammonium formate, and ammonium acetate, have been used for the analysis of pesticides by FIA-MS. A thorough discussion about the characteristics and applicability of this method for multiresidue analysis of pesticides by FIA-MS can be found in the paper published by Nanita and Padivitage (2013). These authors propose the use of a saturated aqueous solution of ammonium chloride as salting-out reagent for the extraction of 14 compounds, within various chemical classes, from several food matrices of plant and animal origin as well as plasma and urine. Based on the validation data obtained, they conclude that the method was appropriate for trace-level quantitation of pesticides in complex matrices. In addition, this method has proved to be suitable for the analysis of pesticides in water meeting the European Union sensitivity requirements (LOQ: 0.1 μg/L) (Nanita et al., 2015). However, it must be indicated that highly polar compounds cannot be salted out, as happened, for instance, for aminocyclopyrachlor, an analyte included among the compounds studied by the authors, which always reminded in the water phase showing recoveries lower than 5%. In these cases the dilute-and-shoot method would work better, as has been demonstrated by Mol and van Dam

(2014) who proposed this strategy for the analysis of highly polar pesticides, such as glyphosate, chlormequat, diquat, and fosetyl, among others, that are not amenable to multiresidue methods because they partition poorly into organic solvents. The results, recoveries, ion suppression, and limits of detection (LODs) obtained after the extraction of the compounds from several matrices (fruits, vegetables, and cereals) with methanol/0.1M HCl and 10-fold dilution showed that for most of the pesticides, the method can be used for verification of MRL compliance for many crops.

Although the general tendency in applying FIA-MS for the analysis of pesticides has been to simplify the sample treatment to reduce analysis time, a new approach has appeared that uses high-performance thin-layer chromatography (HPTLC) for the cleanup before the FIA injection to obtain almost matrix-free extracts. The method known as high-throughput planar solid-phase extraction (HTpSPE) has been tested for the analysis of seven pesticides in several fruit and vegetable matrices (Oellig & Schwack, 2014). Citrate-buffered QuEChERS was used for sample extraction and thin-layer chromatography (TLC) cleanup was performed with an aminopropyl-modified silica aluminum foil and acetonitrile/10 mM ammonium formate as elution solvent. Highly automated planar chromatographic tools were used to apply the samples, develop the TLC separation, and collect the clean extracts into autosampler vials for the subsequent injection in a microliter FIA-MS. Fig. 9.3 shows the method flow chart of an HTpSPE μL-FIA coupled to a nano-ESI TOF MS used for the screening of pesticides.

9.2.3 FIA-MS METHODS PERFORMANCE

In general, quality parameters obtained in FIA-MS methods are quite good. Recoveries higher than 80%—90% with standard deviations between 3% and 21% are obtained, values that meet the current guidelines for multiresidue screening methods. LOQs are between 0.001 and 0.05 mg/kg (Table 9.1) and for most of the compounds they are low enough to verify compliance of products with legal tolerances. However, matrix effects due to the competition between analytes and other ingredients of the sample also present in the source are currently observed when real samples are analyzed by FIA-MS. As a result, there is a decrease in analyte ionization (ion suppression) or an increase in its ionization (ion enhancement). This problem is especially important in FIA-ESI-MS, where the resulting mixtures of compounds of high complexity in the electrospray droplets as a consequence of the simple sample treatments applied, may reduce evaporation efficiency and the ability of the analytes to reach the gas phase. So in the development of an FIA-MS method for a particular application, systems to lessen or to overcome the problems that matrix effect generates must be evaluated. These effects are particularly important when using the "dilute-and-shoot" method for sample treatment and as a result, calibration with pure standard solutions cannot be used. So, other quantitation methods such as matrix matching, standard addition, or isotopic dilution must be applied. Among them, matrix-matched standard calibration is the most popular and generally yields satisfactory results. As an example, Fig. 9.4 shows the FIA-

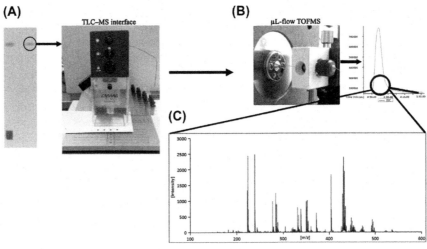

FIGURE 9.3

Flow chart of high-throughput planar solid-phase extraction microliter flow injection analysis time-of-flight mass spectrometry (µL-FIA-TOFMS) screening analysis: (A) elution of pesticides from the target analyte zone after the twofold development (planar solid-phase extraction cleanup) by the thin-layer chromatography mass spectrometry interface; (B) injection of extracts into the µL-FIA-TOFMS system; (C) the obtained full-scan mass spectrum extracted from the entire FIA peak.

Reprinted from Oellig, C., & Schwack, W. (2014). Planar solid phase extraction clean-up and microliter-flow injection analysis-time-of-flight mass spectrometry for multi-residue screening of pesticides in food. Journal of Chromatography A, 1351, 1–11 with permission from Elsevier, Copyright 2014.

MS/MS chronograms of a solvent blank, a control, and a fortified sample, and a matrix-matched standard, obtained for flupyrsulfuron methyl in a rice matrix. The chronograms of the matrix-matched calibration standards prepared at the same concentration level that the fortified samples give an idea of the performance of the method. Matrix-matched calibration curves yielded in general satisfactory linear responses with good correlation coefficients (>0.99%). However, even though most of the published papers used matrix-matched calibration (Table 9.1), in some cases, to better correct the effects of the sample matrix, isotope dilution is proposed, as happens, for instance, in the analysis of highly polar pesticides by the dilute-and-shoot method (Mol & van Dam, 2014). In this case, the recoveries found for some of the compounds when using matrix-matched standards were not acceptable (from 45% to 190%) indicating that when strong suppression occurs, small differences in the composition of the extracts and the matrix-matched standards would result in important different matrix effects. In this particular case, to solve this problem the authors used isotopically labeled internal standards since these standards were commercially available and the other possible approach, standard addition, is a less versatile quantitation method. In this context, it must be mentioned that an advantage of using the

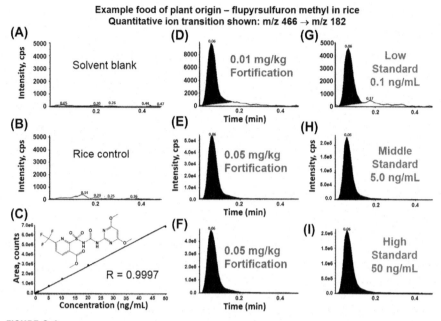

FIGURE 9.4

Representative NH₄Cl salting out FIA/MS/MS ion chronograms and calibration curve obtained for flupyrsulfuron methyl during method validation for a rice matrix: (A) solvent blank, (B) control, (C) calibration curve, (D) control sample fortified at 0.01 mg/kg, (E) control sample fortified at 0.05 mg/kg, (F) control sample fortified at 0.5 mg/kg, (G) 0.1 ng/mL matrix-matched standard, (H) 5.0 ng/mL matrix-matched standard, and (I) 50 ng/mL matrix-matched standard.

Reprinted from Nanita, S. C., & Padivitage, N. L. T. (2013). Ammonium chloride salting out extraction/cleanup for trace-level quantitative analysis in food and biological matrices by flow injection tandem mass spectrometry. Analytica Chimica Acta, *768, 1–11 with permission from Elsevier, Copyright 2014.*

thin-layer cleanup method (HTpSPE) previously commented (Oellig & Schwack, 2014) is that quantitation can be performed using pure standards.

The approach of Nanita is of interest from a theoretical point of view, whereby analyte concentrations were determined at the infinite dilution limit where both the presence and effects of matrix are negligible (Nanita, 2013). The method uses the online real-time dilution achieved in FIA by the diffusion/mixing process that occurs as the sample band travels from the injector to the ion source. Because of matrix effect the concentration measured at the maximum of the peak is inaccurate but accuracy increases when matrix concentration is reduced as happens at both sides of the band. For quantitation the authors proposed to calculate the concentration at infinite dilution limit. Results comparable with those found by LC-MS/MS and FIA-MS/MS with matrix-matched calibration were found indicating that the method would be appropriate for pesticide residue analysis. However, actually the application of this method is not easy and the development of software for automation of data processing is needed to simplify mathematical treatment and reduce data analysis time.

Some authors have performed experiments to evaluate the stability of the FIA-MS system, its ruggedness, and its repeatability simulating the sequential analysis of multiple analytical sets and the results indicate consistent quantitative measurements. For instance, Nanita et al. (2009), in a study where they included a total of 160 injections (blanks, samples, controls, and standard spiked samples), found relative standard deviations (RSDs) between 4% and 11% for all analytes in each sample, with better precision for samples at higher concentrations ($10 \times$ LOQ).

9.2.4 FIA-MS APPLICATIONS

FIA-MS has been applied to the analysis of pesticides since a decade ago when Ito et al. (2003) proposed using this procedure for the analysis of three fungicides in citrus fruits. However, the number of published applications (Table 9.1) is limited and most are devoted to demonstrate the applicability of the technique; to establish its characteristics, such as LODs, analysis time, and matrix effects; to select the optimum sample treatment, but not to apply it for the monitoring of pesticide residues. In addition, the number of researchers working in this field is also limited and most of the studies have been performed by two research groups and always with the main objective of improving the methodology trying to develop a method suitable for the high-throughput analysis of pesticides in a broad scope of matrices.

In general, the number of compounds included in the papers dedicated to FIA-MS is low, from 3 to 14, which is directly related to the MS instrumentation employed. Since MRM with QqQ instruments is the MS mode used in most of the applications and the number of total transitions to be selected is limited by the number of data points recorded and scan times used, the number of compounds cannot be increased too much. Carbamate insecticides, phenyl and sulfonylurea herbicides, fungicides, and also herbicides, as well as plant growth regulators of high polarity such as quaternary ammonium compounds, glyphosate, and maleic hydrazide, among others, have been studied. The applicability of the FIA-MS for the analysis of different matrices such as water, fruits, vegetables, cereals, meat and animal products such as eggs, milk, oil, and even urine and plasma has also been tested. Most of the papers in the literature analyze fortified samples and only few of them apply the optimized method to quantify pesticides in commercial products. Examples are those published at the beginning of the 2000s where thiabendazole and imazalil (Ito et al., 2003) and 2,4-dichlorophenoxy acetic acid (Nakazawa et al., 2004) were found in citrus and grapefruits. In some studies, a comparison of the results obtained when using FIA-MS/MS and LC-MS/MS for the analysis of pesticides in different food matrixes is included. In general, comparable results are obtained, for instance, the amount of chlorantraniliprole in strawberries using a salting-out ($NH_4 OH$) FIA-MS/MS method (Nanita & Padivitage, 2013) or the pesticides found in the screening of several samples (fruits and vegetables) by HTpSPE µL-

FIA-TOF (Oellig & Schwack, 2014). It must be indicated that in this last case, some of the compounds present in the samples could not be identified because of the lower sensitivity of the TOF in front of MRM in QqQ instruments that made difficult the identification of pesticide residues below 0.5 mg/kg. In contrast to what is happening in LC-MS, where the use of HRMS is nowadays almost a routine (García-López et al., 2014; Gómez-Ramos et al., 2013; Zomer & Mol, 2015), only in this paper FIA has been coupled to HRMS. Modern high-resolution instruments such as those equipped with Orbitrap mass analyzers that have been proved to be a good option for LC-MS and direct analysis in real time-mass spectrometry (DART-MS) must be tested.

As commented earlier, most of the FIA-MS methods used for the residue analysis of pesticides employed QqQ instruments, but this instrumentation presents some intrinsic limitations; first, only target analysis can be performed, and second, the number of compounds that can be analyzed in a single run is limited. This means that a multiresidue method based on FIA-MS would require multiple sample injections if hundreds of compounds must be analyzed. The use of HRMS with TOF or Orbitrap mass analyzers providing accurate mass measurements for the subsequent database search would overcome the limitation of MS/MS target analysis. The results obtained when coupling HTpSPE to FIA-MS using a TOF analyzer (Oellig & Schwack, 2014) that improved the cleanup of the extracts and allowed the implementation of a suspected screening approach using pesticide databases, or those obtained in the analysis of psychoactive substances by FIA-HRMS using both full-scan HRMS and data-dependent MS/MS on a quadrupole-Orbitrap system combined with database search (Alechaga, Moyano, & Galceran, 2015), show some of the directions to be followed in further studies.

9.3 AMBIENT MASS SPECTROMETRY

In 2004 a new group of ionization techniques arrived in the field of MS changing the way samples are analyzed and making possible the direct analysis of complex samples. The term "ambient mass spectrometry" was first introduced by Cooks et al. in 2004 to describe ionization sources that operated in open air with ionization taking place in the laboratory environment and without the need of laborious sample treatments and avoiding chromatographic separations. DESI was the first ambient ionization technique (Takáts et al., 2004) and a little later in 2005 DART was developed (Cody, Laramée, & Durst, 2005). These two pioneer ambient mass spectrometric techniques boosted the development and broad applications of a large number of ionization techniques. Today, a wide range of compounds, from low to high molecular weight and from low to high polarity, can be directly analyzed on sample surfaces (or auxiliary surfaces) in open air and in most cases within less than 5 s facilitating high-throughput analysis. In the early days of ambient techniques, it was frequently considered that "no sample preparation" was required. Nowadays, it is more precise to say that "no additional sample preparation" is required since the sample processing takes place in situ during the analysis (Venter et al., 2014). Many ambient ionization techniques have been introduced, which differ in the way that sample processing and ionization mechanisms are combined (Huang et al., 2011; Venter, Nefliu, & Graham Cooks, 2008). A complete discussion of all ambient ionization techniques would exceed the scope of this book chapter and for further in-depth study, a comprehensive review of AMS techniques, from fundamentals to the state-of-the art of applications, can be found in the book published by Domin and Cody (2015).

In the field of pesticide testing and food safety screening, different authors have explored the applicability of AMS as a fast and simple alternative to simplify laboratory workflow and to improve high-throughput analysis. A summary of AMS methods used for the analysis of pesticides is given in Tables 9.2 and 9.3. Among the ambient ionization techniques, the electrospray-based ones, DESI (Cajka et al., 2011; Garcia-Reyes et al., 2009; Mulligan, Talaty, & Cooks, 2006; Schurek et al., 2008), paper spray ionization (PSI) (Soparawalla et al., 2011), probe electrospray ionization (PESI), and extractive electrospray ionization (EESI) mass spectrometry (Chen, Venter, & Cooks, 2006) have been applied to detect pesticides in food and environmental samples. Nevertheless, plasma-based ionization techniques that use thermal desorption/ionization via atmospheric pressure chemical ionization are more popular probably due to the simplicity of their design and operation. DART (Cajka et al., 2011; Crawford & Musselman, 2012; Edison, Lin, Gamble, et al., 2011; Farré, Picó, & Barceló, 2013; Kiguchi et al., 2014; Schurek et al., 2008; Wang et al., 2010), low-temperature plasma (LTP) (Albert et al., 2014; Soparawalla et al., 2011; Wiley et al., 2010), and flowing afterglow atmospheric pressure glow discharge (FAPA) or atmospheric pressure glow discharge (APGD) (Andrade et al., 2008; Jackson & Attalla, 2010) have been used for pesticides. However, it must be pointed out that the majority of the ambient methods developed for pesticide analysis use DESI and DART, probably because there are available commercial platforms.

9.3.1 ELECTROSPRAY-BASED TECHNIQUES

The direct analysis of a sample via electrospray-based ionization techniques combines ESI with sample processing through a substrate-liquid extraction (Venter et al., 2014). DESI (Takáts et al., 2004) and EESI (Chen et al., 2006) are spray desorption methods that use a solvent spray as sample processing agent to remove material from a surface and for subsequent electrospray compound ionization. In contrast, substrate spray methods such as PSI (Wang et al., 2010) and PESI (Hiraoka et al., 2007) use substrates with sharp tips as electrospray emitters. Fig. 9.5 shows the schemes of the electrospray-based interfaces commented in this section.

EESI has been used in the field of pesticide analysis only as proof of concept for detecting atrazine in mouse urine. In this technique, the liquid sample was sprayed into an electrospray beam positioned orthogonally to the sample spray. The ionizing solvent was methanol/water/acetic acid (45:45:10, v/v/v) to favor analyte

Table 9.2 Summary of Electrospray-Based AMS Methods for the Analysis of Pesticides

Compounds	Sample	Sample Handling	AMS Conditions		MS Analyzer (Acquisition Mode)	References
			Technique	Operational Parameters		
Alachlor Atrazine N,N-diethyl-m-toluamide	Cornstalk leaf Vegetables	Direct analysis of peels	DESI (+)	Methanol:water (1:1), 3 μL/min 5 kV, 150 psi	LIT, portable IT (full scan, MS/MS)	Mulligan et al. (2006)
Acetochlor Alachlor Atrazine	Groundwater	Method 1: wet a filter paper Method 2: SPE membrane	DESI (+)	Methanol:water (1:1), 3 μL/min 5 kV, 90–120 psi	LIT (full scan, MS/MS, MS^n)	Mulligan et al. (2007)
2 Organophosphates (dicrotophos, malathion)	Porous inorganic or polymeric organic powder	Affixed 1 mg of powder material to double-side tape on a kapton slide and lightly scrap the surface to remove excess powder	DESI (+)	Methanol:water (1:1), 1.5 μL/min 4.5 kV, 100 psi Incident (collecting) angle: 45° (5°)	IT (full scan, MS/MS)	Hagan et al. (2008)
6 Strobilurins	Wheat	Methanol extraction and C18 solid phase microextraction (pipet tips)	DESI (+)	Methanol:water (1:1); 0.1% formic acid), 2.5 μL/min, 5.5 kV, 120 psi Incident (collecting) angle: 55° (10°)	LIT (full scan, MS/MS, MS^n)	Schurek et al. (2008)
16 Pesticides (insecticides, herbicides, fungicides)	Fruit Vegetables	Method 1: peels Method 2: QuEChERS	DESI (+)	Acetonitrile:water (1:1; 0.1% formic acid), 5 μL/min, 4.5 kV, 150 psi Incident (collecting) angle: 55° (10°)	LIT (full scan, MS/MS, MS^n)	Garcia-Reyes et al. (2009)

Continued

Table 9.2 Summary of Electrospray-Based AMS Methods for the Analysis of Pesticides—cont'd

Compounds	Sample	Sample Handling	AMS Conditions		MS Analyzer (Acquisition Mode)	References
			Technique	Operational Parameters		
2 Dithiocarbamate fungicides (thiram, ziram)	Fruits	QuEChERS	DESI (+)	Methanol:water (1:1, 0.1% formic acid), 5 μL/min, 5 kV, 120 psi Incident angle (collecting distance): 55° (0.5 mm)	LIT (full scan, MS/MS, MS^n)	Cajka et al. (2011)
Atrazine	Mouse urine	Spike urine	EESI (+)	Methanol:water: acetic acid (45:45:10)	LIT (full scan, MS/MS)	Chen et al. (2006)
Atrazine	Standard	Few μL on a chromatographic paper	PSI (+)	Methanol:water (9:1)	LIT (full scan, MS/MS)	Wang et al. (2010)
Thiabendazole	Fruits	Wipe the fruit surface with a methanol wetted filter paper or with a wipe and cut a triangle from the wipe	PSI (+)	Methanol:water (1:1)	Portable IT (full scan, MS/MS)	Soparawalla et al. (2011)
Acephate, acetamiprid, clothianidin, thiophanate-methyl	Living plants	Direct surface analysis	PESI (+)	Acetonitrile:water (1:1) with 0.1% formic acid	TOF (full scan)	Mandal et al. (2013)

AMS, ambient mass spectrometry; DESI, desorption electrospray ionization; EESI, extractive electrospray ionization; IT, ion trap; LIT, linear ion trap; MS, mass spectrometry; MS/MS, tandem mass spectrometry; MS^n, multi-stage mass spectrometry; PESI, probe electrospray ionization; PSI, paper spray ionization; QuEChERS, Quick, Easy, Cheap, Effective, Rugged, and Safe; SPE, solid phase extraction; TOF, time-of-flight.

Table 9.3 Summary of Plasma-Based AMS Methods for the Analysis of Pesticides

Compounds	Sample	Sample Handling	AMS Conditions		MS Analyzer (Acquisition Mode)	References
			Technique	Operational Parameters		
6 Strobilurins	Wheat	Method 1: ethyl acetate extraction Method 2: direct analysis of milled wheat grains placed in a filter paper envelope	DART (+)	Method 1: 200°C Method 2: 300°C 2.9 L/min (He)	TOF (full scan)	Schurek et al. (2008)
132 Pesticides	Fruits	Surface sample swabbing	DART (+)	100–350°C over 3 min	Orbitrap (full scan)	Edison, Lin, Gamble, et al. (2011)
2 Dithiocarbamate fungicides (thiram, ziram)	Fruits	QuEChERS Dip-it scanner autosampler	DART (+)	400°C 5 s (desorption time)	Orbitrap (full scan)	Cajka et al. (2011)
104 Pesticides	Fruits	Surface sample swabbing	DART (+)	100–350°C over 3 min	Orbitrap (full scan)	Edison, Lin, and Parrales (2011)
Carbofuran, ethoprophos, fipronil, phorate	Commercial agrochemicals	Dissolution of solid in acetonitrile and dip a glass rod into the acetonitrile extract	DART (+)	400°C 2.9 mL/min (He) 5 s (desorption time)	Quadrupole (full scan)	Wang et al. (2012)
Dimethoate, methamidophos	Fruits Vegetables	Surface sample swabbing	DART (+)	150–250°C 3 s (desorption time)	Orbitrap (full scan)	Crawford and Musselman (2012)
Imazalil	Fruits	Direct peel surface analysis	DART (+)	400°C 3 L/min (He)	LIT-Orbitrap (full scan, MS^n)	Farré et al. (2013)

Continued

Table 9.3 Summary of Plasma-Based AMS Methods for the Analysis of Pesticides—cont'd

Compounds	Sample	Sample Handling	AMS Conditions		MS Analyzer (Acquisition Mode)	References
			Technique	Operational Parameters		
Acephate, diazinon, EPN, fenitrothion, methamidophos	Dumpling Grapefruit	TLC	DART (+)	260°C 3 L/min (He)	TOF (full scan)	Kiguchi et al. (2014)
13 Pesticides	Fruits Vegetables Water	Direct peel surface analysis Few μL of water sample deposited on a glass slide surface	LTP (+)	400 mL/min (He)	LIT (full scan, MS/MS)	Wiley et al. (2010)
Diphenylamine	Fruits	Direct peel surface analysis	LTP (+)	300 mL/min (He)	Portable IT (full scan, MS/MS)	Soparawalla et al. (2011)
Acetamiprid, cyprodinil, fenhexamid, fludioxonil	Fruits	QuEChERS	LTP (+)	300 mL/min (He)	Orbitrap (full scan)	Albert et al. (2014)
10 Pesticides	Fruit juices Fruits Salad leaves	Few μL of fruit juice deposited on a filter paper Direct peel surface analysis	APGD (+)	He	q-TOF (full scan, MS/MS)	Jecklin et al. (2008)
8 Pesticides	Bulk solutions	Direct analysis of liquid sample	FAPA (+)	600 mL/min (He)	LIT (full scan, MS/MS)	Shelley et al. (2011)

AMS, ambient mass spectrometry; APGD, atmospheric pressure glow discharge; DART, direct analysis in real time; FAPA, flowing afterglow atmospheric pressure glow discharge; IT, ion trap; MS, mass spectrometry; MS/MS, tandem mass spectrometry; MSⁿ, multi-stage mass spectrometry; LIT, linear ion trap; LTP, low temperature plasma; QuEChERS, Quick, Easy, Cheap, Effective, Rugged, and Safe; TLC, thin layer chromatography; TOF, time-of-flight.

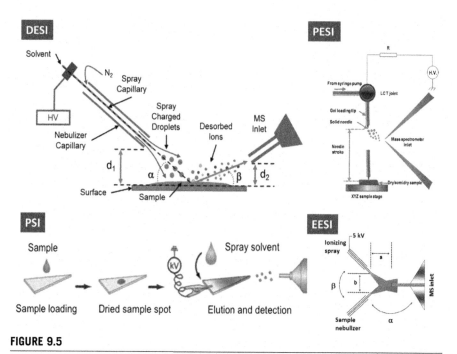

FIGURE 9.5

Schematic of electrospray-based AMS interfaces: (A) DESI, (B) PESI, (C) PSI, and (D) EESI.

(A) Reprinted from Seró, 2016, Copyright 2016 with permission from Elsevier. (B) Reprinted with permission from Mandal, M. K., et al. (2013). Development of sheath-flow probe electrospray ionization mass spectrometry and its application to real time pesticide analysis. Journal of Agricultural and Food Chemistry, 61 (33), 7889–7895, Copyright 2013 American Chemical Society. (C) Adapted with permission from Wang, H., et al. (2010). Paper spray for direct analysis of complex mixtures using mass spectrometry. Angewandte Chemie — International Edition, 49 (5), 877–880, Copyright 2010 Wiley-VCH. (D) Adapted from Chen, H., Venter, A. & Cooks, R. G. (2006). Extractive electrospray ionization for direct analysis of undiluted urine, milk and other complex mixtures without sample preparation. Chemical Communications (Cambridge, England), (19), 2042–2044 with permission from The Royal Society of Chemistry.

protonation. When sample microdroplets interacted with charged microdroplets a liquid extraction/ionization occurred, allowing detection of atrazine $[M+H]^+$ directly in a raw urine sample (Chen et al., 2006). Although DESI involves a more complex desorption/ionization mechanism (Takáts, Wiseman, & Cooks, 2005), more applications have been developed related to the identification of pesticides in food and environmental samples (Kauppila & Vaikkinen, 2014; Nielen et al., 2011). For DESI-MS analysis a solvent, generally methanol:water, is electrosprayed onto a sample surface at an impact angle ($\sim$55 degrees). The first arriving microdroplets wet the sample surface extracting/dissolving sample material (Fig. 9.5). The arrival of subsequent charged droplets cause the splash of the wet surface, which results in the emission of secondary charged microdroplets that contain the extracted/dissolved material. In both ambient techniques, DESI and EESI, compound ionization occurs via ion evaporation from charged microdroplets as in standard ESI.

To initiate the electrospray process in substrate spray methods, PSI and PESI, a high voltage is applied to a sharp tip-shaped substrate, which generates charged droplets. The applicability of PSI has been demonstrated in analyzing thiabendazoles in fruit peels (Soparawalla et al., 2011). A paper wetted with methanol was used to wipe the fruit peel and cut into a triangle shape. After adding a small amount of solvent (methanol:water, 9:1, v/v) to extract the analytes from the sample, a high voltage was applied to the paper substrate generating charge droplets at the apex of the paper triangle by a process similar to ESI and the spray solvent carried the extracted analyte toward the mass spectrometer inlet (Fig. 9.5). Regarding PESI, a sheath-flow version (SF-PESI) has been developed to be applied to direct analysis in real time and it has been tested analyzing pesticides in living plants and leaves (Mandal et al., 2013). In SF-PESI a needle is inserted into a gel-loading pipette tip and a solvent is flowed through the capillary through an LC T-joint (Fig. 9.5). When the needle tip touches the surface of the plant, the components from the plant tissue are extracted by the solvent and picked up by the needle. When the needle was moved up and arrived to the highest position, a high voltage (~ 2 kV) was applied generating a tiny Taylor cone at the tip of the solid needle probe, which yielded analyte ions via conventional electrospray process. Both substrate spray methods demonstrated to be able to produce reasonable ionization efficiency for the real-time pesticide analysis.

9.3.2 PLASMA-BASED TECHNIQUES

The simplicity of design and operation of plasma-based techniques has increased their popularity and nowadays, more than half of the ambient methods developed for pesticide screening rely on this technology. In plasma-based sources an electrical discharge is formed between two electrodes by applying either a direct-current (DC) voltage or an alternating-current (AC) voltage (Domin & Cody, 2015). The discharge is generated in a flowing gas (helium, nitrogen, argon, or air), which produces reactive species such as electrons, radicals, ionized or excited atoms, and molecules. Among the plasma-based sources used for pesticide analysis, DART and FAPA operate with DC voltages, whereas LTP uses AC voltage. In DART, the discharge is located in the enclosed discharge chamber between two electrodes and the ions generated in the plasma are partially filtered on their way to the sample-ionization region (Fig. 9.6). In FAPA sources the discharge is also generated in an enclosed chamber between a pin cathode and a plate (AGPD) or capillary anode (SF-FAPA) (Andrade et al., 2008; Jackson & Attalla, 2010) but the reactive species leave the plasma source through an orifice in the plate or capillary electrode, without previous ion filtering as occurs in the DART source. In the LTP probe (Fig. 9.6), kilohertz AC voltages are applied to one of the electrodes, while the other is grounded (Albert et al., 2014; Harper et al., 2008; Soparawalla et al., 2011; Wiley

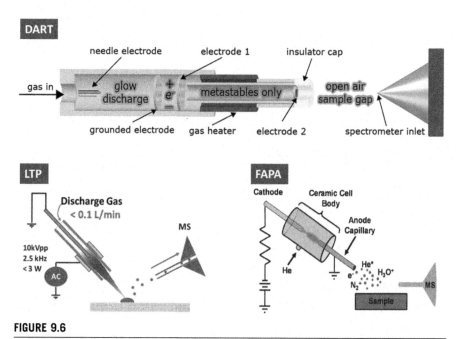

FIGURE 9.6

Schematic of plasma-based AMS interfaces: (A) DART, (B) LTP probe, and (C) FAPA.

(A) Reproduced with permission from R.B. Cody, Wikipedia. https://commons.wikimedia.org/wiki/File:DART_ion_source_schematic.png, from JEOL USA, Inc. (B) Adapted with permission from Harper, J. D., et al. (2008). Low-temperature plasma probe for ambient desorption ionization. Analytical Chemistry. 80 *(23), Copyright 2008 American Chemical Society. (C) Adapted with permission from Shelley, J. T., Wiley, J. S., & Hieftje, G. M. (2011). Ultrasensitive ambient mass spectrometric analysis with a pin-to-capillary flowing atmospheric-pressure afterglow source.* Analytical Chemistry, 83 *(14), 5741–5748, Copyright 2011 American Chemical Society.*

et al., 2010). In this source, a dielectric barrier (a glass or quartz tube) separates the electrodes (a pin electrode and a foil ring electrode) and the plasma is generated within the glass tube and often extends into the ambient environment, the named afterglow discharge region. The electrical discharge in FAPA heats plasma gas, so additional heating is not necessary for the thermal desorption of the compounds as showed by Shelley et al. who used a FAPA source for the analysis of some selected pesticides and herbicides without the need of any external heating source (Shelley, Wiley, & Hieftje, 2011). In contrast, the lower temperature of the DART and LTP plasmas requires the additional heating of the sample to enhance thermal desorption and maximize the signal. For instance, for the identification of dithiocarbamate fungicides in fruit, temperatures ~300°C were needed when using DART (Cajka et al., 2011), so in some applications a temperature gradient to selectively desorb pesticides from fruits is proposed by Edison, Lin, Gamble, et al. (2011) and Edison, Lin, and Parrales (2011). Regarding LTP, some authors manage to detect pesticides without external heating (Albert et al., 2014), but others (Wiley et al.,

2010) recommend the use of a heat gun directly under the sample holder to increase the temperature of the substrate (to $\sim 150°C$), thereby increasing desorption capabilities and enhancing the response.

Regarding ionization, DART plasma mainly contains metastable helium atoms because the high energetic ionic species generated in the glow discharge section are filtered before entering the afterglow discharge region. There, the excited helium atoms interact with atmospheric gases (N_2, O_2, or even H_2O) to generate other reagent ions. However, in FAPA and LTP sources the species in the afterglow discharge already contain the reagent ionic species. Protonated water clusters, $[(H_2O)_n+H]^+$, are the most abundant reagent ions, but other species such as $N_2^{+\bullet}$, NO^+, $O_2^{+\bullet}$, $O_2^{-\bullet}$, NO_2^-, and NO_3^-, as well as thermal electrons, can also be present with different abundances. All these ionic species interact with neutral molecules through chemical reactions in gas phase. Proton transfer, charge transfer, or even electron capture are some of the ionization mechanisms that can take place in gas phase to ionize analyte molecules.

9.3.3 AMBIENT MASS SPECTROMETRY METHOD DEVELOPMENT

9.3.3.1 Ambient Mass Spectrometry Ionization

AMS techniques provide soft ionization generating ions with low internal energy, and so no or very little molecular fragmentation is expected. The generated ions may be both monocharged and multiple-charged in both positive- and negative-ion mode when working with electrospray-based methods (DESI, PSI, PESI), whereas plasma-based techniques (DART, FAPA, LTP) generate only monocharged ions.

Electrospray-based methods produce quasi-molecular ions in positive-ion mode as a result of protonation, $[M+H]^+$, or adduct formation, $[M+NH_4]^+$, $[M+Na]^+$, and $[M+K]^+$. Formic acid and acetic acid are commonly added to the electrospray solvent to favor protonated molecules when working in positive-ion mode. However, there are some pesticides such as isofenphos-methyl, malathion, acetamiprid, thiophanate-methyl, and thiram that yield sodium and/or potassium adducts even when using acidified spray solvents (Cajka et al., 2011; Garcia-Reyes et al., 2009; Mandal et al., 2013). It must be pointed out that these adduct ions can be stable and sometimes do not fragment in tandem MS as happens, for instance, with thiram sodium adduct ion. To solve this problem, the addition of 10 mM of ammonium formate to the spray solvent was proposed to force the generation of thiram ammonium adducts, which can be easily fragmented in MS/MS (Cajka et al., 2011). In general, all ions generated by electrospray-based techniques are even-electron ions and radical ions are not frequently observed, in contrast to what happens in plasma-based ionization techniques. However, water-adduct radical ions $[M+H_2O]^{+\bullet}$ have been observed when analyzing low-volatile organophosphates such as malathion and dicrotophos (Hagan et al., 2008) by DESI. These adducts are more prevalent at mild desolvation conditions, but they readily dissociated upon MS/MS, resulting in the protonated molecular ion $[M+H]^+$. Finally, it must be pointed out that, although electrospray-based ambient techniques can be used in negative-ion mode to generate deprotonated molecules $[M-H]^-$, no examples have been found in the analysis of pesticides.

Although the discharges in different plasma-based techniques are fundamentally different, the reagent ions generated in the plasma source and the ionization mechanisms involved show similarities between them. The mass spectrum is quite similar for most analytes, showing comparable molecular ions, fragment ions, and reactive species, thus indicating a similar ionization mechanism. As commented earlier, reagent ions are produced through the interaction of helium plasma species with atmospheric gases (N_2, O_2, H_2O) and these reagent ions interact with thermally desorbed analyte molecules to generate both positive and negative ions through chemical reactions in gas phase. In positive-ion mode, protonated water and water cluster ions may act as reagent ions for analyte ionization via proton-transfer reactions yielding $[M+H]^+$, whereas molecular radical ions $[M]^{+\bullet}$ can be generated via charge transfer with atmospheric gas-phase ions ($N_4^{+\bullet}$, $O_2^{+\bullet}$). The ratio $[M]^{+\bullet}$ to $[M+H]^+$ mainly depends on the analyte ionization energy (IE) and proton affinity (PA). Analytes with low IE will favor the formation of radical molecular ions, while for those with high PA the generation of protonated molecules is favored. For instance, phorate, carbofuran, and ethoprophos yielded $[M]^{+\bullet}$ instead of $[M+H]^+$ in DART-MS positive-ion mode (Wang et al., 2012). Nevertheless, both ionic species can be simultaneously present, with the consequent overlapping of isotope clusters, which complicates the mass spectrum interpretation. Ammonium adducts have also been observed for some moderate polar analytes as, for instance, in the DART mass spectra of phorate, carbofuran, and ethoprophos where the $[M+NH_4]^+$ is base peak (Wang et al., 2012).

Taking into account that the main gaseous neutral species in the gas phase is nitrogen, the electrons emitted upon $N_2^{+\bullet}$ generation are of relevance. So, in negative mode, electron capture becomes the main ionization mechanism in plasma-based techniques for molecules with a high electron affinity. In addition, atmospheric oxygen yields negative radical reagent ion $O_2^{-\bullet}$, which can interact with neutral molecules to generate adduct ions such as $[M+O]^{-\bullet}$. These adduct ions may either prevail intact or dissociate to yield analyte radical anions $[M]^{-\bullet}$. For instance, Albert et al. (Albert et al., 2014) observed the formation of $[M+O-H]^-$ for fludioxonil when analyzing this pesticide by LTP.

Finally, Edison, Lin, Gamble, et al. (2011) and Edison, Lin, and Parrales (2011) screened a wide range of pesticides by DART-MS and found that many of the classes analyzed could be detected in both positive and negative ionization modes. For instance, pyrimidine, phenylurea, neonicitinoid, benzimidazole, amide, and anilide pesticides were detected in both modes, although the signal was higher (one order of magnitude) in the negative-ion mode, whereas triazine, organophosphorous, and carbamate pesticides showed the best response in the positive-ion mode. Other compounds such as thiocarbamate, tetramic acid derivative, strobilurins, phenylamide, organochlorine, and morpholine pesticides were observed only when using positive-ion mode, whereas for benzoylphenylurea pesticides negative-ion mode had to be used. In addition, the authors pointed out that analytes with a mass greater than 640 Da, such as spinosyns and some macrocyclic lactones, were not detected in either mode, probably due to a molecular weight threshold for the thermal desorption of the molecules.

9.3.3.2 Ambient Mass Spectrometry Working Conditions

AMS techniques can easily be coupled to most types of mass spectrometers equipped with atmospheric pressure interfaces and without any modification of the ion optics or the vacuum interface. Nevertheless, despite the simplicity of AMS techniques, several key parameters that affect the ionization efficiency and signal intensity should be optimized in ambient method development.

A unique feature of electrospray-based techniques compared with plasma-based ones is the ability to tailor the spray solvent for specific analytes. Solubility and acid–base characteristics of analytes should be taken into account when choosing the spray solvent composition, since it may dramatically affect the extraction in DESI and PESI and the partitioning between the paper surface and solvent in PSI, and consequently, the ionization efficiency of pesticides in the electrospray process. Generally, electrospray-based techniques use binary mixtures of water and methanol or acetonitrile (1:1, v/v) as spray solvent and the addition of formic or acetic acid to promote analyte protonation when acquiring data in positive-ion mode. For instance, the use of acetonitrile-based solvents instead of methanol-based ones and the addition of formic acid has been recommended to improve the DESI signal of organophosphorous insecticides (malathion, isofenphos-methyl) (Garcia-Reyes et al., 2009) and methanol:water (9:1, v/v) has been proposed as solvent for a paper spray process to detect thiabendazole, thus diminishing the interaction of compounds with the paper surface and favoring their movement toward the apex of the triangle (Soparawalla et al., 2011).

Other parameters must be taken into account in electrospray-based techniques. For instance, spray solvent flow rate is important in DESI and SF-PESI methods, since both use a flowing solvent. Most of the DESI-MS applications apply spray solvent flow rate values ranging from 2 to 3 μL/min. Nevertheless, higher values may be used to obtain wetter surfaces, which favor analyte extraction and increase the screening sample area, thus improving sensitivity but sacrificing spatial resolution (Cajka et al., 2011). Regarding SF-PESI, the sheath flow rate recommended for a stable electrospray from the tip of the needle is 1.0 μL/min, since lower flow rates generate an electrospray from the terminal end of the plastic capillary rather than from the tip of the needle probably due to the poor wetting of the needle surface. Nevertheless, flow rates higher than 1 μL/min produce a decrease of the analyte signal intensity due to the dilution of the analytes dissolved/extracted from the sample surface (Mandal et al., 2013). Another parameter that must be optimized in DESI is the flow rate of the gas (N_2) used to assist in the formation of the spray of microdroplets. Typical gas pressure values range from 120 to 150 psi for maximal ion signal. Although the increase of gas flow rate can produce smaller but faster charged droplets, too high values may provoke earlier droplet evaporation before impact on the sample surface, which might prevent ionization for some analytes.

Geometrical parameters are a key point of the electrospray-based methods, especially for DESI, where the number of experimental parameters to be optimized is relatively large, hampering their implementation in routine pesticide laboratories. Furthermore, DESI optimal operational parameters found in the literature show a large variation (Douglass et al., 2012) and many of them are interdependent, so different combinations of settings can produce similar results. However, some best practices and a set of standard working conditions can be derived from the literature to be used as starting point for DESI methodology development. The range of values generally applied for some critical geometrical parameters in DESI are 25–80 degrees incident angle (α), 5–10 degrees collection angles (β), 0–2 mm for the sample surface-to-mass spectrometer inlet distance, and 1–10 mm for the spray tip-to-surface distance. A good description of the most important practical aspects and specific features of DESI within pesticide analysis field can be found in the review published by Nielen et al. (2011).

Another feature to be taken into account is the sample surface or sample substrate since surface properties can affect both sensitivity and signal stability. Although the direct analysis of the sample surface is a fast and desirable option for the analysis of fruit and vegetable samples, in DESI-MS sample extraction or cleanup is sometimes required. In these cases, the deposition of an aliquot of the extract (few microliters) on a surface is necessary. Surface materials such as glass, paper, or polytetrafluoroethylene (PTFE) have been used to analyze fruit and vegetable extracts after a liquid–liquid extraction or QuEChERS (Hagan et al., 2008; Garcia-Reyes et al., 2009; Mulligan et al., 2007) but in most cases, the DESI best performance (sensitivity and signal stability) was achieved when using nonpolar PTFE surfaces. This may be due to lower interaction with analytes, which makes easy the extraction/desorption by the sprayed solvent. Nevertheless, signal intensities generally are highly variable during DESI analysis and the introduction of an appropriate internal standard in the spray solvent is recommended to control signal variations resulting in interday variability of <10% RSD (Ifa et al., 2008). In paper spray, the substrate also has an important role, since chromatographic separations can occur for sample components with large differences in distribution coefficients (substrate surface/solvent), which can help to deal with matrix interferences and matrix signal effects. For instance, among the different substrates (filter paper and several commercially available wipes) used to wipe the surface of oranges before analysis by paper spray, the best performance was found for filter paper (Soparawalla et al., 2011), probably because the wipes are markedly less porous than the filter paper reducing the efficiency of analyte transfer onto the wet paper surface. In this technique, it is also important to optimize the relationship between the voltage applied to generate the spray and the spray current. Finally, it should be taken into account that the angle of the triangle tip has a large impact on the generation of the electrospray since the onset voltage increases continuously as the angle increases and the highest signal and spray current generally occurs at a triangle tip angle of 30 degrees, which provides an onset voltage of 3 kV (Yang et al., 2012). In SF-PESI, several interdependent parameters such as the frequency of the needle drive, the protrusion of the needle, and the speed of the moving stage must be optimized. For the analysis of pesticides, it has been recommended (Mandal et al., 2013) to use a repetition rate of 1 Hz, frequency that was enough to allow sample extraction (0.2 s), electrospray acquisition (0.8 s), and to avoid the overlap of the spots touched by the probe. The protrusion of the needle from the tip of the plastic capillary in this method was between 0.1 and 0.2 mm, since at lengths shorter than 0.1 mm, electrospray became unstable and at longer than 0.2 mm the electrospray takes place from the tip of the plastic capillary.

Regarding plasma-based techniques, most of them use helium as ionizing gas since the metastable helium atoms He* generated in the plasma-based sources have internal energies above the IE of atmospheric gases, thus yielding reagent ions. In addition, analyte molecules have to be thermally desorbed to interact with He* and reagent ions in the gas phase. Under this scenario, the key experimental parameters that directly affect the ion intensity are helium beam temperature, He flow rate, and desorption time, since they have an important impact in ionization efficiency and ion transport into MS inlet.

Helium gas temperature is critical to effectively desorb analytes from the sample surface. For the analysis of pesticides with FAPA or APGD the additional heating of the He gas stream is not necessary. As commented earlier, the Joule heat generated in these sources is enough to self-heat the gas and to desorb compounds from the sample surface. In contrast, in DART gas beam has to be heated before impacting onto the sample surface and temperatures generally used in pesticide analysis range from 100°C to 400°C. Kiguchi et al. (Kiguchi et al., 2014) indicated that low temperature settings (250−270°C) are suitable to detect highly volatile and low-molecular-weight pesticides such as methamidophos, acephate, and fenitrothion, whereas high temperature settings (290−300°C) are required to detect semivolatile and higher-molecular-weight pesticides such as diazinon. High temperatures (300 −400°C) were also applied by other authors (Cajka et al., 2011) for an efficient thermal desorption of pesticides such as dithiocarbamate fungicides directly in fruits. Schurek et al. (2008) proposed to reduce helium beam temperature down to 200°C when analyzing strobilurin fungicides in wheat, but it was due to the limited thermal stability of the filter paper in which the sample, milled wheat grains, was placed into the DART source. In fact, the same authors used a higher temperature (300°C) when analyzing the same pesticides in wheat extracts (QuEChERS), thus improving thermal desorption and enhancing the DART-MS signal. However, thermal degradation might occur for thermolabile compounds, which could limit the maximum temperature applicable. For instance, when analyzing dimethoate a relatively low temperature (150°C) was needed to achieve the maximum intensity, since thermal degradation of this compound occurs at higher DART gas temperatures (Andreozzi et al., 1999). In addition, for some applications, where small analyte molecules are immersed in high-mass matrices, the use of low DART temperatures is recommended to decrease mass spectra background noise, preventing the volatilization of matrix components. To deal with a large number of compounds with different volatilities, Edison, Lin, Gamble, et al. (2011) proposed a short temperature gradient (3 min) from 100°C to 300°C with the DART helium stream to achieve some fractionation based on boiling points. Although in many LTP applications external heating has not been applied, different authors recommend increasing the temperature of the sample holder to improve thermal desorption. For instance, Wiley et al. (2010) used a heat gun under the sample holder to screen agrochemicals in foodstuffs and Albert et al. (2014) observed that to detect fungicides (cyprodinil, fenhexamid, and fludioxonil) and insecticides (acetamiprid) in fruit extracts it was necessary to hold the well-plate temperature at 150°C during the analysis using a heating foil.

Helium flow rate has also a prominent impact on the responses in plasma-based techniques. For instance, the DART signal intensity increases when the flow rate increases, but high flow rates provide strong turbulence that might affect the response reproducibility, so a flow rate of ~3 L/min is generally used for most applications (Cajka et al., 2011). Desorption time also has an impact in both sensitivity and selectivity and typical values of 2—5 s are generally applied in static mode to desorb analytes, whereas low sample speeds are needed to guarantee analyte desorption and fast screening of samples when working under dynamic conditions. For instance, in the analysis of thiram and ziram in fruits by DART-MS, it has been reported that 5 s provided enough ion intensity, but a longer desorption time led only to an increased detection of various matrix coextracts present in the sample (Cajka et al., 2011). Farré et al. (2013) indicated that sample speeds between 0.2 and 0.6 mm/s was optimal for the analysis of pesticides in fruit peel using DART-MS, as it provided a rational compromise between the rates of sample analysis and of analyte desorption/ionization to satisfy signal intensity. Finally, the source-to-mass spectrometer inlet distance also affects ion intensities. In DART, a distance of 12—25 mm is generally used as a compromise between the most intense signal and the vacuum fluctuations.

AMS techniques are soft ionization techniques that yield molecular or pseudomolecular ions allowing the assignment of each ion to one compound. Nevertheless, ions with the same m/z values can be produced due to isomeric compounds or isobaric interferences coming from matrix components. Complex samples generate complex ambient mass spectra, since all substances in the sample surface that can be desorbed/ionized will be present in the ambient mass spectrum. In addition, the direct analysis of raw samples without previous sample treatment and chromatographic separation would provide more signals to the background mass spectrum. For these reasons, obtaining reliable information from AMS techniques relies on the capabilities of the mass spectrometric system. Initially, ambient ionization techniques were coupled to low-resolution mass spectrometers (quadrupoles and ion traps), mainly for the target analysis of a small number of pesticides and to prove the applicability of these techniques. Soon tandem MS was used to increase selectivity, to improve confirmation, and to reduce background noise. For instance, most of the pesticide applications developed for electrospray-based methods, LTP and

FAPA, used ion traps, taking advantage of the higher sensitivity of this mass analyzer in front quadrupoles and because of its capability to perform multistage mass spectrometry experiments (MS^n) providing structural information for confirmatory purposes (Table 9.2). The next step is to couple these techniques to high-resolution mass analyzers (TOF and Orbitrap) that are frequently combined with AMS in other applications such as environmental, doping, food characterization, and forensic (Seró, 2016). However, in the pesticide field they have only been used with DART (Table 9.3). It must be taken into account that sometimes the mass resolution of TOF instruments is not enough when dealing with complex food samples, thus making necessary the use of ultrahigh mass resolution instruments such as Orbitrap. For instance Cajka et al. (2011) analyzed thiram in extracts of fruit samples by DART using both TOF and Orbitrap and they concluded that the mass resolution of the TOF instrument was inefficient to separate thiram from matrix coextractants limiting the detection of this pesticide at low concentration levels. A quadrupole-Orbitrap has been used by Edison, Lin, Gamble, et al. (2011) and Edison, Lin, and Parrales (2011) to screen a wide number of multiclass pesticides (up to 240) using DART. The ultrahigh-resolution MS provided reliable isotope patterns to solve some isobaric interferences, accurate mass measurements allowed the determination of elemental compositions narrowing the number of possible pesticide candidates, and the ultrahigh-resolution product ion scan mass spectra offered structural information necessary to confirm the presence of the compounds in samples and to help in nontarget pesticide searching.

9.3.4 SAMPLE HANDLING FOR AMBIENT MASS SPECTROMETRY-BASED METHODOLOGIES

As mentioned earlier, the initial goal of AMS techniques is the direct analysis on a sample with no sample preparation, but in fact, sample processing often takes place during the extraction/desorption/ionization process. The real-time sample preparation leads to a simplified workflow and easier operation that reduce the potential for interface contamination and sample carryover, since samples remain outside of the analytical system. Nevertheless, for accurate results, frequently an additional but little manipulation/preparation of samples is needed. For instance, the analysis of samples with low matrix content such as agrochemical products can be performed directly after dissolving the sample (Wang et al., 2012). However, if quantitative analysis is required for samples with a higher matrix content (pesticides in foodstuffs and plants), some sample treatment is needed to obtain clean extracts before the AMS analysis (Albert et al., 2014; Domin & Cody, 2015).

The most common implementation of AMS techniques for pesticide analysis is the continuous analysis of solid surfaces, such as plant materials, fruit peels, etc. Pesticides that are applied by spraying or dusting onto fruits and vegetables remain on the peel; therefore direct identification and confirmation of these agrochemicals by AMS techniques allows fast screening by MS without any prior sample treatment, using the fruit skin itself as substrate. For instance, imazalil, thiabendazole, DEET

(*N*,*N*-diethyl-*m*-toluamide), and diphenylamine, which are commonly used in agricultural practices, have been directly detected on fruit peels and vegetables with DESI-MS (Garcia-Reyes et al., 2009; Mulligan et al., 2006), DART-MS (Farré et al., 2013), and LTP-MS (Soparawalla et al., 2011; Wiley et al., 2010). To analyze pesticides in fruits by DESI-MS, pieces ($\sim 1\,cm^2$) of peel fruits secured using double-sided tape on a microscope glass slide have been used (Garcia-Reyes et al., 2009). Regarding plasma-based techniques, the direct detection of pesticides in fruit peels can be limited by the thermal stability of the sample surface since high temperatures of the gas stream up to 300—400°C are often required for the thermal desorption of some pesticides. However, since sample surfaces are exposed to the ionizing gas for a relatively short period (scan rate: 0.2 mm/s) no degradation is currently observed even working at 400°C (Farré et al., 2013). Since LTP works at lower temperatures no problems related to the thermal stability of the fruit surface have been observed (Soparawalla et al., 2011; Wiley et al., 2010). For powder samples, some strategies must be considered to prevent the blowing away of the sample from the surface by the gas stream. For instance, Schurek et al. (2008) placed millet wheat grains into a handmade envelope (filter paper) and exposed it to the DART gas stream. The same authors commented that direct analysis of milled grains by DESI was feasible but in practice it is not recommended since the grains are easily blown away from the surface by both the DESI solvent and the gas. Furthermore, powder and dust samples should be handled with particular care since the gas stream might also blow the sample into the laboratory and MS environment and cause contamination (Nielen et al., 2011).

One problem found in the direct analysis of pesticides is to have a representative sample since pesticide contamination is highly variable from unit to unit, due in part to the application methods (spraying or dusting across the crop) as well as to environmental factors such as handling issues or plant growth, among others. So, multiple units must be composited to prepare a sample from each batch in standard analysis protocols. One possibility is compositing using surface-sampling techniques but this procedure presents unique challenges in AMS. In this context, several authors (Crawford & Musselman, 2012; Edison, Lin, Gamble, et al., 2011; Edison, Lin, & Parrales, 2011) used a strategy to screening a wide range of pesticides by a DART-based method that consisted in preparing composites by swabbing multiple pieces of product using one piece of foam. The fruit surface is first misted lightly by spraying a solvent blend and then a foam disk is swabbed across the moistened surface until all of the solvent is absorbed (Fig. 9.7). Finally, the swab is held in the DART heated gas stream for at least 20 s. However, the topography of the commodities is a critical factor when sampling from the surface. For instance, the smooth surfaces of apples or cherry tomatoes are considered easy sampling surfaces since there is little abrasion between the sample surface and the swab material, whereas other less smooth commodities such as peaches and kiwis pose some challenges to the swabbing, resulting in the destruction of the foam. For reproducible results, the use of a polyurethane foam and a protocol that includes dabbing and shaving the haired surface with a razor blade (especially when swabbing kiwis) have been

FIGURE 9.7

Swabbing technique employed to sample pesticides directly from fruit and vegetable surfaces: (A) commercial cotton swab, (B) cotton swab introduction to the DART-SVP source, (C) Alpha polyester cleaning swab, and (D) Alpha swab DART analysis.

Reprinted from Crawford, E., & Musselman, B. (2012). Evaluating a direct swabbing method for screening pesticides on fruit and vegetable surfaces using Direct Analysis in Real Time (DART) coupled to an Exactive benchtop orbitrap mass spectrometer. Analytical and Bioanalytical Chemistry, 403 *(10), 2807–2812 with permission from Springer, Copyright 2014.*

proposed (Edison, Lin, Gamble, et al., 2011; Edison, Lin, & Parrales, 2011). In addition, other nonsmooth surface products such as carrots tend to readily absorb the spraying solvent blend pulling the analytes beneath the surface, thus hindering the detection of pesticides at low concentration levels (Crawford & Musselman, 2012). A similar strategy has been applied for the analysis of thiabendazole in oranges by paper spray (Soparawalla et al., 2011) where a filter paper wetted with methanol was used to wipe over the orange surface to sample any soluble chemical residue and was used as support for PSI.

The AMS analysis of liquid samples such as raw urine, fruit juices, or even liquid extracts requires a different strategy. Chen et al. (2007) used EESI not only to analyze atrazine in raw urine by taking advantage of the double-separate sprayers, one to nebulize the sample solution and the other to produce charged microdroplets of solvent by electrospray, but also to extract atrazine from the bulk solution (raw urine) by liquid—liquid extraction/ionization during the collision of microdroplets. The method allowed direct analysis of liquid samples by infusing them in the microcapillary. Another strategy used with both liquid samples or sample extracts is to wet filter paper strips or to dip glass rods into the liquid samples and place them between the gas stream and the MS inlet. For instance, the screening of pesticides in fruit juices at micrograms per liter concentrations by APGD-MS has been performed by wetting filter paper strips (Jecklin et al., 2008). However, the most popular way to perform the detection of pesticides in fruits and vegetables is by DART taking advantage of the commercially available automatic sampling system (12-Dip-It tip scanner autosampler) able to scan glass-rod surfaces, previously dipped into the liquid sample, through the DART gas stream at a constant speed, thus improving both the reproducibility of the results and the throughput of the analysis (Cajka et al., 2011; Farré et al., 2013; Schurek et al., 2008).

There are two main reasons for applying a more extensive sample treatment before AMS analysis. The first is the high LODs that are often associated with the AMS techniques, which force the application of preconcentration steps to detect pesticides at LODs low enough for real and convenient applications. One example of this treatment is the use of an SPE membrane to extract and preconcentrate pesticides from groundwater (Mulligan et al., 2007). In this case the membrane was directly analyzed by DESI-MS, thus improving LODs. The other reason for requiring a more extensive sample treatment is matrix effect, which might hinder sensitivity and analyte identification. A simple cleanup strategy such as QuEChERS has been applied to reduce coextractant matrix components (Albert et al., 2014; Cajka et al., 2011; Garcia-Reyes et al., 2009) when analyzing fruits and vegetables. For instance, QuEChERS was used for the analysis of ziram and thiram in fruits by both DART-MS and DESI-MS (Cajka et al., 2011). Although matrix components were significantly reduced, some interfering compounds remained in the final extract making difficult the identification of thiram. QuEChERS has also been used to deal with ion suppression problems, which are common in electrospray ionization, since analytes and matrix components compete for desorption/ionization. As an example, Garcia-Reyes et al. (2009) used QuEChERS and then diluted the extracts 1:3 to reduce matrix components when analyzing fruits and vegetables by DESI-MS. However, the authors observed ion suppression in the final extracts and reported LODs 3—15 times higher than the standards depending upon the complexity of the matrix and the MS/MS transition selected. QuEChERS has also been applied to deal with matrix effects when using plasma-based techniques. For instance, important signal variations due to ion suppression have been reported when analyzing dithiocarbamate fungicides by DART-MS (Cajka et al., 2011) and a thorough study of how components of fruit matrices affect the signal intensity of insecticides and fungicides when analyzing them by LTP-MS has also been published (Albert et al., 2014). In this last case, a cleanup step was needed to reduce the amount of plant pigments. However, ion suppression (cyprodinil) and ion enhancement (acetamiprid) were observed, although some compounds such as fluodioxonil did not exhibit matrix effects. Finally, to overcome matrix effects and to improve LODs, Kiguchi et al. (2014) coupled HPTLC to DART-TOFMS. Nevertheless, the method proposed is highly complex since it used accelerated solvent extraction, a precleanup step (C18 cartridges) and a separation by HPTLC with silica gel.

9.3.5 AMBIENT MASS SPECTROMETRY METHOD PERFORMANCE

One of the most frequently emphasized characteristic of AMS techniques is their high throughput. In general, the direct desorption/ionization step is rapid, typically 10 samples in few minutes. For instance, Farré et al. (2013) indicated that an automated DART analysis of 10 samples takes ca. 5 min, which means 30 s per sample. An easy way to take advantage of this fact is to perform the analysis directly onto the surface of the samples without any sample treatment. However, as commented earlier, sometimes an additional sample treatment is needed but even in these cases, a rapid analysis can be performed. For instance, Crawford and Musselman (2012), in the evaluation of a direct swabbing method for the screening of pesticides in fruits, indicated that the analysis of one sample takes less than 1 min, 30 s for swabbing and 20 s for the DART-MS automatic swab analysis.

AMS techniques have some limitations and one of the most important ones is the high fluctuation of signal intensities within repeated measurements of a given sample. For instance, standard deviations up to 60% were found by Edison, Lin, Gamble, et al. (2011) when measuring the normalized level of the protonated molecular ion of several compounds (132 pesticides) in the screening of pesticides using surface swabbing. This relatively poor repeatability can be acceptable for qualitative screening since ion ratios among ions (isotope ions or fragment ions) are maintained; however, this problem is very important in quantitative applications and in this case, the normalization of the responses by using an internal standard is mandatory. This procedure is currently employed by most of the authors, independently of the technique used. It must also be mentioned that sample positioning also affects reproducibility in AMS techniques; for instance, Jecklin et al. (2008) in the optimization of a FAPA source emphasized the need for improving the precision of positioning the samples and Mandal et al. (2013) reported that variations in the sampling depth from spot to spot when using an SF-PESI source deteriorate reproducibility. However, the increasing use of better automated systems as happens with DART is reducing this problem.

In spite of the large amount of matrix material in the ion source and the lack of reproducibility in sampling, the LODs for most of the compounds and techniques ranging from 1 to 90 pg on surface (1–100 µg/kg) are generally below the MRL regulatory levels and good enough for screening purposes. So screening is the main application of AMS techniques in pesticide analysis. These LODs are available in very short times and this characteristic is very useful for field-oriented pesticide testing. Some portable instruments coupled to a DESI source have been developed and used for in situ analysis of pesticides (Garcia-Reyes et al., 2009; Soparawalla et al., 2011), indicating that this application can be a reality in the near future. The coupling of ambient techniques, mainly DART, to HRMS has allowed demonstrating the high capabilities of these techniques for screening. Both TOF and Orbitrap analyzers have been used. Schurek et al. (2008), in one of the first published papers using DART coupled to a TOF for the screening of pesticides, indicated that the mass resolution of the studied compounds (5000–5200 FWHM) was not enough for identification following confirmation requirements. Nowadays, mass analyzers with higher mass resolution have been used, for instance, an Orbitrap working at 60,000 FWHM (Farré et al., 2013) or even at 100,000 FWHM (Edison, Lin, & Parrales, 2011). Working at this last resolution, Edison, Lin, Gamble, et al. (2011) were able to base-line separate a pair of isobaric compounds with differences in theoretical masses of the protonated molecular ions of 25 ppm, such as formothion and barban (Fig. 9.8A), and partially resolve a pair with a 10-ppm mass difference (fenarimol and malathion) (Fig. 9.8B). In addition, the use of instruments that permitted working in MS/MS at high resolution allows increasing identification guarantees in screening. For instance, Kern et al. used a Q-Orbitrap instrument working at a mass resolution of 140,000 FWHM at full scan and 35,000 FWHM in MS/MS for the screening of target pesticides in fruits and vegetables and the method was demonstrated to be able to identify a large number of pesticides (Kern, Lin, & Fricke, 2014). Moreover, several compounds not spiked into the fruits were also identified, showing the potential of the procedure for the analysis of suspected and nontarget compounds.

Quantification of agrochemicals in AMS in complex matrices is not straightforward. Quantitation difficulties are endemic to all AMS sources and arise from a number of factors, including desorption and ionization matrix effects, inconsistences in sample positioning, and variances in analyte transport from the sample surface into the mass spectrometer. Consequently, direct quantification of pesticides on the surface remains one of the most significant issues in AMS. Some authors propose the direct analysis of fruit peels mainly for postharvest agrochemicals since for pesticides that are applied during the growth phase it is complicated because they tend to diffuse from the peel to the flesh of the plant. Farré et al. (2013) indicated that the direct quantitation of fruit peels is possible since they found linear calibration curves and acceptable reproducibility. However, there are several problems related to the direct surface quantitation; one of them is to prepare a representative sample of each batch and the other one is to guarantee that the internal standard is uniformly spiked on the sample, which often is difficult to achieve. The possibility of performing direct analysis of pesticides on fruit peel has also been demonstrated by other authors such as Jecklin et al. (2008), who used APGD-MS for this purpose, achieving LODs on apple skin in the nanogram range (µg/mg) or by Mandal et al. (2013) who used an SF-PESI source to study the distribution of pesticides in different parts of plants. However, to improve quantitation in AMS, sample

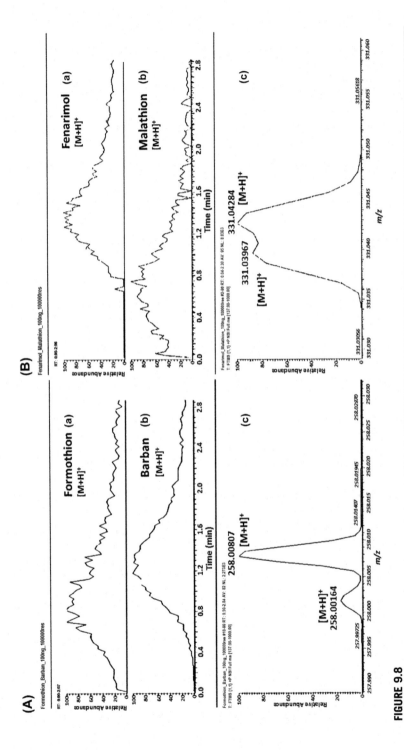

FIGURE 9.8

(A) Ion traces of (a) [M+H]$^+$ ion of formothion (exact mass: 258.00234 m/z); (b) [M+H]$^+$ ion of barban (exact mass: 258.00884 m/z); and (c) mass spectrum of fully resolved ions. (B) Ion traces of (a) [M+H]$^+$ ion of fenarimol (exact mass: 331.04047 m/z); (b) [M+H]$^+$ ion of malathion (exact mass: 331.04387 m/z); and (c) mass spectrum of partially resolved ions.

Reprinted from Edison, S. E., Lin, L. A., & Parrales, L. (2011). Practical considerations for the rapid screening for pesticides using ambient pressure desorption ionisation with high-resolution mass spectrometry. Food Additives & Contaminants: Part A, 28 (10), 1393–1404 with permission from John Wiley and Sons, Copyright 2010.

treatment including an extraction from the homogenized samples with different organic solvents is generally used (Tables 9.2 and 9.3) and in some cases, a cleanup step such as QuEChERS to circumvent matrix effects is also added for the quantitative analysis of pesticides by DESI, DART (Cajka et al., 2011; Garcia-Reyes et al., 2009), and LTP (Albert et al., 2014). For the quantitative analysis of fatty foods and precooked foods a thorough cleanup using C_{18} and the coupling of TLC (silica gel) to DART have been proposed (Kiguchi et al., 2014). In AMS techniques, to provide quantitative reliable results, internal standardization with isotopically labeled standards when available or matrix-matched calibration is used. However, it must be mentioned that with AMS techniques in most of the cases, only semiquantitative results can be obtained. In this context, the primary application of these techniques related to MRL legislation is the rapid screening of the samples to select those of potential interest from the majority of negative samples that would be analyzed by well-established quantitative methods.

9.3.6 AMBIENT MASS SPECTROMETRY APPLICATIONS

Most of the papers published in the literature related to AMS in pesticide analysis are addressed to evaluate the performance of new source designs and also to demonstrate the applicability of the techniques for both identification and quantitation. In these papers, pesticides are used as model compounds probably to take advantage of the availability of standards, the existence of well-established official methods for comparison, and the need for developing fast methodologies for the control of these compounds since strict regulations have been established. It must be indicated that most of the authors used spiked samples and only a few real samples are analyzed. In this section, a revision of these applications is included.

The first papers published in the mid-2000s by Mulligan et al. (2006, 2007) are examples of proof-of-concept approaches since they are oriented to demonstrate the capability of DESI to detect pesticides deposited onto a paper or onto a solid-phase extraction (SPE) membrane; in this last case, contaminated groundwater was analyzed. There are other examples of this kind of applications such as the use of a home-built APGD source for the analysis of pesticides directly from the surface of fruits and vegetables demonstrating that LODs down to the low microgram per kilogram range can be obtained (Jecklin et al., 2008) or the development of a new FAPA source configuration with a pin-to-capillary geometry that is able to analyze bulk solutions and allows improvement of the performance of FAPA sources reducing LODs (Shelley et al., 2011). Additionally, papers focused on the implementation of ambient sources to portable mass spectrometers such as the home-built reduced-size LTP source and the PSI sources can also fall in this category (Soparawalla et al., 2011). The PSI device has been used for the in situ analysis of postharvest agrochemicals.

One of the key characteristics of AMS techniques is their ability to perform direct analysis on native surfaces without sample preparation, which can be very useful for industries handling large quantities of fruits and in agricultural production where it is necessary to have available rapid in situ pesticide methods. So there are several papers orientated to evaluate the possibility of identifying and even quantifying pesticides directly on the surface of fruits and vegetables. For instance, Wiley et al. (2010) evaluated the potential of an LTP-MS for the rapid screening of a fungicide (imazalil) that was applied to fruits and vegetables at the final stage of their production and found that the results obtained by directly analyzing fruit peels with LODs below MRL indicated that LTP-MS can be considered a useful tool. Peels have also been used by other authors for the screening of postharvest fungicides by DART (Farré et al., 2013) and for the direct screening of banned/nonauthorized pesticides by DESI (Garcia-Reyes et al., 2009). An interesting example of the direct analysis of native surfaces can be found in the work of Mandal et al. (2013), where real-time pesticide analysis of living plants by SF-PESI-MS is performed demonstrating a nonhomogeneous distribution of pesticides (organophosphorous) along the plant. However, sometimes the characteristics of the sample make difficult the direct analysis of the surface and in these cases, other strategies must be applied. For instance, Schurek et al. (2008) in one of the first papers evaluating the potential of DART for the analysis of pesticide residues in incurred wheat had to use an envelope to contain the milled wheat grains that were placed directly into the DART gas stream. Other authors, with the prime objective of acquiring quantitative information applied an extraction step. For instance, QuEChERS extraction has been proposed for the analysis of four pesticides that are reported to often exceed legal maximum levels in Germany in products bought in a local grocery store demonstrating that this extraction method combined with LTP-MS is suitable for MRL verification (Albert et al., 2014).

It must be mentioned that, in general, the number of pesticides analyzed in most of the published papers is quite low, from 2 to 16, even if the method is proposed for screening. There are only two works published by Edison, Lin, Gamble, et al. (2011) and Edison, Lin, and Parrales (2011) where a large number of compounds, up to 240, have been included. These authors studied the potential of DART coupled to HRMS (Orbitrap) for the direct screening of pesticides, and several mixtures containing a large range of chemical classes and chemical properties (carbamates, phenylamides, phenylureas, conazoles, benzimidazoles, pyretroids, triazines, organophosphorous, and organochlorine pesticides) were tested to provide a broad idea of the performance of the technique. The authors used spiked samples to evaluate the methodology and propose using foam swabs to collect the pesticides from the skins of fruits and vegetables. To maximize throughput, a composite sample was used and for this purpose swabbing three pieces of product per piece of foam was proposed. The method was able to identify most of the tested compounds of each class, spinosyns being the only exception probably due to their high molecular weight. A custom-made database of pesticides based on the accurate mass of the protonated or deprotonated molecular pesticide ions was used for identification. The results indicate that the method is applicable to a wide range of commodities. Identification capabilities of this method were improved by Kern et al. (2014) using a Q-Orbitrap mass spectrometer, an instrument that can perform data-dependent fragmentation. The authors used a custom-made database with 30 compounds that included accurate masses of precursor molecular ions (two isotopes) and three product ions and were able to identify from 80% to 100% of the compounds spiked in three different matrices (apples, oranges, and broccoli). To generalize this approach, libraries with the accurate masses of precursor molecular ions and product ions of a large number of pesticides must be available. Custom-made databases can be built by analyzing pesticide standards (target compounds) and by including the exact masses of molecular ions and expected product ions obtained from the literature for nonavailable standards (suspected compounds). The identification of nontarget compounds (new pesticides not included in libraries) will require the search in databases of a chemical compound that fits with the formulae obtained from elemental composition, deduced from the accurate mass and isotope pattern, and the use of in silico fragmentation approaches.

9.4 FINAL REMARKS AND FUTURE TRENDS

Method simplicity and time saving, which are directly related to the elimination of chromatographic separation, to the reduction of sample processing, and to the capability of obtaining information from the MS spectrum of the entire sample, are the most important common characteristics of the techniques discussed in this chapter. However, differences exist in the degree of simplicity, throughput, and also applicability of these techniques. A comparison of chromatography-MS, FIA-MS, and AMS methods with respect to analytical figures of merit and analysis throughput can be found in a review about the use of FIA-MS for quantitative analysis (Nanita & Kaldon, 2016). High throughput increases from chromatography-MS to AMS techniques but this improving is counterbalanced by the worsening of analytical figures of merit (selectivity, sensitivity, accuracy, and precision). In this context, FIA-MS occupies a halfway position between these two techniques.

Regarding simplicity, the possibility of performing direct surface analysis with AMS techniques is an unquestionable advantage compared with FIA-MS, which requires the extraction and dilution of the sample. However, in FIA, injection reproducibility is high in contrast to the difficulties that present in most of the AMS techniques, which despite their ability of rapid analysis are prone to large changes in sensitivity because of small changes in sample position and orientation. There is still a need for suitable sample carriers and automatic moving platforms to improve high throughput, although new devices that allow improving automation mainly for DART are available. Interfacing of FIA-MS and AMS to robotics and microfluidics is an interesting research area that could enable automation and integration of these approaches into platforms able to respond to the demands of modern analytical laboratories.

Quantitation capability is of utmost importance in pesticide residue analysis since strict regulations must be followed. In this context, the direct analysis with AMS techniques, without any sample preparation, presents some shortcomings since the method provides only qualitative or semiquantitative information. To improve quantitation, strategies to prepare composite representative samples must be applied and frequently an additional sample preparation step that usually requires the extraction of the analytes from the sample is performed. As a result, the sample treatments used in both FIA and AMS based on a fast extraction are quite similar. Matrix effect is always an important problem in quantitation with both FIA-MS and AMS techniques since the lack of a separation step results in a highly pronounced signal suppression/enhancement. Among the strategies proposed to reduce this effect the dilution of the extracts is a fast and convenient option but matrix-matched or isotopic dilution quantitation methods must always be used to obtain reliable results. In this context, it is interesting to mention that an approach using TLC and semiautomated devices has been proposed as cleanup to prevent matrix effects in both FIA and DART, indicating that the cleanness of the extracts remains to be an unsolved problem for the fast quantitative analysis of complex samples such as foods. Techniques that incorporate sample cleanup into the ionization process such as paper spray could be a useful approach to solve this problem. As regards accuracy and precision, FIA-MS generally provides better results than AMS-based techniques where the use of internal standards to normalize the responses is mandatory. Moreover, to implement these techniques for pesticide analysis, carefully designed quality assurance and quality control (QA/QC) procedures are needed to guarantee high levels of confidence in the process of identification and quantification.

Specificity is another key aspect in both FIA-MS and AMS techniques that traditionally has been improved by using tandem MS and more recently using HRMS. Surprisingly, HRMS analyzers have been scarcely used in FIA and also in DESI in contrast to what happens in DART where they are the most popular mass analyzers. The capability of obtaining full-scan spectra and high-quality accurate mass data with HRMS (TOF and Orbitrap analyzers) combined with databases has permitted rapid screening of target, suspect, and nontarget pesticides by DART-MS. The increasing use of the coupling of HRMS and of hybrid HRMS instruments to the other techniques studied in this chapter (mainly FIA and DESI) is expected to expand their applicability in pesticide analysis. An additional option to increase selectivity that has not been used till now in pesticide analysis is ion mobility spectrometry (May & McLean, 2015), a technique that has suffered a significant growth in recent years and that has also been coupled to DESI (Weston et al., 2005). The ion mobility drift times are not affected by the matrix and can be used to calculate collision cross section (CCS) values that can be included in the library, helping the identification of isomeric and isobaric compounds. The potential of this additional orthogonal dimension in the screening of pesticides in complex samples has not been evaluated yet, but it has a great potential for pesticide screening.

As regards applicability, AMS is the technique of choice for the high-throughput screening of pesticides directly on the surface of food products, whereas FIA-MS seems to be convenient for quantitation purposes. Finally, an interesting area of research and development is the miniaturization and implementation of AMS techniques in portable instruments since it would allow performing in situ applications directly in the field. For this application, well-established analytical methods specified for particular analytes must be developed and validated.

ACKNOWLEDGMENTS

The authors gratefully acknowledge the financial support received from Spanish Ministry of Economy and Competitiveness under the project CTQ2015-63968-C2-1-P and from the Agency for Administration of University and Research Grants (Generalitat de Catalunya, Spain) under the project 2014SGR-539.

REFERENCES

Albert, A., et al. (2014). Rapid and quantitative analysis of pesticides in fruits by QuEChERS pretreatment and low-temperature plasma desorption/ionization orbitrap mass spectrometry. *Analytical Methods, 6*(15), 5463–5471.

Alechaga, É., Moyano, E., & Galceran, M. T. (2015). Wide-range screening of psychoactive substances by FIA-HRMS: Identification strategies. *Analytical and Bioanalytical Chemistry, 407*(16), 4567–4580.

Andrade, F. J., et al. (2008). Atmospheric pressure chemical ionization source. 1. Ionization of compounds in the gas phase. *Analytical Chemistry, 80*(8), 2646–2653.

Andreozzi, R., et al. (1999). The thermal decomposition of dimethoate. *Journal of Hazardous Materials, 64*(3), 283–294.

Botitsi, H. V., et al. (2011). Current mass spectrometry strategies for the analysis of pesticides and their metabolites in food and water matrices. *Mass Spectrometry Reviews, 30*, 907–939.

Cajka, T., et al. (2011). Direct analysis of dithiocarbamate fungicides in fruit by ambient mass spectrometry. *Food Additives & Contaminants: Part A, 28*(10), 1372–1382.

Chen, H., et al. (2007). Differentiation of maturity and quality of fruit using noninvasive extractive electrospray ionization quadrupole time-of-flight mass spectrometry. *Analytical Chemistry, 79*(4), 1447–1455.

Chen, H., Venter, A., & Cooks, R. G. (2006). Extractive electrospray ionization for direct analysis of undiluted urine, milk and other complex mixtures without sample preparation. *Chemical Communications (Cambridge, England)*, (19), 2042–2044.

Cody, R. B., Laramée, J. A., & Durst, H. D. (2005). Versatile new ion source for the analysis of materials in open air under ambient conditions. *Analytical Chemistry, 77*(8), 2297–2302.

Crawford, E., & Musselman, B. (2012). Evaluating a direct swabbing method for screening pesticides on fruit and vegetable surfaces using Direct Analysis in Real Time (DART) coupled to an Exactive benchtop orbitrap mass spectrometer. *Analytical and Bioanalytical Chemistry, 403*(10), 2807–2812.

Domin, M., & Cody, R. B. (2015). *Ambient ionization mass spectrometry*. Cambridge, UK: Royal Society of Chemistry, ISBN 978-1-84973-926-9.

Douglass, K. A., et al. (2012). Deconstructing desorption electrospray ionization: Independent optimization of desorption and ionization by spray desorption collection. *Journal of the American Society for Mass Spectrometry, 23*(11), 1896–1902.

Edison, S. E., Lin, L. A., Gamble, B. M., et al. (2011). Surface swabbing technique for the rapid screening for pesticides using ambient pressure desorption ionization with high-resolution mass spectrometry. *Rapid Communications in Mass Spectrometry, 25*(1), 127−139.

Edison, S. E., Lin, L. A., & Parrales, L. (2011). Practical considerations for the rapid screening for pesticides using ambient pressure desorption ionisation with high-resolution mass spectrometry. *Food Additives & Contaminants: Part A, 28*(10), 1393−1404.

EU Legislation on MRLs. (2016). *European Commision: Food safety and plants, pesticides.* Available at http://ec.europa.eu/food/plant/pesticides/max_residue_ levels/eu_rules/index_en.htm.

EU-SANTE. (2015). *Guidance document on analytical quality control and method validation procedures for pesticide residues analysis in food and feed.* /11945/2015. European Commission, SANTE.

Farré, M., Picó, Y., & Barceló, D. (2013). Direct peel monitoring of xenobiotics in fruit by direct analysis in real time coupled to a linear quadrupole ion trap-orbitrap mass spectrometer. *Analytical Chemistry, 85*(5), 2638−2644.

García-López, M., et al. (2014). Evaluation and validation of an accurate mass screening method for the analysis of pesticides in fruits and vegetables using liquid chromatography-quadrupole-time of flight-mass spectrometry with automated detection. *Journal of Chromatography A, 1373*, 40−50.

Garcia-Reyes, J. F., et al. (2009). Desorption electrospray ionization mass spectrometry for trace analysis of agrochemicals in food. *Analytical Chemistry, 81*(2), 820−829.

Geerdink, R. B., Kienhuis, P. G., & Brinkman, U. A. T. (1993). Fast screening method for eight phenoxyacid herbicides and bentazone in water Optimization procedures for flow-injection analysis-thermospray tandem mass spectrometry. *Journal of Chromatography A, 647*, 329−339.

Gómez-Ramos, M. M., et al. (2013). Liquid chromatography-high-resolution mass spectrometry for pesticide residue analysis in fruit and vegetables: Screening and quantitative studies. *Journal of Chromatography A, 1287*, 24−37.

Goto, T., et al. (2003). Simple and rapid determination of *N*-methylcarbamate pesticides in citrus fruits by electrospray ionization tandem mass spectrometry. *Analytica Chimica Acta, 487*(2), 201−209.

Hagan, N. A., et al. (2008). Detection and identification of immobilized low-volatility organophosphates by desorption ionization mass spectrometry. *International Journal of Mass Spectrometry, 278*(2−3), 158−165.

Harper, J. D., et al. (2008). Low-temperature plasma probe for ambient desorption ionization. *Analytical Chemistry, 80*(23), 9097−9104.

Hiraoka, K., et al. (2007). Development of probe electrospray using a solid needle. *Rapid Communications in Mass Spectrometry, 21*(18), 3139−3144.

Huang, M. Z., et al. (2011). Ambient ionization mass spectrometry: A tutorial. *Analytica Chimica Acta, 702*, 1−15.

Ifa, R. D., et al. (2008). Quantitative analysis of small molecules by desorption electrospray ionization mass spectroemtry from polytetrafluoroethylene surfaces. *Rapid Communications in Mass Spectrometry, 22*(4), 503−510.

Ito, Y., et al. (2003). Simple and rapid determination of thiabendazole, imazalil, and *o*-phenylphenol in citrus fruit using flow-injection electrospray ionization tandem mass spectrometry. *Journal of Agricultural and Food Chemistry, 51*(4), 861−866.

Jackson, P., & Attalla, M. I. (2010). Fast analysis of high-energy compounds and agricultural chemicals in water with desorption electrospray ionization mass spectrometry. *Rapid Communications in Mass Spectrometry, 24*(24), 3567–3577.

Jecklin, M. C., et al. (2008). Atmospheric pressure glow discharge desorption mass spectrometry for rapid screening of pesticides in food. *Rapid Communications in Mass Spectrometry, 22*(18), 2791–2798.

John, H., et al. (2010). Simultaneous quantification of the organophosphorus pesticides dimethoate and omethoate in porcine plasma and urine by LC-ESI-MS/MS and flow-injection-ESI-MS/MS. *Journal of Chromatography B: Analytical Technologies in the Biomedical and Life Sciences, 878*(17–18), 1234–1245.

Kaufmann, A. (2012). The current role of high-resolution mass spectrometry in food analysis. *Analytical and Bioanalytical Chemistry, 403*(5), 1233–1249.

Kauppila, T. J., & Vaikkinen, A. (2014). Chapter 7: Ambient mass spectrometry: Food and environmental applications. In O. Núñez, H. Gallart-Ayala, C. P. B. Martins, & P. Lucci (Eds.), *Fast liquid chromatography–ambient mass spectrometry: Food and environmental applications* (pp. 271–323). Imperial College Press, ISBN 978-1-78326-493-3.

Kern, S. E., Lin, L. A., & Fricke, F. L. (2014). Accurate mass fragment library for rapid analysis of pesticides on produce using ambient pressure desorption ionization with high-resolution mass spectrometry. *Journal of the American Society for Mass Spectrometry, 25*(8), 1482–1488.

Kiguchi, O., et al. (2014). Thin-layer chromatography/direct analysis in real time time-of-flight mass spectrometry and isotope dilution to analyze organophosphorus insecticides in fatty foods. *Journal of Chromatography A, 1370*, 246–254.

Mandal, M. K., et al. (2013). Development of sheath-flow probe electrospray ionization mass spectrometry and its application to real time pesticide analysis. *Journal of Agricultural and Food Chemistry, 61*(33), 7889–7895.

May, J. C., & McLean, J. A. (2015). Ion mobility-mass spectrometry: Time-dispersive instrumentation. *Analytical Chemistry, 87*(3), 1422–1436.

Mol, H. G. J., & van Dam, R. C. J. (2014). Rapid detection of pesticides not amenable to multi-residue methods by flow injection-tandem mass spectrometry. *Analytical and Bioanalytical Chemistry, 406*(27), 6817–6825.

Mol, H. G. J., et al. (2015). Identification in residue analysis based on liquid chromatography with tandem mass spectrometry: Experimental evidence to update performance criteria. *Analytica Chimica Acta, 873*, 1–13.

Mulligan, C. C., et al. (2007). Fast analysis of high-energy compounds and agricultural chemicals in water with desorption electrospray ionization mass spectrometry. *Rapid Communications in Mass Spectrometry, 21*(22), 3729–3736.

Mulligan, C. C., Talaty, N., & Cooks, R. G. (2006). Desorption electrospray ionization with a portable mass spectrometer: In situ analysis of ambient surfaces. *Chemical communications (Cambridge, England)*, (16), 1709–1711.

Nakazawa, H., et al. (2004). Rapid and simultaneous analysis of dichlorvos, malathion, carbaryl, and 2,4-dichlorophenoxy acetic acid in citrus fruit by flow-injection ion spray ionization tandem mass spectrometry. *Talanta, 64*(4), 899–905.

Nanita, S. C. (2011). High-throughput chemical residue analysis by fast extraction and dilution flow injection mass spectrometry. *The Analyst, 136*(2), 285–287.

Nanita, S. C. (2013). Quantitative mass spectrometry independence from matrix effects and detector saturation achieved by flow injection analysis with real-time infinite dilution. *Analytical Chemistry, 85*(24), 11866–11875.

Nanita, S. C., & Kaldon, L. G. (2016). Emerging flow injection mass spectrometry methods for high-throughput quantitative analysis. *Analytical and Bioanalytical Chemistry, 408*(1), 23–33.

Nanita, S. C., Kaldon, L. G., & Bailey, D. L. (2015). Ammonium salting out extraction with analyte preconcentration for sub-part per billion quantitative analysis in surface, ground and drinking water by flow injection tandem mass spectrometry. *Analytical Methods, 7*(6), 2300–2312.

Nanita, S. C., & Padivitage, N. L. T. (2013). Ammonium chloride salting out extraction/cleanup for trace-level quantitative analysis in food and biological matrices by flow injection tandem mass spectrometry. *Analytica Chimica Acta, 768*, 1–11.

Nanita, S. C., Pentz, A. M., & Bramble, F. Q. (2009). High-throughput pesticide residue quantitative analysis achieved by tandem mass spectrometry with automated flow injection. *Analytical Chemistry, 81*(8), 3134–3142.

Nanita, S. C., et al. (2011). Fast extraction and dilution flow injection mass spectrometry method for quantitative chemical residue screening in food. *Journal of Agricultural and Food Chemistry, 59*(14), 7557–7568.

Nielen, M. W. F., et al. (2011). Desorption electrospray ionization mass spectrometry in the analysis of chemical food contaminants in food. *TrAC Trends in Analytical Chemistry, 30*(2), 165–180.

Oellig, C., & Schwack, W. (2014). Planar solid phase extraction clean-up and microliter-flow injection analysis-time-of-flight mass spectrometry for multi-residue screening of pesticides in food. *Journal of Chromatography A, 1351*, 1–11.

Ružička, J., & Hansen, E. H. (1975). Analyse d'echantillons de cobalt par spectrometrie gamma directe apres irradiation au moyen de protons de 10 mev. *Analytica Chimica Acta, 78*, 145–157.

Schug, K. A. (December 12, 2013). The LCGC blog: Flow injection analysis can be used to create temporal compositional analyte gradients for mass spectrometry-based quantitative analysis. *LC-GC Europe.*

Schurek, J., et al. (2008). Control of strobilurin fungicides in wheat using direct analysis in real time accurate time-of-flight and desorption electrospray ionization linear ion trap mass spectrometry. *Analytical Chemistry, 80*(24), 9567–9575.

Seró, R., Núñez, Ó., & Moyano, E. (2016). Chapter 3: Ambient ionisation-high-resolution mass spectrometry: Environmental, food, forensic and doping analysis, *Vol. 71. Comprehensive analytical Chemistry, applications of time-of-flight and Orbitrap mass spectrometry in environmental, food, doping, and forensic analysis (comprehensive analytical Chemistry)* (pp. 51–88). Elsevier, ISBN 978-04446-357-23.

Shelley, J. T., Wiley, J. S., & Hieftje, G. M. (2011). Ultrasensitive ambient mass spectrometric analysis with a pin-to-capillary flowing atmospheric-pressure afterglow source. *Analytical Chemistry, 83*(14), 5741–5748.

Soparawalla, S., et al. (2011). In situ analysis of agrochemical residues on fruit using ambient ionization on a handheld mass spectrometer. *The Analyst, 136*(21), 4392.

Takáts, Z., Wiseman, J. M., & Cooks, R. G. (2005). Ambient mass spectrometry using desorption electrospray ionization (DESI): Instrumentation, mechanisms and applications in forensics, chemistry, and biology. *Journal of Mass Spectrometry, 40*(10), 1261–1275.

Takáts, Z., et al. (2004). Mass spectrometry sampling under ambient conditions with desorption electrospray ionization. *Science, 306*(5695), 471–473.

USEPA. (2016). *Pesticides*. US Environmental Protection Agency. Available at https://www.epa.gov/pesticides.

Venter, A., Nefliu, M., & Graham Cooks, R. (2008). Ambient desorption ionization mass spectrometry. *TrAC Trends in Analytical Chemistry, 27*(4), 284–290.

Venter, A. R., et al. (2014). Mechanisms of real-time, proximal sample processing during ambient ionization mass spectrometry. *Analytical Chemistry, 86*(1), 233–249.

Wang, H., et al. (2010). Paper spray for direct analysis of complex mixtures using mass spectrometry. *Angewandte Chemie - International Edition, 49*(5), 877–880.

Wang, L., et al. (2012). Direct analysis in real time mass spectrometry for the rapid identification of four highly hazardous pesticides in agrochemicals. *Rapid Communications in Mass Spectrometry, 26*(16), 1859–1867.

Weston, D. J., et al. (2005). Direct analysis of pharmaceutical drug formulations using ion mobility spectrometry/quadrupole-time-of-flight mass spectrometry combined with desorption electrospray ionization. *Analytical Chemistry (Washington), 77*(23), 7572–7580.

Wiley, J. S., et al. (2010). Screening of agrochemicals in foodstuffs using low-temperature plasma (LTP) ambient ionization mass spectrometry. *The Analyst, 135*(5), 971–979.

Yang, Q., et al. (2012). Paper spray ionization devices for direct, biomedical analysis using mass spectrometry. *International Journal of Mass Spectrometry, 312*, 201–207.

Zomer, P., & Mol, H. G. J. (2015). Simultaneous quantitative determination, identification and qualitative screening of pesticides in fruits and vegetables using LC-Q-Orbitrap™-MS. *Food Additives & Contaminants: Part A, 32*(10), 1628–1636.

Identification of Pesticide Transformation Products in Food Applying High-Resolution Mass Spectrometry

10

Imma Ferrer[1], Jerry A. Zweigenbaum[2], E. Michael Thurman[1]

University of Colorado, Boulder, CO, United States[1]; Agilent Technologies Inc., Wilmington, DE, United States[2]

10.1 INTRODUCTION

Pesticides are usually found in food commodities such as fruits and vegetables. The source of pesticides in food often comes from spraying directly onto crops and/or leaves to combat plagues (insecticides) or to preserve the commodity for transportation (fungicides). But a major part of pesticides (fungicides, herbicides, insecticides) are usually applied to soils to clear any undesired fungus, weeds, or bugs around the plant. It is well known that a plant can uptake these pesticides and assimilate them in their metabolism. However, not many studies of transformation of pesticides in plants have been published to date. Most of these studies refer to mass balances between the applied pesticide and the plant uptake, but only a few show the potential metabolites that the soil or plant can form.

High resolution using time-of-flight mass spectrometry techniques have become popular in the last 15 years, because it gives full-spectrum data at all times. A large number of compounds (virtually no limit) can be analyzed in a single run while obtaining valuable accurate mass information for each compound that ionizes. Furthermore, extra information on metabolites or transformation products can be achieved by exploring the accurate mass spectra of unknown peaks in the chromatograms. Pesticide metabolites have been often identified in environmental water samples using these types of techniques.

In a publication from our group (Ferrer & Thurman, 2015), a summary of accurate mass tools was reviewed for their use in the identification of pesticides and their transformation products. The tools include the use of accurate mass databases, diagnostic ions, generation of molecular features, application of isotope filters, and mass profiling. Therefore, instead of focusing on the generalities of how these tools operate (which were discussed in Ferrer & Thurman, 2015), we will give examples of identification of different metabolites by using these unique features. Three different

Applications in High Resolution Mass Spectrometry. http://dx.doi.org/10.1016/B978-0-12-809464-8.00010-5

pesticides were studied: imidacloprid (insecticide), imazalil, and propiconazole (fungicides). These pesticides were applied to and studied on onions and lettuce to represent a root vegetable and a leafy commodity, respectively.

10.2 EXPERIMENTAL
10.2.1 CHEMICALS AND REAGENTS

Individual pesticide stock solution ($\sim$1000 µg/mL) were prepared in methanol and stored at $-18°$C. From these solutions, working standard solutions were prepared by dilution with acetonitrile and water. High-performance liquid chromatography (HPLC)-grade acetonitrile, methanol, and water were obtained from Burdick and Jackson (Muskegon, MI, USA). Formic and acetic acids were obtained from Sigma-Aldrich (St. Louis, MO, USA). The guanidine metabolite was synthesized by hydrolysis of imidacloprid in 1.0 N HCl for 1 h at 45°C. The yield was low at $\sim$1%, which was sufficient for the tandem mass spectrometry (MS/MS) analysis needed for metabolite identification and for diagnostic ion studies.

10.2.2 GREENHOUSE STUDY

Onions (*Allium cepa* L.) and lettuce (*Lactuca sativa*) were grown in a greenhouse environment for 3 months from March to May 2012. Plants were raised from certi-fied seeds and planted in greenhouse soil in 2-inch pots and grown as seedlings for 1 month. Plants were placed in a well sunlit area of the greenhouse with tempera-tures that varied from 68°F at night to 90°F in the day. Plants were watered daily or as needed for 4 weeks until plants reached a height of $\sim$6 inches. At that time, plants were repotted again in greenhouse soil and raised for 2 subsequent months to full size before the addition of the pesticides. A single application of imidacloprid, imazalil, and propiconazole was applied at 5 mg/kg in the top layer of soil on May 1, 2012 and samples of the plants and soil were taken at three periods (28, 38, and 53 days after application). Some plants did not receive the pesticide application (used as controls). The sampling design included three replicates at each of the three sampling periods for a total of 36 samples for onions and 36 samples for lettuce (control samples of soil and plant were taken from the control pots at the same time). Furthermore, water samples (leachate) were collected once a week for 3 weeks.

10.2.3 PLANT EXTRACTION

The harvested plant was representatively sampled, rinsed, and shaken of soil gently with deionized water and then grounded up with either a blender or by hand with mortar and pestle. In both cases, 6 mL of extractant was added per 3 g of plant matter or 1 g of soil. The extractant consisted of methanol/water (80/20%) with 0.1% acetic acid. The sample was spiked with 100 μL d-5 imidacloprid (100 mg/L in methanol) and rotated for 30 min in a 40 mL glass vial with Teflon cap. The tube was then centrifuged for 15 min at 3500 rpm. The methanol/water extractant was carefully decanted and pipetted leaving all plant debris in the vial. The extractant was evaporated to 0.5 mL of final volume by weight and transferred to vials for analysis by liquid chromatography/quadrupole-time-of-flight mass spectrometry (LC/QTOF-MS).

10.2.4 LIQUID CHROMATOGRAPHY/QUADRUPOLE-TIME-OF-FLIGHT MASS SPECTROMETRY ANALYSIS

The ultrahigh-performance liquid chromatograph consisted of the Agilent model 1290 pump, autosampler, and column compartment (Agilent Technologies, Inc., Santa Clara, CA, USA) with a Zorbax C-8 reverse phase column (4.6 mm × 150 mm, 3.5 μm) having 120,000 theoretical plates per meter (Agilent Technologies, Inc., Santa Clara, CA, USA). The mobile phase began with 90% water/0.1% formic acid (A) and 10% acetonitrile (B), the gradient was held for 5 min, then over 25 min the mobile phase changed linearly to 100% B at a constant rate and held for 1 min before returning to starting conditions. The flow rate was 0.6 mL/min. This is a generic gradient that has been used for many applications of pesticides in food and found to be useful to reach high peak capacities for large pesticide mixtures (Thurman, Ferrer, & Fernandez-Alba, 2005). The ultrahigh-performance liquid chromatography (UHPLC) system was connected to an ultrahigh-definition quadrupole time-of-flight mass spectrometer Model 6540 Agilent (Agilent Technologies, Inc., Santa Clara, CA, USA) equipped with electrospray Jet Stream Technology, operating in positive ion mode, using the following operation parameters—capillary voltage: 3500 V; nebulizer pressure: 45 psig; drying gas: 10 L/min; gas temperature: 250°C; sheath gas flow: 11 L/min; sheath gas temperature: 350°C; nozzle voltage: 0 V in positive ion mode, fragmentor voltage: 190 V; skimmer voltage: 65 V; octopole RF: 750 V. LC/MS accurate mass spectra were recorded across the range 50−1000 m/z at 2 GHz. The data recorded were processed with MassHunter software (Agilent Technologies, Santa Clara, CA, USA). Accurate mass measurements of each peak from the extracted ion (EI) chromatograms were corrected with internal reference masses by means of a reference solution delivered by an external quaternary pump. This solution contains the internal reference masses—purine ($C_5H_4N_4$) at m/z 121.0509 and HP-921 [hexakis-(1H,1H,3H-tetrafluoro-pentoxy)phosphazene] ($C_{18}H_{18}O_6N_3P_3F_{24}$) at m/z 922.0098. The instrument provides a mass resolving power of 35,000 ± 500 (m/z 1522). Stability of mass accuracy was daily checked, and if values went above 2-ppm error, then the instrument was recalibrated. The instrument was operated in both single MS full-spectrum mode and tandem (MS/MS) mode. The isolation width of the quadrupole was set at medium (~ 4 m/z) and collision energies of 10, 20, and 40 eV were used for MS/MS experiments.

10.2.5 MASS PROFILER SOFTWARE

This software was used to detect differences in the control and pesticide-applied samples. The software program operates by extraction of the molecular features of the chromatogram. A molecular feature consists of the monoisotopic mass, its isotopic pattern, and any adducts that may be present, such as sodium, ammonium, or potassium. These features are statistically compared between a control and sample for differences. The molecular features that are different are compiled for the user to examine. Insignificant differences may be removed by a process called recursion, which corrects inconsistencies of the molecular feature finding algorithm.

10.3 IMIDACLOPRID METABOLITES IN PLANTS

Imidacloprid is a neonicotinoid pesticide that was introduced to the market in the late 1990s for the control of homopteran pests, such as aphids, planthoppers, and whiteflies, as well as certain beetles (Kagabu, 1997; Nauen, Tietjen, Wagner, & Elbert, 1998), where it now shares 20% of the insecticide market (Kagabu, 2011). Imidacloprid acts as an agonist of the nicotinoid acetylcholine receptor (Tomizawa & Casida, 2003), which is highly specific to insect receptors. Insecticides acting on the nicotinoid acetylcholine receptor are not common; thus, imidacloprid is a valuable pesticide when combined with others such as organophosphates, carbamates, and pyrethroids. Imidacloprid has been studied on many crops including sugar beets, tomatoes, cotton, sunflowers, spinach, potatoes, and onions (Alsayeda, Pascal-Lorber, Nallanthigal, Debrauwer, & Laurent, 2008; Ford & Casida, 2008; Laurent & Rathahao, 2003; Mandic, Lazic, Okresz, & Gaal, 2005; Nauen, Reckmann, Armborst, Stupp, & Elbert, 1999; Olson, Dively, & Nelson, 2004; Westwood, Bean, Dewar, Bromilow, & Chamberlain, 1998). It is often applied as a systemic insecticide in a drench mode (Nauen et al., 1998), where it is affixed to soil and seed. Imidacloprid has also been implicated in toxicity to beneficial insects, such as that to honeybees (Iwasa, Motoyama, Ambrose, & Roe, 2004). Thus, it is an important insecticide to investigate in plant fate and metabolism studies.

There are a few studies of imidacloprid in a number of plant applications often using C-14 imidacloprid to follow the metabolism of the pesticide (Alsayeda et al., 2008; Laurent & Rathahao, 2003; Nauen et al., 1999; Westwood et al., 1998). This includes studies of sugar beets, tomatoes, sunflowers, and cotton. Ford and Casida (2008) and Casida (2011) used HPLC DAD and HPLC/MS (single quadrupole analysis) to identify plant metabolites of imidacloprid. In this study, we use both UHPLC and high-resolution mass spectrometry for the analysis of imidacloprid and its plant metabolites. In particular, we use UHPLC and LC/QTOF-MS in MS/MS mode. We chose this methodology because of the power of the method to determine accurate mass and the molecular formulas of various degradation products. Furthermore, we can apply MS/MS analysis with accurate mass to identify new or hypothesized metabolites of imidacloprid.

10.3.1 ACCURATE MASS DATABASES

An accurate mass database, built from the published known structures of imidacloprid metabolites, was used to search for potential metabolites. The database was compiled from references Casida (2011) and Ford and Casida (2008). The chemical structure of each metabolite was drawn in ChemDraw and exact calculated masses and formulas were determined for a total of 12 metabolites (Table 10.1). This database was used to examine each of the plant extracts for the parent and metabolites of imidacloprid. These compounds were searched in the total ion chromatogram (TIC) from all onion extracts. Fig. 10.1 shows the TIC for a complex onion extract with the nominal mass of the extracted ion chromatogram (EIC) for imidacloprid (m/z 256) and metabolite-1 (m/z 211) from Table 10.1. The EIC at m/z 256 (Fig. 10.1) had a measured accurate mass of m/z 256.0594, which was within 1 ppm of the exact calculated mass of imidacloprid (Table 10.1), and had the same retention time as the standard of imidacloprid. MS/MS analysis of the ion at m/z 256 confirmed that it was imidacloprid.

Next, the large major extracted peak (m/z 211) in Fig. 10.1 with a retention time of 6.6 min had a measured accurate mass of m/z 211.0743, which is within 1 ppm of the calculated exact mass of the 211.0745 for the guanidine metabolite of imidacloprid (metabolite-1, Table 10.1). To verify the identity of the proposed guanidine metabolite, MS/MS analysis of the m/z 211 was carried out. Fig. 10.2 shows the MS/MS analysis of the m/z 211 at a collision energy of 20 eV, medium energy. The spectrum is rich with major fragment ions at m/z 84.0550, 126.0096, and 175.0965. The putative structures for each of these fragment ions, based on the guanidine structure, are also shown in Fig. 10.2. Finally, the guanidine metabolite was synthesized and analyzed by MS/MS and the same fragments and retention time of the peak at 6.6 min were obtained. The guanidine metabolite has been reported as a nontoxic metabolite of imidacloprid in onions, tomatoes, spinach, potatoes, cotton, and sunflowers (Alsayeda et al., 2008; Ford & Casida, 2008; Laurent & Rathahao, 2003; Nauen et al., 1999; Westwood et al., 1998) for insects; however, mammalian toxicity does exist (Tomizawa & Casida, 2005). In the study of tomato plants (Westwood et al., 1998), the m/z 211 guanidine metabolite was the major plant metabolite found in leaves, which agrees with the results reported here, where the m/z 211 was the major metabolite in all of the nine onion plants extracted and in none of the control extracts.

The exact mass database of search for metabolites showed detections of two other compounds of the 12 shown in Table 10.1. They are 4-hydroxyimidacloprid (metabolite 2) m/z 272.0545 and the urea analogue of imidacloprid (metabolite 3) m/z 212.0585. Metabolite 3 was present in trace quantities while metabolite 2 (4-hydroxyimidacloprid) was verified as the correct putative structure by MS/MS analysis. None of the other nine metabolites reported in Table 10.1 were present in the onion extracts based on accurate mass and EIC chromatograms within a 5-ppm accurate mass window.

Table 10.1 List of Reported Plant Metabolites (Spinach, Cotton, Tomatoes, Sunflowers, and Sugar Beets) of Imidacloprid and Their Calculated Exact Masses of Protonated Molecules (Casida, 2011; Ford & Casida, 2008; Westwood et al., 1998).

Compound Name	Elemental Composition	Retention Time (min)	Calculated Exact Mass [M + H]⁺	Chemical Structure
Imidacloprid 1-((6-Chloropyridin-3-yl) methyl)-4,5-dihydro-N-nitro-1H-imidazol-2-amine	$C_9H_{10}ClN_5O_2$	14.1	256.0596	
Metabolite-1: Guanidine analogue imidacloprid 1-((6-Chloropyridin-3-yl) methyl)-4,5-dihydro-1H-imidazol-2-amine	$C_9H_{11}ClN_4$	6.6	211.0745	
Metabolite-2: 4-Hydroxyimidacloprid 1-((6-Chloropyridin-3-yl) methyl)-4,5-dihydro-2-(nitroamino)-1H-imidazol-4-ol	$C_9H_{10}ClN_5O_3$	12.5	272.0545	
Metabolite-3: Urea analogue of imidacloprid 1-((6-Chloropyridin-3-yl) methyl)imidazolidin-2-one	$C_9H_{10}ClN_3O$	6.1	212.0585	
Metabolite-4: 5-Hydroxyimidacloprid 1-((6-Chloropyridin-3-yl) methyl)-4,5-dihydro-2-(nitroamino)-1H-imidazol-5-ol	$C_9H_{10}ClN_5O_3$	—	272.0545	

Continued

Metabolite-5: 4,5-Deshydro-imidacloprid 1-((6-Chloropyridin-3-yl)methyl)-N-nitro-1H-imidazol-2-amine		254.0439	—	$C_{18}H_{19}NO_4$
Metabolite-6: 4,5-Dihydroxyimidacloprid 1-((6-Chloropyridin-3-yl)methyl)-4,5-dihydro-2-(nitroamino)-1H-imidazole-4,5-diol		288.0494	—	$C_9H_{10}ClN_5O_4$
Metabolite-7: Reduced NO analogue of imidacloprid 1-((6-Chloropyridin-3-yl)methyl)-4,5-dihydro-N-nitroso-1H-imidazol-2-amine		240.0647	—	$C_9H_{10}ClN_5O$
Metabolite-8: Imidacloprid-urea-olefin 3-((6-Chloropyridin-3-yl)methyl)-1H-imidazol-2(3H)-one		210.0429	—	$C_9H_8ClN_3O$
Metabolite-9: Hydroxyimidacloprid-glucoside 3-(1-((6-chloropyridin-3-yl)methyl)-4,5-dihydro-2-(nitroamino)-1H-imidazol-4-yloxy)-tetrahydro-6-(hydroxymethyl)-2H-pyran-2,4,5-triol		434.1073	—	$C_{15}H_{20}ClN_5O_8$

Table 10.1 List of Reported Plant Metabolites (Spinach, Cotton, Tomatoes, Sunflowers, and Sugar Beets) of Imidacloprid and Their Calculated Exact Masses of Protonated Molecules (Casida, 2011; Ford & Casida, 2008; Westwood et al., 1998).—cont'd

Compound Name	Elemental Composition	Retention Time (min)	Calculated Exact Mass [M+H]+	Chemical Structure
Metabolite-10: Urea-imidacloprid-glucoside 1-((6-Chloropyridin-3-yl)methyl)-3-(tetrahydro-2,4,5-trihydroxy-6-(hydroxymethyl)-2H-pyran-3-yl)imidazolidin-2-one	C₁₅H₂₀ClN₃O₆	—	374.1113	
Metabolite-11: Urea-imidacloprid-gentiobioside 1-((6-Chloropyridin-3-yl)methyl)-3-(tetrahydro-2,4,5-trihydroxy-6-(hydroxymethyl)-2H-pyran-3-yl)imidazolidin-2-one6-(tetrahydro-2,4,5-trihdyroxy)-2H-pyran-2,4,5-triol	C₂₁H₃₀ClN₃O₁₁	—	536.1642	
Metabolite-12: N-((6-chloropyridin-3-yl)methyl)-N-nitromethanetriamine	C₇H₁₀ClN₅O₂	—	232.0596	

All measured masses were between 1 and 3 ppm measured exact masses.

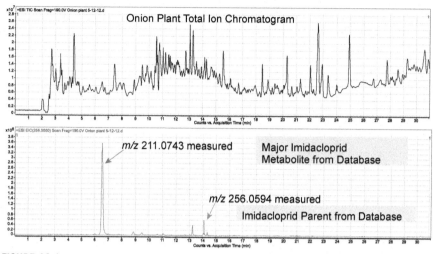

FIGURE 10.1

Onion extract showing a complex total ion chromatogram. The *red peak* (dark gray in print version) is a guanidine metabolite of imidacloprid *m/z* 211.0745 exact mass found using a simple database search and the *green arrow* (light gray in print version) is the imidacloprid parent at *m/z* 256.0596 exact mass.

The fragments obtained for the guanidine metabolite were key to the identification of newer metabolites of imidacloprid that had never been reported before. Table 10.2 shows the names, accurate masses, and chemical structures of all the metabolites identified by accurate mass. A paper published by our group discusses in detail all the approaches and MS/MS accurate mass information used for the elucidation of all the metabolites of imidacloprid (Thurman, Ferrer, Zavitsanos, & Zweigenbaum, 2013b).

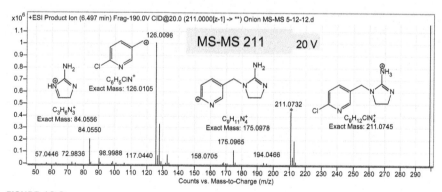

FIGURE 10.2

MS/MS analysis of the guanidine metabolite of imidacloprid *m/z* 211 at a collision energy of 20 V. Metabolite-2 (Table 10.1) with putative fragment ion structures. The measured mass is shown along with the calculated exact mass of the fragments.

Table 10.2 List of Newly Discovered Plant Metabolites (Not Confirmed by Standards) of Imidacloprid Found in This Study of Onions and Their Calculated Exact Masses of Protonated Molecules (Thurman et al., 2013b).

Name	Elemental Composition	Retention Time (Min)	Calculated Exact Mass $[M+H]^+$	Putative Chemical Structure (Not Confirmed by Standards)
Metabolite-1: Methylated imidacloprid 1-(1-((6-Chloropyridin-3-yl)methyl)-4,5-dihydro-1H-imidazol-2-yl)-2-methylhydrazine	$C_{10}H_{14}ClN_5$	8.3	240.1010	
Metabolite-2: Isomer of guanidine imidacloprid 3-((6-Chloropyridin-3-yl)methyl)-2,3-dihydro-1H-imidazol-2-amine	$C_9H_{11}ClN_4$	7.1	211.0745	
Metabolite-3: Olefin of guanidine imidacloprid 3-((6-Chloropyridin-3-yl)methyl)-1H-imidazol-2(3H)-imine	$C_9H_9ClN_4$	6.4	209.0589	
Metabolite-4: Imidacloprid-amine analogue 1-(1-((6-Chloropyridin-3-yl)methyl)-4,5-dihydro-1H-imidazol-2-yl)hydrazine	$C_9H_{12}ClN_5$	6.0	226.0854	
Metabolite-5: Olefin-imidacloprid-amine analogue 1-(1-((6-Chloropyridin-3-yl)methyl)-1H-imidazol-2-yl)hydrazine	$C_9H_{10}ClN_5$	4.8	224.0697	

Metabolite-6: Hydroxy-guanidine analogue 1-((6-Chloropyridin-3-yl)methyl)-2-iminoimidazolidin-4-ol	$C_9H_{11}ClN_4O$	4.7	227.0694
Metabolite-7: Glutamic acid conjugate of imidacloprid amine 1-(1-((6-Chloropyridin-3-yl)methyl)-4,5-dihydro-1H-imidazol-2-yl)hydrazine-glutamic acid	$C_{14}H_{16}ClN_5O_4$	10.7	354.0964

The new metabolites were found in nearly all of the nine onion extracts.

10.3.2 MASS PROFILER PROFESSIONAL

Mass Profiler Professional (MPP, Agilent Technologies) is a multivariate statistical program. In this study, it was applied to complex mass spectral chromatograms from 18 combined samples (plants that received pesticide application and control plants that were not treated) to tease out differences that arise from the metabolism of the pesticide imidacloprid, when applied to onions. The purpose of MPP, in general, is to assist in analyzing the complex mass spectral data that arise when high-resolution accurate mass is gathered from samples of different treatments containing thousands of accurate mass ions.

MPP was used to interrogate the samples as described next. First, the data were filtered by flags, keeping all entities in each sample that were present or marginal. Next, the data were filtered by variability, again keeping entities in each sample that had a coefficient of variation of 25% or less. This filtered data set was then processed by principal component analysis. Finally, the same filtered data were examined using find-by-similarities to discover new metabolites.

10.3.2.1 Principal Component Analysis

Principal component analysis reveals complex data variation. This data set clearly shows differences in the samples treated with imidacloprid versus those not treated. Fig. 10.3 displays these differences in a three-dimensional graph. This graph accounts for 71% of the total variance in the data set. The separation between the treated and untreated samples is clearly seen in Fig. 10.3, where the red points

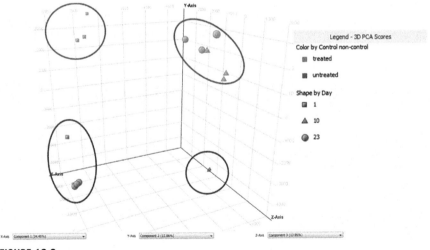

FIGURE 10.3

Principal component analysis of imidacloprid treated and untreated onion plants taken on 3 separate sampling days. Three individual plants were sampled on each day and each plant was only sampled once (18 separate plants in total).

are treated samples and the blue points are untreated samples. Interestingly, day 1 of the treated samples is quite different from day 10 and day 23. These are the three different sampling periods of the study. In addition, day 10 is different from days 1 and 23 of the untreated samples. One simple explanation for this variance for untreated samples is the senescence of the onion plant, which is a natural process of browning or wilting of the plant due to different water uptake and sun exposure in the greenhouse. These living organisms have variations in the natural organic compounds present in the plant, which are important for photosynthesis. These compounds are measured quite easily by accurate mass analysis.

An explanation for the treated samples could be the stress that the plant receives after application of the insecticide imidacloprid. This stress or release of stress (we cannot be sure) changes the plant metabolism. This is reflected in the ion chromatograms measured in the extracted plant materials. Our previous reports (Thurman, Ferrer, & Zweigenbaum, 2013a; Thurman et al., 2013b) show that imidacloprid is primarily transformed from the insecticide to its metabolites. It was found that the parent resides primarily in soil and the metabolites reside primarily in the onion plant.

10.3.2.2 Find Similar Entities

The second process used in MPP is called "Find Similar Entities," found in the analysis tab. To use this process, we applied the major guanidine metabolite previously found (Ferrer, Thurman, & Zweigenbaum, 2014; Thurman et al., 2013b) at 6.6 min at a neutral mass of 210.0668. Fig. 10.4 shows the five entities that have similar patterns to the guanidine metabolite (including this metabolite there are six) for both the treated and untreated sample set. The x-axis shows the nine untreated and the nine

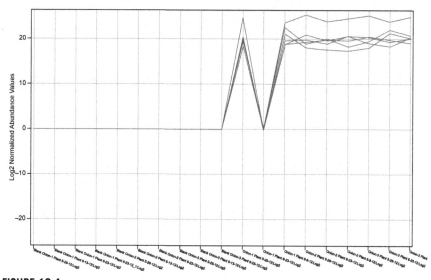

FIGURE 10.4

Plot of similar entities to the guanidine metabolite (Thurman et al., 2013a).

treated samples in this study. The y-axis shows the normalized data set. A zero value shows that all the samples are identical. A negative value would show that those samples are lower than the zero (median) value. A positive value shows that those samples are higher than the zero (median) value with respect to referenced entity at the mass of 210.0668. There was one treated sample which was of zero value. This means that it was the same as the untreated samples. However, the soil showed the parent compound only, therefore it means that the plant did not uptake imidacloprid.

Fig. 10.5 shows the list of six entities displayed as a positive value (+20) in Fig. 10.4. A metabolite of imidacloprid not previously identified was found at a neutral mass of 295.0817 and a retention time of 11.01 min. The molecular formula given by the software was $C_{12}H_{14}ClN_5O_2$. From our previous experience with the metabolism of imidacloprid in onions (Thurman et al., 2013b), a putative structure is proposed in Fig. 10.5. The guanidine metabolite, which was used for this correlation, is the basic structure that is proposed for conjugation. In this case, the amino acid, alanine, has the correct mass and formula for the new conjugated metabolite. The entity at neutral mass, 208.0516, at 5.93 min, was examined and found to be a fragment of a previously identified metabolite of imidacloprid amine (Thurman et al., 2013a). The remaining three entities did not contain chlorine in their molecular formulas and were not metabolites of imidacloprid, probably other natural components in the plants that may have been produced by stress or removal of stress by the pesticide.

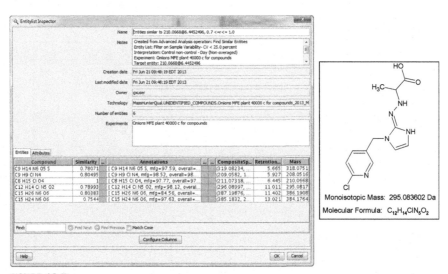

FIGURE 10.5

Screenshot of list of entities from find similar entities with the guanidine metabolite as the reference (entity with similarity of 1 in table). Putative structure of imidacloprid metabolite discovered using find similar entities in Mass Profiler Professional (MPP) is also shown.

10.3.3 METABOLITE DISTRIBUTION AND MASS BALANCE FOR IMIDACLOPRID METABOLITES

The metabolite distribution for imidacloprid metabolites in plants was studied. Fig. 10.6 shows a pie diagram showing the different metabolites identified in onion and lettuce plants. As it can be seen in this figure, the guanidine metabolite at m/z 211 was the major metabolite found, followed by the olefin of the same compound at m/z 209. Both plants showed similar characteristics when comparing the distribution of metabolites. However, more metabolites were identified in onion plants as compared to lettuce plants. This could be due to the intrinsic characteristics of the onion plant itself in terms of a more diverse metabolism.

A mass balance comparing the concentrations applied and the concentrations found for imidacloprid in the soils and in the plants was carried out. The data are shown in Fig. 10.7. Interestingly, the parent compound imidacloprid was not found in the plant but rather in the soil, whereas the main metabolite at m/z 211 was mostly found in plants and not in soil. This suggests that this important metabolite is formed in the plant itself after it has absorbed the parent compound, rather than forming in the soil before the plant uptakes it. For some reason plant number 7 did not show up any metabolite in the plant; we think this is due to witting or drying of the plant. On the other hand, after a certain period of time (a month) the metabolite starts forming in the soil as it is shown in plant numbers 7–9. No significant concentrations were found in the analyzed water leachates, indicating that the parent compound and their metabolites remain either in the soil or are uptaken by the plant.

10.4 IMAZALIL METABOLITES IN PLANTS AND SOIL

Imazalil is a fungicide widely used in agriculture, particularly in the postharvesting stage. We chose this pesticide for plant application mainly because of the high sensitivity that it exhibits under high-resolution mass spectrometry using an electrospray source. Moreover, the chemical structure contains two chlorine atoms, which makes the molecule very amenable to be detected after degradation (assuming one or two

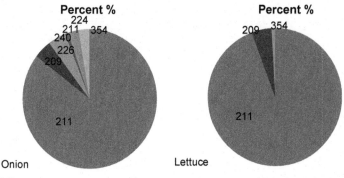

FIGURE 10.6

Pie diagram of plant metabolites in onion and lettuce as a percentage of the total imidacloprid metabolites.

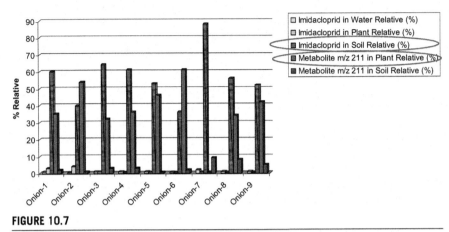

FIGURE 10.7

Percentage of imidacloprid and main metabolite in water leachate, plant, and soil.

chlorines remain in the chemical structure). Imazalil was applied to the soil in the pots of both onion and lettuce plants. Degradation products were identified following two different approaches, which are discussed in the following sections.

10.4.1 CHLORINE FILTER APPROACH

A useful tool that can be applied for metabolite discovery is the chlorine filter. The chlorine filter was first published by Ferrer and Thurman (2010) as a tool for the environmental analysis of pharmaceuticals in wastewater samples. Basically, this tool first extracts all of the ions in the chromatogram above an ion intensity of 10,000 counts, which generates accurate mass formula for the monoisotopic ion and its A + 1 and A + 2 isotopes. Then the program compiles molecular formulas that are forced to contain one or two chlorine atoms, because that is the basic structure of the pesticide studied here. When this program was run on the onion and lettuce extracts, there were detections of several new metabolites. Fig. 10.8 shows a table of potential metabolite hits that contain formulas with chlorine atoms

Show/Hid	Cpd	Formula	RT	m/z	Score	Diff (	Polarity	Height	Ions	Algorithm
☑	1387	C9 H10 Cl N5 O2	14.094	256.0578	96.24	4.78	Positive	601240	5	Find by Molecular Feature
☑	336	C8 H15 Cl O4	6.544	211.0738	95.94	-2.92	Positive	3575138	4	Find by Molecular Feature
☑	314	C8 H13 Cl O4	6.011	209.0582	93.65	-2.87	Positive	145349	3	Find by Molecular Feature
☑	337	C6 H4 Cl N	6.544	126.0104	98.08	1.05	Positive	393210	3	Find by Molecular Feature
☑	773	C14 H16 Cl N5 O4	10.964	354.0955	91.98	1.78	Positive	254645	5	Find by Molecular Feature
☑	1413	C14 H14 Cl2 N2 O	14.257	297.0552	82.87	1.63	Positive	99487	3	Find by Molecular Feature
☑	625	C12 H14 Cl N5 O2	10.192	296.0891	90.18	4.44	Positive	101477	3	Find by Molecular Feature
☑	785	C12 H14 Cl N5 O2	11.045	296.0903	91.83	2.25	Positive	87287	3	Find by Molecular Feature
☑	1135	C11 H10 Cl2 N2 O	12.814	257.0241	98.99	1.97	Positive	390498	5	Find by Molecular Feature
☑	1700	C10 H9 Cl2 N3 O	17.054	258.0189	85.64	2.35	Positive	218712	3	Find by Molecular Feature
☑	1		2.019	146.1649			Positive	225109	2	Find by Molecular Feature

FIGURE 10.8

Table of potential metabolites containing at least one chlorine atom in the chemical structure.

for an onion extract. In this table, the previous imidacloprid metabolites identified in the last section were also found. However, we were now looking for potential metabolites of imazalil, so we focused on formulas that contained two chlorine atoms. One example is the peak at 12.8 min (imazalil elutes at 16.0 min), which contains the formula $C_{11}H_{10}Cl_2N_2O$ and an accurate mass at m/z 257.0241. By comparing this formula to the imazalil chemical structure, one realizes that the group propene is missing, thus giving rise to a hydroxylated metabolite (metabolite 1) as shown in Fig. 10.9. This metabolite was called R14832 and it was previously reported in citrus

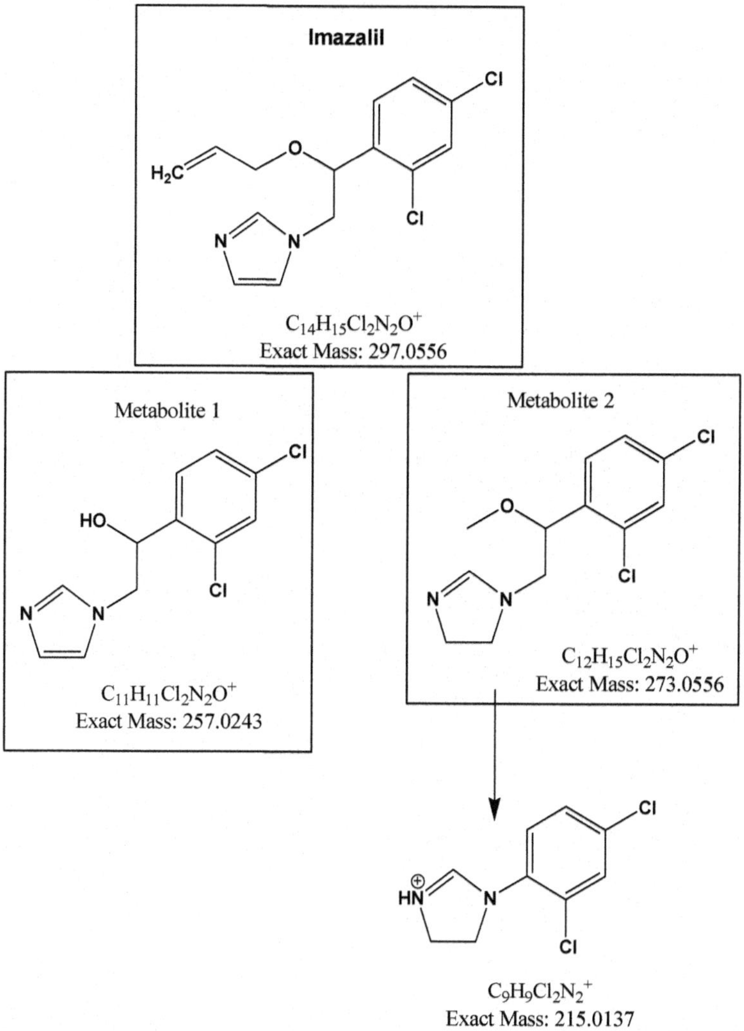

FIGURE 10.9

Chemical structures and accurate mass for imazalil and two of its metabolites. Also showing a fragment ion of Metabolite 2.

fruits (Thurman et al., 2005). An MS/MS of this metabolite was performed and all the fragments matched well with previously reported data, and were consistent with the structure.

10.4.2 SOIL METABOLITES

A soil metabolite of imazalil was found at 15.4 min with m/z 273.0549 (Metabolite 2). This metabolite is interesting from the point of view of molecular rearrangement under MS/MS conditions. The formula obtained for this accurate mass is $C_{12}H_{14}Cl_2N_2O$, which is a methylated form from the previous reported metabolite (Metabolite 1). Under MS/MS conditions a main peak is obtained at m/z 215.0137, which is a loss of C_3H_6O from the metabolite (see Fig. 10.9). For this to happen, a chemical structure rearrangement occurs: the propylene oxide moiety breaks apart and the two rings (the aromatic one with the two chlorines and the cyclopentane ring) attach to each other.

The two metabolites reported for imazalil are formed in the soil. However, only one of them (m/z 257.0243) was found in onion plants, and none in lettuce. This means that the onion plants uptake the metabolite directly from the soil, rather than forming it in their metabolism. The lettuce plant, on the other hand, does not uptake these metabolites from the soil. The reason for this difference between onions and lettuces could be the nature of the vegetable itself; onions have extensive roots that go into the soil, whereas lettuce is a leafy vegetable that sits mostly on the surface of the pot. An interesting fact was also that the parent compound, imazalil, was not found in either plant, thus indicating that it remains in soil, probably because of its hydrophobicity and affinity for the soil matrix.

10.5 PROPICONAZOLE METABOLITES IN PLANTS AND SOIL

Propiconazole is a triazole molecule and it is used agriculturally as a systemic fungicide on turfgrasses grown for seed and aesthetic or athletic value, mushrooms, corn, wild rice, peanuts, almonds, sorghum, oats, pecans, apricots, peaches, nectarines, plums, and prunes. Similar to imazalil we chose this pesticide because of the presence of chlorine atoms in the chemical structure. Propiconazole was applied to the soil in the pots of both onion and lettuce plants. Degradation products were identified as follows.

10.5.1 CHLORINE FILTER APPROACH

Once again we used the chlorine filter approach to rapidly detect any metabolites that could be present in the plants. As observed in Fig. 10.8, one of the metabolites was observed at 17.0 min at m/z 258.0189 with two chlorines and with a molecular formula of $C_{10}H_9Cl_2N_3O$. The proposed metabolite is shown in Fig. 10.10. This metabolite was mostly present in onion plants only; lettuce did not show up the

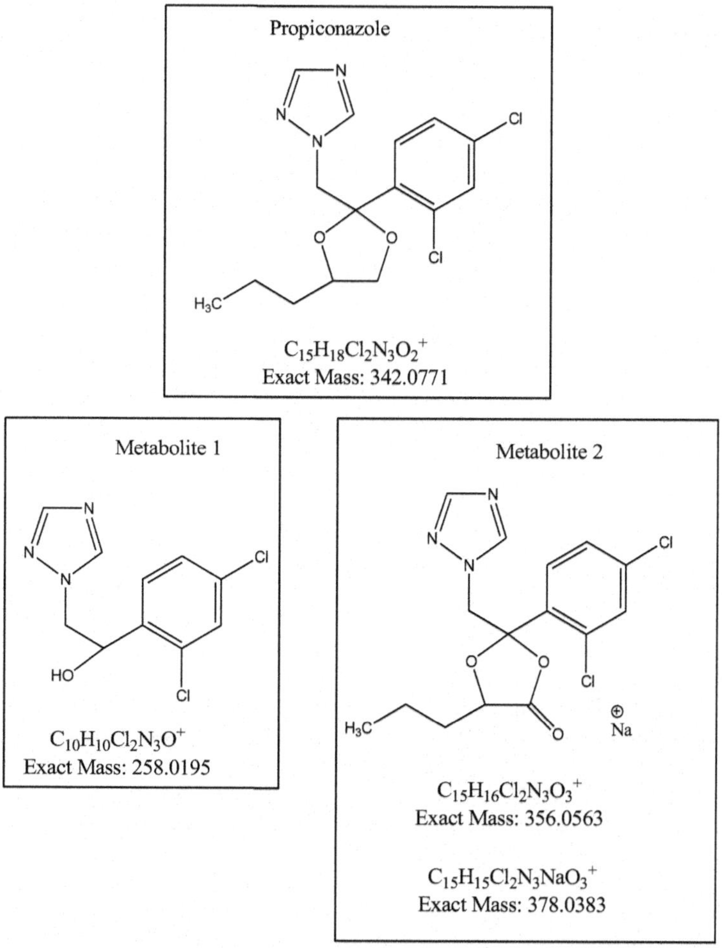

FIGURE 10.10

Chemical structures and accurate mass for propiconazole and two of its metabolites.

presence of this compound. Likewise, the concentrations of this metabolite in soil were not significant, thus suggesting this metabolite is formed in the onion plant metabolism.

Another metabolite identified was at 19.3 min at m/z 356.0563 with a formula of $C_{15}H_{15}Cl_2N_3O_3$, which is the addition of one oxygen atom and the loss of two hydrogens to the propiconazole molecule (Fig. 10.10). Interestingly, this molecule exhibits a strong sodium adduct under electrospray conditions, which generates from the presence of a keto group in the molecule, as it has been reported in some adduct ionization studies. This metabolite, similar to metabolite 2 from imazalil, was only found in soil and not in the plants.

10.6 CONCLUSIONS

High-resolution mass spectrometry has powerful tools for identifying and characterizing metabolites of pesticides in plants, such as onion and lettuce. These tools enable detailed studies of the fate of the pesticides imidacloprid, imazalil, and propiconazole in plants. For example, a previously overlooked metabolite, the guanidine—alanine conjugate, was discovered using the MPP tool set. Furthermore, this tool is useful for portraying changes in both the treated and untreated samples for future studies. Another conclusion from this study is the importance of soil organic matter to retain the more hydrophobic pesticides, which are typically the parent compounds while the more water-soluble pesticides or transformation products are present in the plants. This result emphasizes the importance of using accurate mass with high resolution to identify transformation products and plant metabolites. The toolbox described here does an excellent job of carrying out this process.

REFERENCES

Alsayeda, H., Pascal-Lorber, S., Nallanthigal, C., Debrauwer, L., & Laurent, F. (2008). Transfer of the insecticide [C-14] imidacloprid from soil to tomato plants. *Environmental Chemistry Letters, 6,* 229—234.

Casida, J. E. (2011). Neonicotinoid metabolism: Compounds, substituents, pathways, enzymes, organisms, and relevance. *Journal of Agricultural and Food Chemistry, 59,* 2923—2931.

Ferrer, I., & Thurman, E. M. (2010). Identification of a new antidepressant and its glucuronide metabolite in water samples using liquid chromatography/quadrupole time-of-flight mass spectrometry. *Analytical Chemistry, 82,* 8161—8168.

Ferrer, I., & Thurman, E. M. (2015). Application of LC—MS/MS and LC—TOF-MS for the identification of pesticide residues and their metabolites in environmental samples. In D. Tsipi, H. Botitsi, & A. Economou (Eds.), *Mass spectrometry for analysis of pesticide residues and their metabolites.* Hoboken, NJ: Wiley.

Ferrer, I., Thurman, E. M., & Zweigenbaum, J. A. (2014). *Discovery of Imidacloprid metabolites in onion using mass profiler professional.* Agilent Technologies Publication, 5991—3931EN.

Ford, K. A., & Casida, J. E. (2008). Comparative metabolism and pharmacokinetics of seven neonicotinoid insecticides in spinach. *Journal of Agricultural and Food Chemistry, 56,* 10168—10175.

Iwasa, T., Motoyama, N., Ambrose, J. T., & Roe, R. M. (2004). Mechanism for the differential toxicity of neonicotinoid insecticides in the honey bee, *Apis mellifera. Crop Protection, 23,* 371—378.

Kagabu, S. (1997). Discovery of the chloronicotinyl insecticides. *Abstracts of Papers of the American Chemical Society, 214,* 4-Agro.

Kagabu, S. (2011). Discovery of imidacloprid and further developments from strategic molecular designs. *Journal of Agricultural and Food Chemistry, 59,* 2887—2896.

Laurent, F. M., & Rathahao, E. (2003). Distribution of [C-14]imidacloprid in sunflowers (*Helianthus annuus* L.) following seed treatment. *Journal of Agricultural and Food Chemistry, 51,* 8005–8010.

Mandic, A. I., Lazic, S. D., Okresz, S. N., & Gaal, F. F. (2005). Determination of the insecticide imidacloprid in potato (*Solanum. tuberosum* L.) and onion (*Allium cepa*) by high-performance liquid chromatography with diode-array detection. *Journal of Analytical Chemistry, 60,* 1134–1138.

Nauen, R., Reckmann, U., Armborst, S., Stupp, H. P., & Elbert, A. (1999). Whitefly-active metabolites of imidacloprid: Biological efficacy and translocation in cotton plants. *Pesticide Science, 55,* 265–271.

Nauen, R., Tietjen, K., Wagner, K., & Elbert, A. (1998). Efficacy of plant metabolites of imidacloprid against *Myzus persicae* and *Aphis gossypii* (Homoptera: Aphididae). *Pesticide Science, 52,* 53–57.

Olson, E. R., Dively, G. P., & Nelson, J. O. (2004). Bioassay determination of the distribution of imidacloprid in potato plants: Implications to resistance development. *Journal of Economic Entomology, 97,* 614–620.

Thurman, E. M., Ferrer, I., & Fernandez-Alba, A. R. (2005). Matching unknown empirical formulas to chemical structure using LC/MS TOF accurate mass and database searching: Example of unknown pesticides on tomato skins. *Journal of Chromatography A, 1067,* 127–134.

Thurman, E. M., Ferrer, I., & Zweigenbaum, J. A. (2013a). *Identification of imidacloprid metabolites in onions using high-resolution mass spectrometry and accurate mass tools.* Agilent Technologies Publication, 5991–3397EN.

Thurman, E. M., Ferrer, I., Zavitsanos, P., & Zweigenbaum, J. A. (2013b). Identification of imidacloprid metabolites in onion (*Allium cepa* L.) using high-resolution mass spectrometry and accurate mass tools. *Rapid Communications in Mass Spectrometry, 27,* 1891–1903.

Tomizawa, M., & Casida, J. E. (2003). Selective toxicity of neonicotinoids attributable to specificity of insect and mammalian nicotinic receptors. *Annual Review of Entomology, 48,* 339–364.

Tomizawa, M., & Casida, J. E. (2005). Neonicotinoid insecticide toxicology: Mechanisms of selective action. *Annual Review of Pharmacology and Toxicology, 45,* 247–268.

Westwood, F., Bean, K. M., Dewar, A. M., Bromilow, R. H., & Chamberlain, K. (1998). Movement and persistence of [C-14]imidacloprid in sugar-beet plants following application to pelleted sugar-beet seed. *Pesticide Science, 52,* 97–103.

Index

'*Note*: Page numbers followed by "f" indicate figures and "t" indicate tables.'

A

Accurate mass, 2, 12
  absolute value, 8
  databases, 319–323, 320t–322t, 323f, 324t–325t
  error minimization, 9
  flonicamid pesticide, fragmentation pattern, 7f, 8
  isotopic composition, 7
  mass defect, 7
Acetic acid, 292–293
Acquisition modes
  data-dependent acquisition (DDA), 41f. *See also*
    Data-dependent acquisition (DDA)
  data-independent acquisition (DIA), 41f. *See also*
    Data-independent acquisition (DIA)
  postacquisition approaches, 44
  types, 39
Alternating-current (AC) voltage, 290–292
Ambient mass spectrometry (AMS)
  applications, 305–307
  electrospray-based techniques, 284–290, 289f
  method development
    ionization, 292–294
    sample handling, 298–302, 300f
    working conditions, 294–298
  method performance, 302–305, 304f
  plasma-based techniques, 290–292, 291f
Analog-to-digital converter (ADC), 248
Analytical Quality Assurance Cycle (AQAC), 100
Association of Analytical Communities
  (AOAC), 139
Atmospheric pressure chemical ionization
  (APCI), 16, 234–235
Atmospheric pressure gas chromatography
  background, 236–237
  carrier gas flow rate, 246–247
  current design, 237–238, 238f
  high-resolution mass spectrometry, 239–245
  ionization mechanisms, 238–239
  ion mobility separation. *See* Ion mobility
    separation
  overview, 236
  pesticide screening, 239–245
  pesticides, ionization trends for, 239, 240t–243t
  selectivity, 245–246
  sensitivity, 245–246
  time-of-flight mass spectrometry (TOF MS),
    247–248
Atmospheric pressure glow discharge (APGD),
  284

Atmospheric pressure ionization (API), 268
Atomic mass, 2
Automatic gain control (AGC), 67
Average mass, 2, 3t

C

Chemical Abstracts Service (CAS), 83
ChemSpider, 45, 76, 77f
Chlorine filter approach, 330–333, 330f–331f,
  333f
Chlorpyrifos, 222–223, 223f
Codex Alimentarius Commission (CAC), 88
Codex Committee on Pesticide Residues
  (CCPR), 88
Collision cross section (CCS), 250–251, 254f
Collision-induced dissociation (CID), 63

D

Dalton (Da), 1
Data analysis
  automated component detection, 70
  DG SANTE, 69–72
  extraction mass, 70–71
  fragmentation patterns, 70–71
  full scan-MS and MS/MS chromatograms, 68
  nontarget analysis, 69
    accurate mass error, 76
    ChemSpider, 76, 77f
    comprehensive molecular databases, 78
    endogenous matrix components, 75–76
    MS/MS fragmentation prediction tools, 76–78
    pesticide residue analysis, 75
  qualitative screening method validation, 73–75
  relative isotope abundances (RIA), 72
  retrospective analysis, 69–70
  RF window tolerance, 72
  screening methods, 72
  suspect screening, 69
  target screening, 69
  transformation products (TPs), 70–71
  workflows, 69, 69f, 73, 74f
Data-dependent acquisition (DDA), 29, 63
  ion intensity-dependent, 41
  isotope pattern–dependent, 42
  list dependent, 42
  MS/MS analysis, 39–40
  parallel reaction monitoring, 42–43
  precursor ion scan (PIS), 40–41
  pseudo neutral loss-dependent acquisition, 41

Data-independent acquisition (DIA), 29, 63
  full scan, 43
  nonselective MS/MS spectra, 44
  PIS without precursor ion isolation, 43
Desorption electrospray ionization (DESI), 267,
    283−284, 294
Differential ion mobility (DMS), 30
Dilute-and-shoot (DS), 137−138, 143t−146t
Direct analysis in real time (DART), 267,
    283−284
Direct analysis in real time-mass spectrometry
    (DART-MS), 282−283
Dispersive solid-phase extraction (d-SPE), 139,
    211
Drift time ion mobility spectrometry (DTIMS),
    249−250

E

Electron capture detection (ECD), 236−237
Electron ionization (EI), 9−10
Electrospray-based methods, 292−293
Electrospray ionization (ESI), 15, 234−235
EU multiannual coordinated control program
    (EU-MACCP), 90
EU proficiency tests (EUPTs), 107−108
European Analytical Chemistry group
    (EURACHEM), 92
European Food Regulation, 89−90
European Food Safety Authority (EFSA), 89
Exactive-Orbitrap, 204−208
Exact mass, 2
Exponentially weighted moving average
    (EWMA) control, 97−98
Extractive electrospray mass spectrometry (EESI),
    284−290
Extraneous maximum residue limits
    (EMRLs), 88

F

FDA Foods and Veterinary Medicine (FVM),
    115−116
FDA Foods Program Guidelines, 106
Fiehn Library, 47−48
Field-asymmetric waveform ion mobility
    spectrometry (FAIMS), 249−250
Flow injection analysis-mass spectrometry
    (FIA-MS)
  applications, 282−283
  defined, 268−283, 269t−271t, 272f
  method development, 273−276
    ionization efficiency, 273−274
    MS working conditions, 274−276, 275f

methods performance, 279−282
  sample preparation, 277−279, 280f
Food and Agricultural Organization (FAO), 88
Food and Drug Administration (FDA), 91
Food Emergency Response Network, 91
Formic acid, 292−293
Fourier transformion cyclotron resonance
    (FT-ICR), 16, 19−21, 20f
Fungicides, 84

G

Gas chromatography (GC), 265−266
Gas chromatography−high-resolution mass
    spectrometry (GC−HRMS), 16
Gas chromatography−Q-Orbitrap, 39, 40f
Gaussian-like distribution, 268−273
Geometrical parameters, 295

H

Helium flow rate, 297
Helium gas temperature, 296−297
Herbicides, 84−85
High-field asymmetric waveform ion mobility
    (FAIMS), 30
High-field (LTQ) Orbitrap (Orbitrap Elite),
    37−38, 37f
High-performance thin-layer chromatography
    (HPTLC), 279
High-throughput planar solid-phase extraction
    (HTpSPE), 279
Human Metabolome Database (HMDB), 47
Hybrid ion trap time-of-flight, 32
Hybrid linear ion trap Orbitrap (LTQ-Orbitrap), 37
Hybrid quadrupole Orbitrap (Q-Exactive), 35−36,
    36f, 36t
Hybrid quadrupole time-of-flight instrumentation,
    22−24, 25f

I

Imazalil metabolites, 329−332
  chlorine filter approach, 330−332, 330f−331f
  soil metabolites, 332
Imidacloprid metabolites, 318−329
  defined, 318−329
  mass balance, 329, 329f−330f
  metabolite distribution, 329, 329f−330f
Insecticides, 85
Internal quality control (IQC) system, 107
International Union of Pure and Applied
    Chemistry (IUPAC), 1, 83
International Vocabulary of Metrology (VIM), 93
Ionization energy (IE), 293

Ion mobility quadrupole time-of-flight
  accurate mass measurements, 29
  Agilent 6560, 30, 31f
  data-dependent acquisition (DDA), 29
  data-independent acquisition (DIA), 29
  differential ion mobility (DMS), 30
  gas-phase IM, 29
  LC–IMS-Q-TOF, 31
  Synapt high-definition mass spectrometry system, 29, 30f
  traveling wave—enabled stacked ring ion guides (TWIGs), 29
Ion mobility separations (IMSs), 235
  background, 249–251, 250f
  spectral selectivity enhancement, 253–255, 255f, 257f, 259f–260f
  theory, 249–251, 250f
  traveling wave ion mobility spectrometry, 251–253, 252f–254f
Isotopic mass, 3t
Isotopic mass average (IMA), 12

L

Limit of detection (LOD), 165–167, 192
Limit of quantitation (LOQ), 192
Liquid chromatography (LC) technique, 317. *See also* Pesticide residue analysis
Liquid—liquid partition, 277–278
Lock-plus-cali mass technique, 10–11
Low-resolution mass spectrometry (LRMS)
  vs. high-resolution mass spectrometry
    characteristics, mass analyzer, 4, 5t
    double-focusing magnetic-sector mass instruments, 4
    efficiency, 4
    linear dynamic range, 4
    sensitivity, 4
    structure elucidation, 3–4
  nominal mass, 3–4
Low-temperature plasma (LTP), 284

M

Mass accuracy, 2
Mass balance, 329, 329f–330f
MassBank, 45
Mass calibration, 2
  electron ionization (EI), 9–10
  external mass calibration, 9–10
  FT-ICR mass analyzers, 10
  Lock-plus-cali mass technique, 10–11
  MALDI, 9–10
  mass reference compounds, 9

  multiple ion detection (MID) mode, 10–11, 11f
  orthogonal acceleration TOF analyzers, 10
  selection of, 9–10
Mass defect, 2
Mass defect filter (MDF), 44
Mass limit, 2
Mass number, 2
Mass Profiler Professional (MPP), 326–328
  defined, 326–328
  find similar entities, 327–328, 327f–328f
  principal component analysis, 326–327, 326f
Mass profiler software, 318
Mass spectrometry (MS), 1
Matrix-assisted laser desorption/ionization (MALDI) experiments, 17–18
Matrix solid-phase extraction (MSPD), 141–142, 143t–146t
Maximum residue limits (MRLs), 83, 165
  acceptable daily intake (ADI), 89
  Codex Alimentarius Commission (CAC), 88
  Codex Committee on Pesticide Residues (CCPR), 88
  N-desmethyl-acetamiprid, 102
  environmental sources, 89
  European Food Regulation, 89–90
  European Food Safety Authority (EFSA), 89
  extraneous maximum residue limits (EMRLs), 88
  Food and Agricultural Organization (FAO), 88
  metabolite, definition, 88
  OECD, 90
  residue, definition, 88
  Spirotetramat, 101, 102f
Metabolite distribution, 329, 329f–330f
METLIN Metabolomics Database, 45, 46f, 47
Metrology of Qualitative Chemical Analysis (MEQUALAN), 106
Microwave-assisted extraction (MAE), 149–150
Monoisotopic mass, 2
MPP. *See* Mass Profiler Professional (MPP)
Multiple ion detection (MID) mode, 10–11, 11f
Multiple-reaction monitoring (MRM) mode, 59, 234–235, 265–266
mz Cloud, 48, 49f

N

National Metrology Institute of Japan (NMIJ), 108
Nematicides, 84
NIST MS/MS database, 48

Nitrogen, 237–238
Nominal mass, 2, 3t, 12

O

Office of Regulatory Affairs (ORA) laboratories, 112–114
Orbital ion trapping analyzer (Orbitrap), 15–16, 21, 22f, 23t, 34t, 67
  advantages, 68
  GC–Q-Orbitrap, 39, 40f
  high-field (LTQ) Orbitrap (Orbitrap Elite), 37–38, 37f
  hybrid linear ion trap Orbitrap (LTQ-Orbitrap), 37
  hybrid quadrupole Orbitrap (Q-Exactive), 35–36, 36f, 36t
  limitations, 68
  stand-alone Orbitrap (Exactive), 33–35
  tribrid (LTQ/Q-Orbitrap), 38–39, 38f

P

Paper spray ionization (PSI), 284
Peak-matching mode, 12
Pesticide Data Program (PDP), 91–92
Pesticide identification, transformation products
  accurate mass databases, 319–323, 320t–322t, 323f, 324t–325t
  chemicals and reagents, 316
  greenhouse study, 316
  imazalil metabolites, 329–332
  imidacloprid metabolites, 318–329
  liquid chromatography, 317
  Mass Profiler Professional (MPP), 326–328
  mass profiler software, 318
  overview, 315–316
  plant extraction, 316–317
  propiconazole metabolites, 332–333
  quadrupole, 317
  time-of-flight mass spectrometry analysis, 317
Pesticide residue analysis, 59
  advantages, 60
  Analytical Quality Assurance Cycle (AQAC), 100
  animal farming, 203
  calibration levels, 104
  CCPR, 98–99
  chromatographic techniques, 133, 152t–153t
    cleanup procedures, 135–136
    dilute-and-shoot (DS), 137–138, 143t–146t
    matrix solid-phase extraction (MSPD), 141–142, 143t–146t
    microwave-assisted extraction (MAE), 149–150

organochlorine compounds, 136–137
    pressurized liquid extraction (PLE), 150–151
    QuEChERS method, 143t–146t. See also QuEChERS method
    sample preparation, 135–136
    solid-phase extraction (SPE), 142–147
    solid-phase microextraction (SPME), 147–148
    stir bar sorptive extraction (SBSE), 148–149
  commodity group, 101
  N-desmethyl-acetamiprid, 102
  extraction and chromatographic conditions
    chromatographic separation, 214–215
    dilute and shoot, 211, 212t–213t, 214
    dispersive SPE (d-SPE), 211
    flow rate and injection volume, 215, 216t–219t, 220f
    nonpolar pesticides, 210–211
    QuEChERS, 211, 212t–213t
  FDA Foods and Veterinary Medicine (FVM), 115–116
  FDA Foods Program Guidelines, 106
  fragment ions, 112
  fruit and vegetable matrices
    accuracy, 192
    carbendazim and methomyl, 196–197
    carbon, chlorine and bromine isotopes, 195, 195f
    chlorine isotope determination, 192, 194f
    imidacloprid, 195–196, 196f
    LC-HRMS, 168–186, 169t–185t
    limit of detection (LOD), 192
    limit of quantitation (LOQ), 192
    matrix effect, 187–189
    matrix interference reduction, 189, 190f
    matrix-matched calibration, 189, 191f
    precision, 192
    quantitation process, 192, 193f
    QuEChERS method, 195–196
    sample preparation and chromatographic conditions, 186–187, 190f
    strawberry samples, 197
  GC–QTOF MS, 119
  guidance documents, 98–99
  identification requirements, 112, 113t, 115–116, 116t
  instrumental requirements
    characteristics, 210
    Exactive-Orbitrap, 204–208
    food from animal origin, 204–208, 205t–207t
    mass extraction window, 208, 209f
    resolution power, 204–208
    SANTE guidelines, 210

ionization suppression phenomenon, 104
ion ratio criteria, 114, 114t
ion selection criteria, 114
IPs assignment, ORA-LAB.010, 112—114, 114t
LC—ESI-QTOF-MS system, 117—118
LC—MS and GC—MS identification criteria, 111
LC-Q-Orbitrap MS detection, 117
legislative identification criteria, 115
limit of quantification (LOQ), 105—106
low-resolution mass spectrometry vs.
   high-resolution mass spectrometry
   analytical methods, 226—228, 227t
matrix effects and coextracted components,
   147—149
matrix-induced enhancement effect, 104, 105f
matrix-matched calibration, 104
maximum residue limits (MRLs), 165
method development, 100
Metrology of Qualitative Chemical Analysis
   (MEQUALAN), 106
MS spectra, 111
multiresidue methods (MRMs), 99
Office of Regulatory Affairs (ORA) laboratories,
   112—114
operational parameters, 118
Orbitrap-based mass analyzers, 165—167
organic contaminants, 131
PDP standard operation procedures, 99
physicochemical properties, 131
qualitative binary response, 132
quality assurance, 107—109
quantitative and qualitative applications
   chlorpyrifos, 222—223, 223f
   coumaphos metabolites, 225—226, 225f
   2,6-dichlorobenzamide identification,
     225—226, 226f
   honey samples, 222
   linear ion trap-Orbitrap, 222
   maximum residue limit (MRL), 222—223
   meat and meat-related products, 222
   narrow window-extracted ion chromatograms,
     224—225, 224f
   positive samples, 215—221, 221t
   Q-Orbitrap, 221
   retrospective analysis, 224—225
   triple quadrupole (QqQ), 215—221
retention time, 111
SANTE/11945/2015 document, 99
screening criteria, 115
screening detection limit (SDL), 167
selected reactions mode (SRM), 116—117
selectivity

accurate mass, 60
automatic search algorithms, 62
chromatographic gradients, 60—61
mass window width, 61
matrix components, 60—61
Orbitrap systems, 61
retention time, 61
tandem mass spectrometry. See Tandem mass
   spectrometry
Spirotetramat, 101, 102f
statistical tests, 104
target analysis, 65—67
Tepraloxydim, 101, 103f
time-of-flight (TOF), 165—167
toxic effects, 131—132
trace organic analysis, 110
triple quadrupole (QqQ), 132
ultrahigh-pressure liquid chromatography
   (UHPLC) systems, 165—167
uncertainty, 109—110
USDA criteria, 112
validation parameters and criteria, 102, 103t
Pesticides
analytical quality control, method validation
   accuracy, 94—95
   analytical sensitivity, 94
   detection limit/limit of detection (LOD), 94
   European Analytical Chemistry group
     (EURACHEM), 92
   exponentially weighted moving average
     (EWMA) control, 97—98
   external QC, 98
   interlaboratory approach, 93
   internal QC, 98
   International Vocabulary of Metrology
     (VIM), 93
   limit of quantitation (LOQ), 94
   measurement results quality, 97, 97f
   precision, 95, 96f
   QC samples, 97—98
   residue analysis. See Pesticide residue analysis
   selectivity, 93
   Shewhart control, 97—98
   total error, 97
   trueness, 95
   working range, 93
biotransformation pathways, 85—86
Chemical Abstracts Service (CAS), 83
classification, 84—85, 84t
identification, 83—84
legislation, 86, 87t
   authorization, 86—88

Pesticides (*Continued*)
    maximum residue limits (MRLs). *See*
      Maximum residue limits (MRLs)
    monitoring programs, 90–92
   metabolites, 86
   physicochemical properties, 83–84
   transformation products (TPs), 85–86
Plant extraction, 316–317
Plasma-based techniques, 296
DL-Polyalanine, 250–251
Polytetrafluoroethylene (PTFE), 295–296
Precursor ion scan (PIS), 40–41
Pressurized liquid extraction (PLE), 150–151
Principal component analysis, 326–327, 326f
Probe electrospray ionization (PESI), 284, 290, 294
Propiconazole metabolites, 332–333
   chlorine filter approach, 332–333, 333f
   defined, 332–333
Protomers, 255–258
Proton affinity (PA), 293

Q

Quadrupole linear ion trap (QqLIT) analyzers, 59
Quadrupoles, 265–266, 283, 317
Qualitative screening method validation, 73–75
Quantitative trace analysis, 15
QuEChERS method, 195–196, 211, 212t–213t
   Association of Analytical Communities (AOAC), 139
   dispersive solid-phase extraction (d-SPE), 139
   pesticide characteristics, 139–140
   screening analysis, 141
   signal suppression, 140
   solid-phase extraction (SPE), 140

R

Recursion, 318
Relative isotope abundances (RIA), 72
Relative isotopic mass defect, 2
Relative standard deviations (RSDs), 282
Resolution/mass resolving power, 2, 6, 7f
Retention times (RTs), 254f

S

SANTE/11945/2015, 246–247
SANTE guidelines, 210
Sciex TripleTOF 5600 Q-TOF, 26, 27f
Screening detection limit (SDL), 167
Selected reaction mode (SRM), 116–117
Soil metabolites, 332

Solid-phase extraction (SPE), 140, 142–147, 277–278
Solid-phase microextraction (SPME), 147–148
Spectral Data Base for Organic Compounds (SDBS), 47
Spectral selectivity enhancement, 253–255, 255f, 257f, 259f–260f
Stand-alone electrospray ionization, 22–24
Stand-alone mass spectrometry
   ambient mass spectrometry (AMS). *See* Ambient mass spectrometry (AMS)
   atmospheric pressure glow discharge (APGD), 284
   atmospheric pressure ionization (API), 268
   desorption electrospray ionization (DESI), 267, 283–284, 294
   direct analysis in real time (DART), 267, 283–284
   direct analysis in real time-mass spectrometry (DART-MS), 282–283
   electrospray-based methods, 292–293
   extractive electrospray mass spectrometry (EESI), 284–290
   flow injection analysis-mass spectrometry (FIA-MS). *See* Flow injection analysis-mass spectrometry (FIA-MS)
   formic acid, 292–293
   gas chromatography (GC), 265–266
   Gaussian-like distribution, 268–273
   geometrical parameters, 295
   helium flow rate, 297
   helium gas temperature, 296–297
   high-performance thin-layer chromatography (HPTLC), 279
   high-throughput planar solid-phase extraction (HTpSPE), 279
   ionization energy (IE), 293
   liquid–liquid partition, 277–278
   low-temperature plasma (LTP), 284
   multiple reaction monitoring (MRM), 265–266
   paper spray ionization (PSI), 284
   plasma-based techniques, 296
   polytetrafluoroethylene (PTFE), 295–296
   probe electrospray ionization (PESI), 284, 290, 294
   proton affinity (PA), 293
   quadrupoles, 265–266, 283
   relative standard deviations (RSDs), 282
   solid-phase extraction (SPE), 277–278
Stand-alone Orbitrap (Exactive), 33–35

Stir bar sorptive extraction (SBSE), 148—149
Synapt high-definition mass spectrometry system, 29, 30f

T

Tandem mass spectrometry, 63f
  collision-induced dissociation (CID), 63
  data-dependent acquisition (DDA), 63
  data-independent acquisition (DIA), 63
  diagnostic fragment ions, 62—63
  extracted ion chromatograms, 65, 66f
  fragmentation spectra, 64
  in-source fragmentation, 62—63
  information-independent acquisition mode, 64
  isotopic pattern, 62
  LC-QqQ-MS/MS, 63—64
  ultrahigh-performance liquid chromatography, 64
  variable data-independent acquisition (vDIA), 64—65
Tandem quadrupole (TQ), 247
Thin-layer chromatography (TLC), 279
Time-of-flight (TOF), 165—167
  accurate mass measurements, 26—27
  analog-to-digital conversion (ADC) detectors, 18—19
  continuous calibration systems, 27—29, 28f
  flight path length, 26
  gas chromatography (GC), 32—33, 33f
  hybrid ion trap time-of-flight, 32
  hybrid quadrupole time-of-flight instrumentation, 22—24, 25f
  ion detection, 26
  ion mobility quadrupole time-of-flight. *See* Ion mobility quadrupole time-of-flight
  ion sampling, 24—26
  ionization step, 24
  mass calibration, 26—27
  Micromass Q-TOF-1, 22
  MS/MS mode, 27

orthogonal acceleration TOF, 17—18, 17f
pulsed ionization source, 18
resolving power, 26
Sciex TripleTOF 5600 Q-TOF, 26, 27f
space focusing, 18
stand-alone electrospray ionization, 22—24
time-to-digital conversion (TDC) detectors, 18—19
velocitron, 17—18
Time-of-flight mass spectrometry (TOF MS), 247—248
Time-of-flight mass spectrometry analysis, 317
Time-to-digital converter (TDC), 248
Total Diet Study Program (TDS), 91
Traveling wave ion mobility spectrometry, 251—253, 252f—254f
Traveling wave ion mobility spectrometry (TW-IMS), 249—250
Tribrid (LTQ/Q-Orbitrap), 38—39, 38f
Triple quadrupole (QqQ) analyzers, 59

U

Ultra performance liquid chromatography (UPLC), 235
Ultrahigh-pressure liquid chromatography (UHPLC) systems, 165—167, 188f
Unified atomic mass unit, 1
United States Environmental Protection Agency (EPA), 222—223
USDA Food Safety and Inspection Service (FSIS), 91—92
USDA's Agricultural Marketing Service (AMS), 92
USDA's National Residue Program (NRP), 91—92

W

Wiley Registry MS/MS, 48

Printed in the United States
By Bookmasters